ECHINODERM STUDIES

VOLUME 4

ECHINODERM STUDIES

Edited by
MICHEL JANGOUX
Université Libre de Bruxelles, Belgium

JOHN M. LAWRENCE
University of South Florida, Tampa, USA

VOLUME 4

CRC Press is an imprint of the
Taylor & Francis Group, an **informa** business
A BALKEMA BOOK

CRC Press
Taylor & Francis Group
6000 Broken Sound Parkway NW, Suite 300
Boca Raton, FL 33487-2742

First issued in hardback 2017

ISBN-13: 978-90-5410-117-8 (pbk)
ISBN-13: 978-1-138-44073-9 (hbk)

Visit the Taylor & Francis Web site at
http://www.taylorandfrancis.com

and the CRC Press Web site at
http://www.crcpress.com

Contents

How to study evolution in echinoderms?

BRUNO DAVID

URA CNRS 157, Sciences de la Terre, Université de Bourgogne, Dijon, France

Final manuscript acceptance:March 1991

CONTENTS

1 INTRODUCTION

How to study evolution in echinoderms? Such a question might seem surprisingly odd after an entire century of very intensive research on this topic. But the recent increase of conceptual and methodological tools for the study of evolution in the last few years has brought new issues to bear on the problem. Indeed, during the last two decades, the introduction of new principles has greatly increased the set of possible methods of investigation. Although sometimes expressed in controversial or provocative terms, these principles have addressed fundamental questions: 1) the emerging theories of macroevolution and the reconsideration of the mode and tempo of evolution by punctuated equilibria (Eldredge & Gould 1972, Stanley 1975, 1979, Gould & Eldredge 1977, Gould 1980b); 2) the development of phylogenetic systematics and historical biogeography (Hennig 1965, Rosen 1978, Nelson & Platnick 1981, Wiley 1979, 1981, Myers & Giller 1988b); 3) the use of molecular methods in the construction of phylogenetic trees (Zuckerkandl & Pauling 1962, Sibley & Ahlquist 1983); 4) the great renewal of interest in the role of ontogeny (Gould 1977, Alberch et al. 1979, Bonner 1982, Dommergues et al. 1986). All of these have gained a pervasive influence in the field of evolution. Faced with all of these approaches, and considering the characteristics of echinoderms, it is pertinent to ask whether these methods are equally effective and suitable for analysing evolutionary problems in echinoderms. The goal of the present paper is to evaluate the stengths and weaknesses of the various approaches in establishing phylogenies, using echinoderms as a model. Much of this paper will be devoted to questions of method, of how different ways of analysis have been used to reconstruct echinoderm phylogenies.

Now the basic question is: what is evolution? A synthetic answer could be: evolution is change. Evolution corresponds to change at different scales, from that of molecules to those of cells, organisms or ecosystems; evolution also corresponds to change within different regimes, from that of genetic to those of physiology, anatomy or ethology. Therefore, what kinds of data and what kinds of methods best allow the detection of these changes and lead to the clearest understanding of evolution? To answer this question, I would like to propose a twofold comparison between a phylogeny and a theatrical play. When looking at a play or at an evolutionary example, two main types of data are available to reconstruct the scenario of the play or the phylogeny. The first type corresponds to the characters of the play, i.e., to the evolving organisms or the evolving taxa; the second type corresponds to the scenery and the timing of the play, i.e., to the spatio-temporal frame of evolution: stratigraphy, palaeogeography or ecology. Since one cannot understand the part of the players in complete independence from the set of the play, an accurate study of evolution must take into account information given both by the taxa and by

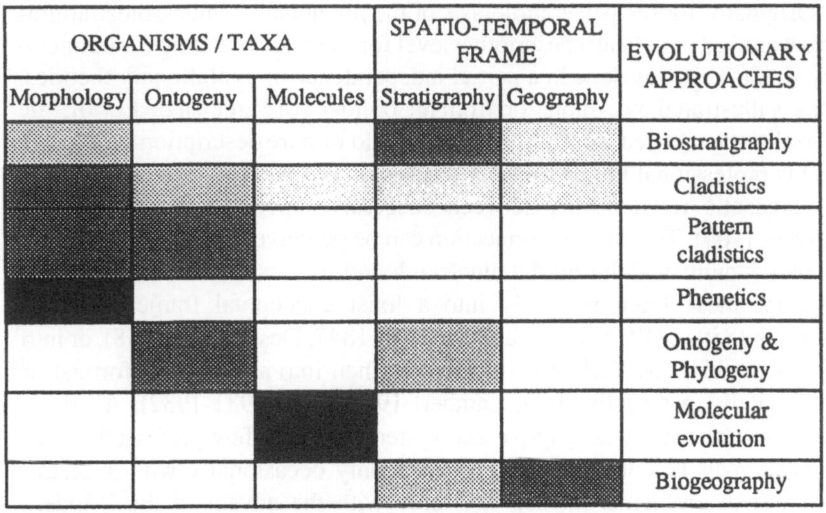

ORGANISMS / TAXA			SPATIO-TEMPORAL FRAME		EVOLUTIONARY APPROACHES
Morphology	Ontogeny	Molecules	Stratigraphy	Geography	
					Biostratigraphy
					Cladistics
					Pattern cladistics
					Phenetics
					Ontogeny & Phylogeny
					Molecular evolution
					Biogeography

Figure 1. Data sources (top lines) and their relative importance according to some of the main approaches to evolution (right column). The darker the rectangles are, the more reliant an approach is on that type of data (e.g. the cladistic approach to evolution relies a great deal on morphology, less on ontogeny, and only slightly on stratigraphy and geography; the phenetic approach relies uniquely on morphology).

the spatio-temporal frame. Thus evolution can be studied in several ways combining to varying degrees the two kinds of data (Fig. 1): e.g., the cladistic approach to evolution gives priority to data taken from the organism's morphology and ontogeny and much less importance to the stratigraphy. Given this choice, and considering the characteristics of echinoderms, is it appropriate to give a greater priority to certain approaches when analysing evolution in this phylum? Details and specific illustrations of these questions will be given in the following sections.

2 THE STRATIGRAPHIC APPROACH TO EVOLUTION

2.1 Historical framework

Because of the irreversibility of life's history, the fossil record possesses an intrinsic value as a source of information. It thus becomes possible to reconstruct phylogeny from the stratigraphic succession observed. Although palaeontologists have always relied first and foremost on morphology to group organisms, stratigraphic order has also been an important guide in the interpretation of evolutionary trends.

3

Originally, for the palaeontologists of the nineteenth century, biostratigraphy was used mainly at a descriptive level for the relative dating of sedimentary strata. This was done in a partial independence of evolutionary theories. This is illustrated by a quotation from the obituary of Perceval de Loriol: ' de Loriol always stayed carefully within the field of pure description (...) during all his professional life he avoided discussion of character evolution and of phylogenetic relationships between successive species' (translated from Sarasin 1909). This lack of connection can be perceived from the succession of works published during the nineteenth and the first part of the twentieth century: these fitted primarily into a fixist conceptual framework (e.g. Agassiz 1839, 1840, Agassiz & Desor 1846-1847, Desor 1855-1858), or into a catastrophist one (Orbigny 1853-1855), then into a more transformist or even vitalist perspective (e.g. Lambert 1902-1907, 1931-1932). All these studies were mainly descriptive and systematic, with few professed evolutionary goals (Lambert 1920), and dealt only occasionally with ancestor-descendent problems. In fact, it is only with the advent of the 'Modern Synthesis' (Simpson 1953) that biostratigraphy became strongly linked with evolutionary studies and became a real tool for phylogenetic reconstruction (e.g. Kermack 1954, Devriès 1960, Kier 1962, 1965, Durham 1966 for echinoids).

During the last 40 years, the stratigraphic approach to evolution has produced a number of 'classical' studies in palaeontology. Calibrated against the geological time scale, evolutionary changes can be investigated in terms of modes and tempos. According to the current conceptual credo, they can be thus recognized respectively as cladogenetic events and anagenetic trends (Modern Synthesis), punctational events and stasis (punctational model, Eldredge & Gould 1972), or constriction of lineages (bottleneck effect, Stanley 1975). However, such changes were traditionally viewed as largely oriented by directional selection and adaptation, the two main tenets of the neo-Darwinian theory. In this perspective, the evolving organisms are almost passively submitted to external contingencies through an omnipotent selection. It could be defined as an externalist point of view which will be compared in the following sections with more internalist approaches.

To illustrate these different and successive aspects of the use of stratigraphy in echinoderms it is impossible to avoid the classic story of the spatangoid *Micraster*, one of the most frequently related examples of a trend in evolutionary literature.

2.2 *The Micraster lineage: History of an evolutionary paradigm*

2.2.1 *The first steps*
The first part of the story represents a phase of accumulation of basic data (Fig. 2): the oldest studies correspond to rough sketches of siliceous molds

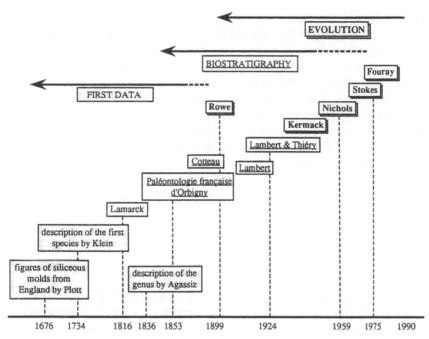

Figure 2. Chronological review of the studies dealing with the *Micraster* lineage. Three parts are distinguished in the history of the studies: 1) accumulation of the basic data; 2) biostratigraphic use of the genus; 3) evolutionary studies. The successive boxes make reference to Plott (1676), Klein (1734), Lamarck (1816), Agassiz (1836), Orbigny (1853-1855), Cotteau (1883), Rowe (1899), Lambert (1895, 1898), Lambert & Thiéry (1909-1925), Kermack (1954), Nichols (1959a), Stokes (1975), Fouray (1981), and Fouray & Pomerol (1985).

from the Chalk of England (Plott 1676, Breynius 1732, Bruguières 1791); description of the first species *M. coranguinum* by Klein (1734); and description of the genus by Agassiz (1836). After a critical amount of knowledge had been acquired about species of *Micraster* (Lamarck 1816, Goldfuss 1826, Des Moulins 1835-1837, Forbes 1850), a second phase began corresponding to a biostratigraphic use of the genus (Wright 1878, Bucaille 1883, Gauthier 1886, Lambert 1895). It was at this time, with the publications of the French palaeontologist Alcide d'Orbigny that the use of stratigraphy as a guideline for evolution reached a climax. D'Orbigny was a 'catastrophist,' and he divided the sedimentary strata into twenty-seven stages, each characterized by a particular fauna. He considered the chronological order of strata as the sole reliable data when interpreting the fossil succession, as illustrated by this quotation: 'If we find in nature forms without any conspicuous difference, although separated by a stratigraphic gap of several stages, we shall without

5

any doubt consider them as distinct' (translated from d'Orbigny, 1850). Morphology was explicitly excluded from phylogenetic considerations. The third phase of the story is the most important and marks the development of attempts to decipher the phylogeny of *Micraster*. It begins at the turn of the century with Rowe (1899), who advanced the first modern phylogenetic interpretation of *Micraster* from the Chalk of England, grounded both on stratigraphy and morphology. Rowe divided the *Micraster* lineage into four branches or groups descending one from each other and constituting a continuous trend from *M. corbovis* to *M. coranguinum*. Subsequent studies included the contribution of the Modern Synthesis and several proposed improvements over the initial attempt of Rowe (1899). Illustrating different approaches, each emphasizes a particular aspect of the problem: geographical distribution, ecology, or evolutionary tempo, but they all share a stratigraphic background which remains very important in the interpretation of evolutionary trends.

2.2.2 *The Micraster lineage: Chronological study*
Ernst (1972) proposed an improved phylogeny of the *Micraster* lineage based on samples collected with great stratigraphical accuracy from northern Germany, and covering the whole stratigraphic range of the lineage from the Turonian to the Maestrichtian (Upper Cretaceous) (Fig. 3). This attempt was founded on direct stratigraphic superposition to document the evolution of the species, the justification being the density of the fossil record. In this new framework, ideas about the evolution of *Micraster* become more complex and the resulting phylogeny consists of three main branches and some secondary stems. The splitting of the lineage into these different branches is deduced from morphological analysis of the plastronal area and of ambulacral zones. Although morphology is accurately included in the phyletic interpretation of Ernst, the temporal distribution of species still provides the most tangible evidence of the evolution of the genus. The resulting pattern of relationships thus appears as an empirical compromise between morphological similarities and data drawn from the fossil record.

2.2.3 *The Micraster lineage: Geographical study*
Stokes (1975) introduced a spatial dimension by establishing several faunal provinces as a framework for understanding the evolution of *Micraster*. He separated a European realm from a north-African realm on the basis of the general distribution of many genera, arguing that *Micraster* and such holasterids as *Infulaster, Hagenowia, Echinocorys* or *Cardiaster* are almost completely restricted to the European realm. At another scale Stokes used the geographical distribution of allopatric species of *Micraster* to propose distinctions between faunal provinces or subprovinces across the European realm (Fig. 4). He considered that the faunal differences probably originated

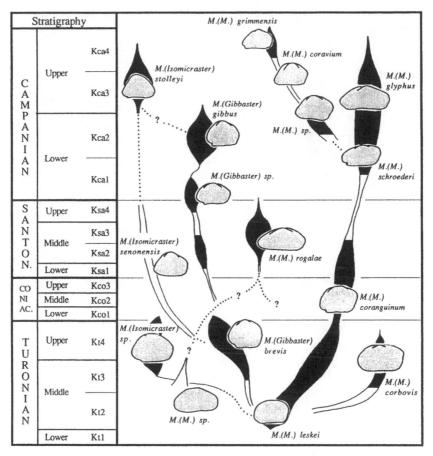

Figure 3. Evolution and stratigraphy of the *Micraster* lineage in the Upper Cretaceous from northern Europe. There are three major branches: the main lineage leading from *M.(M.) leskei* to *M.(M.) glyphus,* the *Gibbaster* lineage, and the *Isomicraster* lineage. The phylogeny is grounded on stratigraphic order; stratigraphic subdivisions according to German nomenclature; ranges of occurrence of the different species are in solid black (modified from Ernst 1972).

from variations in temperature and oceanographic conditions. The southward expansion of the northern province during the Coniacian and Campanian could thus be related to a temperature drop which induced a displacement of the cold waters toward the South. From his comprehensive palaeobiogeographic study Stokes (1975) inferred a phylogeny for the whole *Micraster* lineage where both stratigraphic boundaries and the recognition of geographic provinces had great importance for understanding evolution in *Micraster.*

7

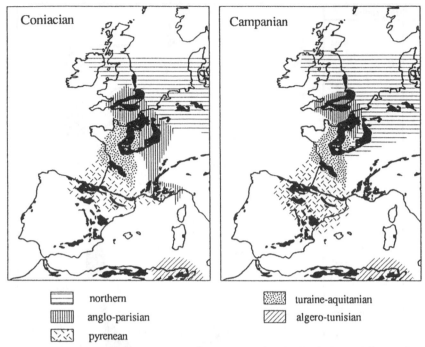

	northern		turaine-aquitanian
	anglo-parisian		algero-tunisian
	pyrenean		

Figure 4. Evolution and biogeography of *Micraster* during the Coniacian and Campanian (Upper Cretaceous). Faunal provinces are defined according to the distribution of the species of *Micraster*. Areas in solid black correspond to the present-day outcrops (redrawn from Stokes 1975, and Smith 1984a).

2.2.4 *The Micraster lineage: Ecological study*

An ecological dimension can be added to the former morpho-stratigraphic or spatio-stratigraphic approaches by taking into account information from the ecology of extant forms. Such an approach was pioneered by Nichols (1959a, b), who compared the functional biology of the species in the *Micraster* lineage with that of living species of *Echinocardium* and *Spatangus*. Later Smith (1984a) revisited this approach and provided a well documented ecological interpretation for part of the lineage leading from *Micraster leskei* to *M. coranguinum* in southern England (Fig. 5). In northwestern Europe, the replacement of sandy-coarse sediments by fine-grained chalk facies during the Cenomanian and Turonian caused a crisis among infaunal dwellers and triggered the replacement of *Epiaster,* well adapted for life in coarse sediments, by *Micraster.* The subsequent evolution of *Micraster* can be interpreted as a succession of adaptive responses to these newly developed environments. The morphological changes that occurred between *M. leskei* and *M. coranguinum* can be viewed as the result of improved burrowing

8

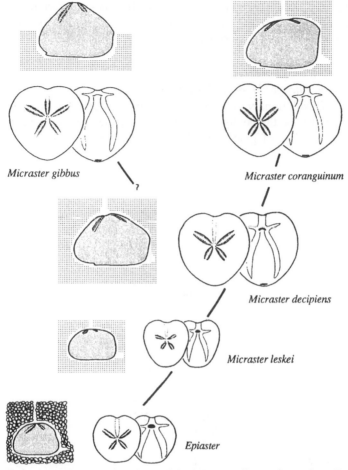

Figure 5. Evolution and ecology of the *Micraster* lineage in southern England. Shaded profiles illustrate the position of the echinoids in their burrows. The illustrated phylogeny corresponds to the earliest part of the lineage on Figure 3 (redrawn from Smith 1984a).

technique in a more cohesive sediment: elevation of the test led to a general body shape that probably allowed a better division between respiratory and nutritional functions; the broadening of the subanal fasciole led to an improvement in the discharge of wastes out of the burrow; the enlargement of the petals is related to an increase in the number of respiratory tube feet; and the development of a dense miliary-spine canopy protects the test from being smothered by fine sediment. Furthermore, the deepening of the anterior groove, the anterior shift of the mouth and the increased projection of the labrum over the peristome correspond to an improvement of the feeding technique. As stated by Smith (1984a), the success of *Micraster* in colonizing the chalk seas corresponds thus to the modification of a 'sand-dwelling' morphotype to a 'mud-dwelling' morphotype along a progressive evolutionary trend.

2.2.5 *The Micraster lineage: Mode and tempo of evolution*

The elucidation of evolutionary patterns was not the primary goal of the studies outlined above and the phylogenies proposed by these workers were grounded on the principle of gradual evolution involving anagenesis and

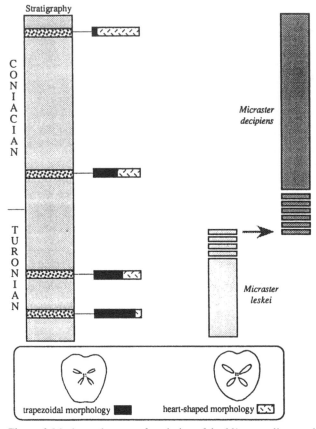

Figure 6. Modes and tempo of evolution of the *Micraster* lineage during the Turonian and Coniacian (Upper Cretaceous) in North-West France. Two simultaneous evolutionary events occur across the lineage: 1) the gradual replacement of the trapezoidal morphologies by heart-shaped forms; 2) the sudden replacement of *M. leskei* by *M. decipiens*. The *Micraster* were sampled in four beds of indurated chalk (black dots on the stratigraphic column) (modified from David & Fouray 1984).

cladogenesis. David and Fouray (1984) proposed a complementary approach documenting evolutionary modes and tempos, supported by analyses of the morphological variation of the oldest part of the lineage leading from *Micraster leskei* to *M. decipiens*. Morphological variation was analysed in samples collected in four stratigraphically superimposed beds from the chalk cliffs of Picardy (NW France), *M. leskei* occuring in the two lower beds and *M. decipiens* in the two upper beds. The study was based on statistical analyses of numerous morphological characters. Factor analyses performed on both quantitative and qualitative parameters demonstrate that the transition between these two species was achieved without intermediary steps (Fig. 6). The distinction between *M. leskei* and *M. decipiens* is very sharp and concerns the general shape, the number of respiratory tube feet, and the ornamentation of petals and periplastronal areas, i.e. on features related to burrowing and feeding techniques, as stated in the interpretation of Nichols (1959a, b) and Smith (1984a). Superimposed on this major evolutionary step between *M. leskei* and *M. decipiens* is the gradual development of a heart-shaped morphology. Each of the four beds contains both trapezoidal and heart-shaped sea urchins, statistically well discriminated and with few intermediary forms. But while the *M. leskei* of the lower beds are almost exclusively trapezoidal, *M. decipiens* shows a well-balanced dimorphism in the third bed, and becomes predominantly heart-shaped in the upper bed. This trend involves features of the ambitus shape as well as of the petals and can be interpreted as a progressive drift between the two morphotypes. Thus the evolution of *Micraster* across the Turonian and Coniacian from Picardy brings together two main modes of variation: 1) the gradual shift towards heart-shaped morphotypes, a trend which continued as far as the Santonian; 2) a rapid speciation process which could be a punctational event. This contrast between the sudden replacement of *M. leskei* by *M. decipiens* and the gradual development of an heart-shaped morphology attests to two different speeds of evolution which seem related to the functional significance of the features involved.

2.3 *Limits of application*

Because the fossil record represents the first line of evidence when reconstructing phylogenies, the stratigraphic approach to evolution, as illustrated by the different attempts dealing with *Micraster,* has been the most widely used in constructing phylogenies of fossil echinoderms. In the last decades, this approach has been used to investigate the two competing models of evolutionary processes: phyletic gradualism (e.g. Tintant 1963, Kellogg 1983) and punctuated equilibria (e.g. Eldredge 1971, Bonis 1983). High confidence in the reliability of the fossil record has led to the development of the so-called stratophenetic method (Gingerich 1979) which emphasizes the importance of a continuous fossil record for understanding phylogenetic relationships. Stratophenetics is an empirical approach explicitly combining stratigraphy and morphology. Even though the stratophenetic method was first established to study phyletic gradualism in Tertiary mammals (Gingerich 1985, 1987) it provides an accurate picture of how numerous phylogenies have been reconstructed in palaeontology since the advent of the Modern Synthesis.

All the echinoderm classes appeared in the initial radiation of the Cambrian and Ordovician (Sprinkle 1983). The phylum possesses a long and continuous fossil record throughout the Phanerozoic, and this fossil record offers an inherent potential for generating phylogenetic hypotheses. Actually, the stratigraphic approaches to evolution has been utilized to varying degrees in different classes of echinoderms. Some of the numerous Palaeozoic extinct classes have been closely examined from a phylogenetic point of view: blastoids (e.g. Breimer & Macurda 1972, Sprinkle 1973, Waters et al. 1982, Waters et al. 1985, Horowitz et al. 1986) or edrioasteroids (e.g. Bell 1976, Smith 1983) because of their relative abundance and diversity and because of their potential stratigraphic value; cystoids (e.g. Bockelie 1981, Sprinkle 1982) and in early studies, the stylophoran 'carpoids' (e.g. Chauvel 1941, Ubaghs 1968, 1975) due to their puzzling morphologies. Of the five classes which survived the end Palaeozoic crisis and are still extant, not all are equally suitable for the type of approach used by 'evolutionary systematists' (Schoch 1986). On the one hand, holothuroids, which possess an insufficiently dense fossil record, have been poorly studied from an evolutionary point of view particularly because of the extreme scarcity of complete fossils (they have a greatly reduced calcareous skeleton and fossils usually consist of isolated spicules, Frizzell et al. 1966). Asteroids and ophiuroids are rarely preserved as complete specimens, but discrete separated ossicles are abundant and carry much more information than do holothurian spicules, allowing these two classes to be employed to some extent in evolutionary studies (e.g. Spencer 1913, Fell 1963a, Schulz & Weischat 1975, 1981, Breton 1990). On the other hand, echinoids are both diverse and abundant as fossils, particularly during Mesozoic and Cenozoic times. Moreover, phylogenetic hypotheses have been primarily developed in relation to stratigraphic and taxonomic works in which echinoids have been the most frequently involved (see bibliographic reviews in Péron 1895, and in Pajaud et al. 1976 for Cotteau and Lambert's works respectively). More recent stratophenetic studies of echinoids are gathered in the 'Treatise of invertebrate paleontology' (Durham et al. 1966) and in the papers of Kier (e.g. 1965, 1974, 1982, 1984), but also in numerous other works (e.g. Devriès 1960, Roman 1965, Durham 1966, Ernst 1970). The phylogeny of crinoids has also been intensively studied, involving both Palaeozoic (Moore & Laudon 1943, Ausich &

Lane 1982, Brower 1982), and post-Palaeozoic forms (Macurda & Meyer 1977, Roux 1979, 1987). Because they are generally rather well done and well argued, these works dealing with echinoderm phylogeny at various taxonomic scales have gained a pervasive influence in our understanding of the evolution the phylum, and still often represent the most accurate account of groups, at least up to the family level (e.g. see above the quite satisfactory understanding of *Micraster*'s evolution).

But all of these works depend on the reliability of the fossil record which is itself correlated with the steadiness of sediment accumulation. It is well known that sedimentation rates vary (Sadler 1981) and that removal of sediment by synsedimentary erosion can produce gaps in the record. For the majority of the sections, the record is highly discontinuous and the preserved sediments represent generally less than 10% of the elapsed time (Schindel 1982). Because of the incompleteness of stratigraphic sections and correlatively of the fossil record, geological sequences often provide ambiguous answers (see discussion in Novacek 1987). Stratigraphy has thus failed to resolve some of the most interesting problems of relationships, for instance the relationships between the five extant classes. Ambiguous evolutionary patterns or idiosynchratic scenarios have sometimes emerged because of the absence of a clear methodological approach linking morphological analysis and stratigraphy. Thus, in reaction, some palaeontologists have chosen to emphasize other types of data, and have advocated the use of a strictly morphological approach to evolution.

3 THE CLADISTIC APPROACH TO EVOLUTION

3.1 *Theoretical basis*

Because the morphology of Recent and fossil organisms can be understood as the result of their history, the investigation of the relationships between taxa can be grounded in formal analyses of character distribution. Characters of organisms themselves (e.g. morphology) here gain priority over extrinsic parameters (e.g. stratigraphy or biogeography) for tracing phylogenetic pathways. In this field of research, and as a general opposition to the approach of the evolutionary systematists, several different techniques have been developed to deduce phylogenies. They range from the strict use of characters to measure morphological distances (phenetics) to the distinction between primitive and derived characters, and the acceptance of palaeontological data (phylogenetic systematics, i.e. classical Hennigian cladistics), or to the dismissal of the palaeontological argument as uncertain and the use of the ontogenetic criterion (i.e. so-called pattern cladistics). These different approaches have given rise to a conflicting debate largely set out in the journal

Systematic Zoology during the 1970's and 1980's, a debate mainly running through academic papers with few concrete examples (see some historical and theoretical points in Farris 1980, Forey 1982, Dupuis 1984, and a review of the most controversial aspects in Janvier 1984). From an evolutionary point of view, the result of this debate has been the acceptance of the 'logic of apomorphies' as the best rule (Wiley 1981), and the retention of Hennigian cladistics as the most appropriate method for phylogenetic purposes involving both extant and fossil forms. The phenetic technique has to be rejected as it postulates taxonomic similarity without historical background, and the more extreme aspects of the pattern cladistics can be discarded (e.g. those that do not take into account evolutionary implications of palaeontological data).

Phylogenetic systematics was introduced by Hennig as early as 1950, but the method reached a worldwide success only at the end of the 1960's after the German master's book was translated into English (1965). Basically, phylogenetic systematics is rooted in the principles of relationship and resemblance: all the species known at a given time are more or less related, and their level of relationship is a function of their morphological resemblance. The key to the use of this function is that 'resemblances between organisms do not always have the same bearing on relationships. Organisms, or groups of organisms, can resemble each other because they share either advanced or primitive characters. Hennig called the advanced, or derived characters apomorphies, and the primitive characters plesiomorphies. When such characters are shared by several organisms, they are synapomorphies or symplesiomorphies. Hennig argued that synapomorphies alone indicate relationship between organisms.' (Janvier 1984, p. 43). As a prerequisite, a Hennigian approach to phylogeny is grounded on two tenets: the tenet of irreversibility of evolution (the so called Dollo's law), and the tenet of differentiation of characters during evolution. Subsequently, the cladistic method is not without ambiguity because of exceptions to these two tenets upon which it is grounded. These exceptions correspond to reversions, which concerns character states (a derived character returns secondarily to a state similar to that of the ancestor); and convergence, which concerns the nature of characters (two apparently identical derived characters result from two different evolutionary ways). In order to overcome these limitations, one may analyse a large number of features and use parsimony to recognize the apomorphic patterns when establishing cladograms (i.e. to choose the hypothesis which implies the least number of reversion and convergence events). As an alternative to parsimony, some authors have proposed an *a priori* weighting of characters (e.g. Hecht & Edwards 1977), but this generally leads to more or less putative assumptions in the ranking of characters. Furthermore this cannot be a substitute for parsimony as the weighted characters must be reintroduced in a parsimony procedure for the computation of cladograms (Farris 1982). In short, the search for congruence in terms of probability within a large set of features appears without doubt to be the least hazardous method. With parsimony it is thus possible to perform an auto-weighting which is beyond the bounds of chance (Patterson 1982).

Another approach to detect reversion and convergence is to use data independent of the morphological characters as test of congruence. Among them, chronological data derived from the fossil record appear well suited for checking the cladograms. Indeed, the palaeontological argument offers a stratigraphic control of cladistic hypotheses which can be usefully performed (as exemplified by Smith, 1981), even though largely criticised as uncertain (Novacek 1987), or clearly rejected by pattern cladists as a polarizing criterion ('the relevant factor is not the age of the fossil (...) and the palaeontological argument is fallacious,' Nelson 1978, p. 329). Besides chronological data, fossils provide complementary morphological data unanimously accepted by systematists of every conceptual leaning. However, the way in which they are used is controversial: pattern cladists restrict their use (Patterson 1981), while phylogenetic cladists believe that morphological and chronological data can be associated in the same analyses (see a methodological account of the use of 'stem' and 'crown' groups in Jefferies 1979, and a defence of the use of fossils in Smith 1984b). Fossils may play several roles in phylogenetic analyses: they provide complementary data; they allow dating of the divergence events; and because they are close to divergence events they preserve the morphologies typical of groups at the time of those events.

3.2 Cladistics and echinoderms

The impact of cladistics in deciphering echinoderm evolution is both intense and widespread. It is intense because the skeletal morphology of echinoderms bears numerous features easy to observe, often clearly related to biological functions, and available both in fossil and Recent forms. Furthermore important anatomical or embryological data are available in living forms. The impact is widespread because most major groups of echinoderms have been subject to cladistic analyses which have led to important restatements of classification and evolution for groups with a very poor fossil record, for example the asteroids (Gale 1987, Blake 1987), as well as for groups with a great range of diversity, for example the Palaeozoic echinoderms (e.g. Jefferies 1979, Donovan & Paul 1985) and the post-Palaeozoic echinoids (e.g. Jensen 1981, David 1985, Harold & Telford 1990, Mooi 1990a).

3.2.1 New phylogenetic evidence for echinoderms

The foremost use of cladistics has been to present phylogenetic hypotheses of relationship within groups formerly insufficiently known. Such an impact has been particularly important among Palaeozoic echinoderms whose fossil record is not complete enough to always provide reliable evidence of phylogenies. The early Palaeozoic was a time of explosive morphological diversification for echinoderms: about 20 echinoderm classes appeared between the early Cambrian and the middle Ordovician (Sprinkle 1983). Such a rapid evolution has given rise to numerous important morphological innovations making the echinoderms highly suitable for cladistic analyses as illustrated by works involving a wide range of groups: exploration of the early evolution of crinoids (Donovan 1988); investigation of the phylogeny of cystoids *sensu lato* resulting in the separation and reorganization of diploporites and rhombiferans (Paul 1988); study of the first steps of the diversification of the edrioasteroids (Smith 1985); and reappraisal of the radiation and classification of echinoderms (Paul & Smith 1984, Smith 1984b). The last point makes reference to wider attempts. Paul and Smith (1984) proposed a cladogram of relationships focusing first on the early

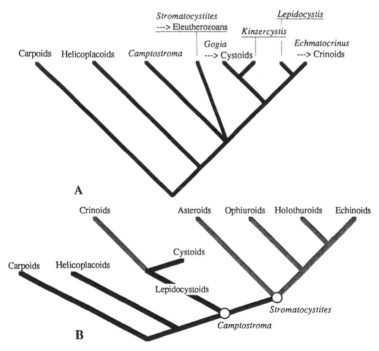

Figure 7. Early radiation and relationships of echinoderms: A) cladogram of relationships for the early radiation of Cambrian echinoderms; B) phylogenetic pattern illustrating how the five extant classes (shaded branches) are rooted in the Cambrian radiation of the phylum (modified from Paul & Smith 1984).

16

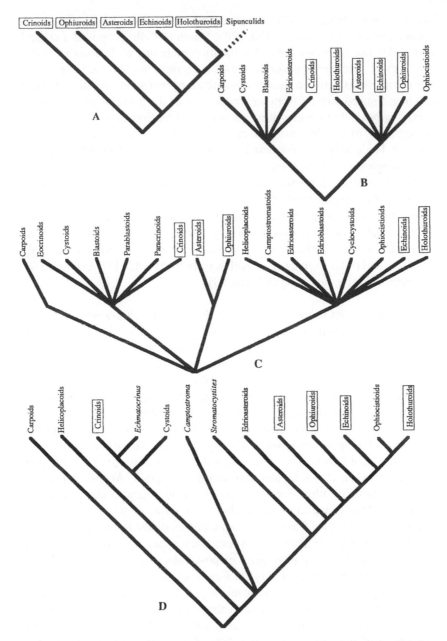

Figure 8. Evolution of echinoderm classification as expressed in cladograms (the five extant classes are highlighted by rectangles). Relationships of echinoderm groups according to: A) Forbes (1841); B) Hyman (1955); C) Moore & Teichert (1978); D) Smith (1984b). The more greater the number of polytomies, the more ambiguous the proposed classifications are (modified from Smith 1984b).

radiation of echinoderms (Fig. 7A), and then explaining the phylogeny to encompass the entire phylum (Fig. 7B). Smith (1984b) proposed a critical reconsideration of echinoderm classification including an instructive historical survey. This survey illustrates how cladograms may be used to perform fair comparisons between former classifications whose information content can be estimated directly from the importance of the polychotomies. Therefore, it appears that the classification proposed by Forbes as early as 1841 corresponds to a fully resolved cladogram (Fig. 8A). Further attempts (Hyman 1955, Moore & Teichert 1978) included new palaeontological and embryological discoveries, but despite this increasing data set, the information content of the proposed classifications decreased (Fig. 8B-C). Smith argued that this paradoxical situation probably results from an inappropriate taxonomic methodology that led to poor or inadequate utilization of palaeontological data, and he proposed a highly informative classification scheme (Fig. 8D) where relationships between the five extant classes are fully resolved, and differ from those of Forbes only by a switch between asteroids and ophiuroids (compare Figs 8A and 8D).

3.2.2 Cladistics and chronological data: Former hypotheses revisited

Another use of cladistics is to deal with groups whose fossil record has allowed classical ancestor-descendant relationships to be formerly proposed: phylogenetic analyses provide testing and reinterpreting for previous hypotheses. Because many echinoderms exhibit a rather dense fossil record, examples of the use of chronological and morphological data in combination within a cladistic framework are not rare. This is shown by several studies which have led to reappraisals of previous phylogenies (e.g. Lewis & Jefferies 1980, Smith 1981, Smith & Wright 1989 for fossil echinoids).

An adequate fossil record also offers the opportunity to check for congruence between chronological data and a cladogram. At this step of the study, stratigraphic information should provide additional evidence to corroborate the morphological data. It also allows the scope of the cladistic analysis to be extended from the cladogram of relationships to a phyletic tree, and to be further developed into evolutionary scenarios. This can be illustrated by a study of the phylogeny of post-Palaeozoic crinoids by Simms (1988). Simms used an explicitly cladistic approach to propose a pattern of relationships close to the family level (Fig. 9A). Assuming that 'the fossil record is reasonably representative of the history of the group' (p. 269) he undertook a comparison of the cladogram with the known stratigraphic occurence of the taxa studied (Fig. 9B). This comparison highlights that stratigraphy is in good agreement with morphology for almost all groups: the timing of appearance of groups matches the succession of the nodes along the cladogram. For instance, the suborder Isocrinina is subdivided into three groups which occur on the cladogram in a series Isocrinidae – Cainocrinidae

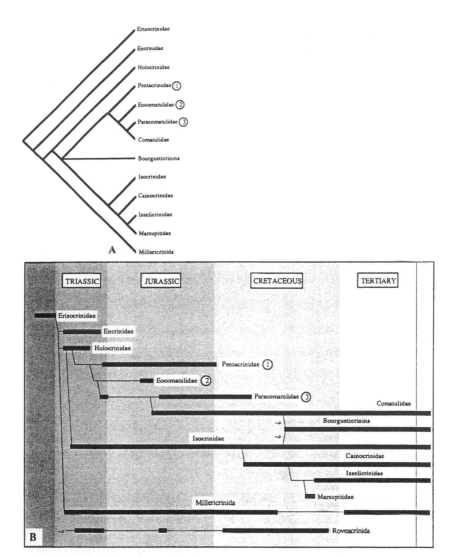

Figure 9. Cladogram and stratigraphic data. A) Cladogram of relationships for the post-Palaeozoic crinoids. B) Stratigraphic distribution and inferred phylogeny; the length of the horizontal thin lines gives an estimate of the discrepancy between the cladogram and the fossil record. The relative positions of the three numbered groups (shaded lines) on A and B provide an illustration of such a discrepancy (modified from Simms 1988).

– Isselicrinidae, from the less to the more derived. This array is fully supported by stratigraphic data, the earliest representative of each group being from the Triassic, early Cretaceous and late Cretaceous respectively. But the morphological hypothesis can also indicate some gaps in the fossil

19

record. For instance, the three taxa 1) Pentacrinitidae, 2) Eocomatulidae and 3) Paracomatulidae are related in the order 1-2-3 on the cladogram, whereas they occur in the order 3-1-2 according to their earliest stratigraphic occurrence. In this particular case, the conflict concerns chiefly the ancestry of Eocomatulidae and the discrepancy can be explained by invoking the poor fossil record of this monospecific family. Thus, the comparison between the cladogram and the known fossil record allows the cladistic hypothesis of relationships to be corroborated, and provides an idea of the degree of discrepancy. The longer the thin lines above the nodes (Fig. 9B), the greater is the degree of discrepancy between morphology and stratigraphy. At this point, it becomes possible to propose a 'story' implementing evolutionary scenarios.

More generally, where incongruency occurs between stratigraphy and morphology the conflict must be discussed. It comes either from an incorrect sister-group relationship or from incompleteness of the fossil record (see Novacek 1987 for discussion). In this conflict, it is often concluded that the fossil record is erroneous. In echinoderms this can be partly explained by the rarity of complete specimens of some groups. Such an argument has been used to justify the position of Echinothurioida (a sea urchin group which possesses a very thin and flexible test with imbricated plates) at the origin of other euechinoids despite the fact that they have not been recorded before the Middle Jurassic (Smith 1981). Moreover it is possible to compute intervals to 'assess the reasonability of large gaps in the fossil record' (Marshall 1990, p. 402). In the case of echinothurioids, computation of confidence intervals are consistent with a 50 Myr gap in their fossil record (Marshall 1990).

3.3 Conclusions about cladistics

After a long domination by the stratigraphic approach to evolution, the cladistic approach has noticeably changed our familiar landscape concerning echinoderm evolution. This has led to major advances in our understanding of the phylogenetic history of echinoderms, particularly concerning Palaeozoic groups whose fossil record is often imprecise. But whatever its success, the cladistic approach gives only a more or less probable picture of evolution since the reliability of morphological information remains limited on two accounts. First, its reliability depends on an ability to recognize homologies, indeed 'synapomorphies are homologies by definition' (Eldredge 1979, p. 181). Homology appears thus without question as the most important principle on which the cladistic method is grounded. This concept has been discussed at length (e.g. Cracraft 1967, Wiley 1975, Bock 1977, Riedl 1979), and masterly reviewed by Patterson (1982). Second, it is subordinated to the frequency of reversals which are fortuitous events, not generally detectable

20

from morphological methods, and which we must assume to be rare. The difficulty in recognizing homology becomes particularly clear when derived features evolve independently to produce convergence. Contrary to reversion, convergence is detectable from morphological methods at least in theory, but unfortunately seldom in practice. In the case of conflict between two sets of apparent homologies, only one is due to true relationships while the other must result from convergence. Faced with such a choice, the sole reasonable answer must be expressed, as for reversion, in statistical terms and is obtained by the application of parsimony. The statistical value of the results primarily depends on the number of characters taken into account. Rigorous applications have been developed through the use of computers and powerful software. The ability of parsimony to arrive at robust and reliable solutions is also dependent on the taxonomic group under consideration. As a result the complex morphology of echinoderms makes them among the best suited invertebrate groups for cladistic analyses.

One way to overcome these weaknesses, and to improve the reliability of results is to supplement them with results given by other methods. This has been done by comparison with stratigraphic data, as done above with crinoids, but can also be done by using molecules instead of morphology.

4 THE MOLECULAR APPROACH TO EVOLUTION

4.1 *The molecular 'clock'*

For some years, the investigation of evolutionary pathways has been expanded into molecular analyses using molecular structure instead of morphological characters. The molecular approach to evolution shares with the cladistic approach the common belief that organisms possess in themselves some evidence of their evolution.

The basic idea was that the rate of change of a given molecular structure is approximately constant for more or less large groupings of taxa over time. This is known as 'the molecular evolutionary clock hypothesis,' first suggested by Zuckerkandl and Pauling (1962) and later related to the neutral theory (Kimura 1983). This has led to the conclusion that the molecular difference between two taxa is proportional to their phyletic gap. Although this proportionality is not absolutely constant and the molecular clock does not beat the same time in all the groups (e.g. Jukes & Holmquist 1972, Goodman 1981, Wu & Li 1985, Britten 1986, Catzeflis et al. 1987), particularly since the effect of generation time makes the clock inaccurate (Kimura, 1987), it remains true that a functional class of molecules to some degree carries a record of evolution and permits phylogenetic inference. This idea generated great interest among biologists and molecular biologists who have elaborated different methods for constructing phylogenetic trees from amino acid and nucleotide sequences. However, as important as the notion of a molecular clock has been to establish the importance of molecular methods for phylogenetic reconstruction, the *a priori* assumption of a molecular clock is not required to retrieve phylogenetic information from molecular data (Marshall pers. comm.).

4.2 *Molecular methods for studying evolution*

The molecular approaches can be artificially classified into two main categories. The first category, 'electrophoretic polymorphism' involves comparison of the amino-acid composition or specific enzymatic activities of proteins. The second category deals with nucleic acids and concerns data from nuclear DNA, mitochondrial DNA, or ribosomal RNA .

4.2.1 *Amino-acid composition and protein sequencing*
Historically, the first attempts to gain phylogenetic information from molecules were analyses based on the amino acid composition of proteins (Hill et al. 1963). In particular cytochrome C has been studied in a variety of animal taxa (Margoliash 1963, Fitch & Margoliash 1967).

In echinoderms, collagen has been the most widely analyzed. This protein offers the advantage of being present in many body tissues of echinoderms and of having an appropriate rate of substitution (Matsumura et al. 1979). The molecular structure of collagen readily allows neutral changes since the substitution of amino acids at many sites has few effects on the chemical properties of the molecule. The use of the amino-acid composition of collagen to derive phylogenetic trees in echinoderms was pioneered by Matsumura et al. 1979. Results have been obtained both at the order-level and at the class-level. However, the amino acid composition is not always in accordance with systematic data, particularly in camarodont echinoids. This suggests 'that the substitution of amino acids in echinoderm collagens cannot be regarded as being due solely to neutral mutation, and that there are substitutions due to Darwinian pressure during evolution' (Matsumura & Shigei 1988, p. 48). This conclusion is probably related to the fact that their method does not take into account variation in the rate of molecular evolution within different parts of the collagen molecule. In addition to collagen studies, comparative analyses have been performed on the free amino-acid composition in eggs and ovaries of several species of regular echinoids (Nagaoki 1985).

4.2.2 *Specific enzymes*
The use of selective staining techniques combined with electrophoresis allows the detection of specific enzymatic activities that can be used to

22

propose biochemically grounded evolutionary hypotheses. Electrophoretic analyses of enzymes permit a wide range of comparisons, from that of different species (e.g. Bullimore & Crump 1982 for asteroids, Liddell & Ohlhorst 1982 for crinoids) including the discovery of sibling species (e.g. Manwell & Baker 1963 for two holothuroids of the genus *Thyonella*), to that of the structure of populations including the characterisation of hybrids (Schopf & Murphy 1973) or of sub-species, and the study of polymorphism. Into this prospect, several studies that evaluate the polymorphism of invertebrates in deep sea environments, regarded as stable, involve echinoderms (Gooch & Schopf 1973, Murphy et al. 1976 for echinoderms in general, Doyle 1972, Ayala & Valentine 1974 for ophiuroids, Ayala et al. 1975 for asteroids, Costa et al. 1982, Bisol et al. 1984 for holothuroids).

Since they are abundant and widespread, regular echinoids have been investigated biochemically more than other echinoderms to clarify the relationships among genera of diverse families. For instance, Matsuoka successively investigated the Toxopneustidae (Matsuoka 1985), Diadematidae (Matsuoka 1989), and Echinometridae (Matsuoka & Suzuki 1989a) by electrophoretic analyses of various enzymes (Fig. 10). The genetic distances expressed on the biochemical dendrograms are consistent with the generally accepted relationships inferred from morphology. According to Nei (1975), they can also be used to compute a divergence time between groups, and thus are directly comparable with data from the fossil record (Fig. 10). But Nei's distances have given unrealistic divergence times when applied to camarodont echinoids (Matsuoka 1987), and they should be used carefully. Dealing with the influence of geographic isolation in speciation processes, Lessios (1981) compared morphological and enzymatic differences between geminate species of *Eucidaris, Diadema* and *Echinometra* located on the two sides of the Isthmus of Panama. The magnitude of divergence between each pair of species shows 'striking differences' amongst the genera and enzymatic characters studied: major molecular differences can be associated with little morphological differentiation.

At another scale, the analysis of the echinoids *Temnopleurus toreumaticus* and *T. hardwickii* illustrates the use of electrophoretic methods to compare morphologically very closely related species (Matsuoka 1984, Matsuoka & Suzuki 1987 for another example). The genetic distance observed between these two species is comparable to that reported between sibling species, and supports the palaeontological hypothesis of a relatively recent speciation event (Nisiyama 1966). But the widest use of enzyme electrophoresis concerns the study of the genetic variation between conspecific populations. Echinoderms are well suited for such analyses, and many studies have been done on echinoids and asteroids (Marcus 1977, 1980, Rosenberg & Wain 1982, Matsuoka & Suzuki 1989b for a review). For the most part, these studies discussed the electrophoretic differences between populations in

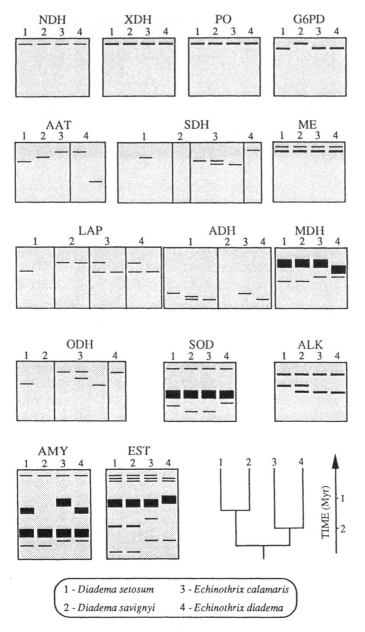

Figure 10. Phylogenetic relationships between four species of the family Diadematidae (regular echinoids) inferred from electrophoretic analyses of 15 enzymes. The biochemical dendrogram confirms the former taxonomic distinction between *Diadema* and *Echinothrix*, and Nei's (1975) technique provides an estimate of the divergence times between each pair of species at 2.4 and 1.2 Myr respectively (modified from Matsuoka 1989).

terms of genetic drift and evolutionary divergence. In doing so, they did not take into account the possibility that other factors have affected the enzymatic polymorphism. For instance, marine animals, and particularly echinoderms, are directly affected by environmental parameters whose precise effects on the organisms are often undocumented, and that can greatly influence evolutionary interpretations (Hochachka 1971, Penniston 1971). The results may also depend on other non-evolutionary factors directly linked to the organisms. For example, in the spatangoid *Brissopsis lyrifera* (Féral et al. 1989, 1990) a discriminant factor analysis performed on zymograms of the female gut showed a rather good discrimination of five gonadal states according to their enzymatic activities (Fig. 11). That is to say, the enzymatic activities of females appear to be partly seasonal dependent according to the physiological state of the specimens. This emphasizes that care must be taken in the use of enzymatic polymorphism analyses in taxonomy and evolution. Since marine invertebrate populations are generally widespread, and environmental parameters insufficiently known (Gooch & Schopf 1973), further studies must be always associated with accurate controls of the homogeneity of samplings, especially of sex, age and physico-chemical conditions. Other ways to overcome these limitations are to use data which may be less disturbed by non-evolutionary factors. This can be done by working directly on sequences obtained from nucleic acids, which may also be more informative since they are a more direct indication of genetic content.

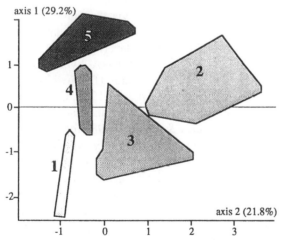

Figure 11. Discriminant factor analysis showing the enzymatic polymorphism of females of the spatangoid *Brissopsis lyrifera* according to their maturation stages (based on gut extracts). Projection on factors F1 and F2 (percentage variance explained by each factor in parentheses); states 1 to 5 are from gonadal inactivity to post-spawning (modified from Féral et al. 1990).

4.2.3 Nuclear DNA

The most common analyses of nuclear DNA interspecies sequence divergences in echinoderms have used the technique of the DNA-DNA hybridization, pioneered by Hoyer et al. (1964) and by Britten & Kohne (1968). This method is based on the formation of heteroduplexes between two species by reassociation of previously separated strands of DNA after removal of repeated sequences. The formed duplexes are then heated to determine their thermal stability which is used as an index of DNA sequence homology and of phylogenetic distance between each pair of compared species. This method has been criticized in a recent controversy concerning primates, which focuses on the accuracy of the different metrics used to determine the thermal stability of the duplexes (see a summary in Lewin 1988). On the other hand, the DNA-DNA hybridization offers the advantage that it utilizes a large part of the enormous complexity of the nuclear DNA (comparisons involve several millions of bases). However, this method could be limited by variation of the rate of evolution of DNA between taxa (Britten 1986), as well as within species (Britten et al. 1978, Grula et al. 1982). In fact, the rate of variation itself does not affect the method (the clock hypothesis is not required, Marshall 1992), but overall rates do affect how deeply in the geological time the method can be used. As a consequence, DNA-DNA hybridization works well for taxa that have diverged relatively recently (i.e. from the species to the family level), but does not provide evidence about distant relationships such as between classes or phyla. Another limitation could be the horizontal transfer of DNA via the incorporation of viral genes (see a review for mammals in Benveniste 1985) or via the mitochondrial DNA (Jacobs et al. 1983). However, this concerns very small quantities of DNA which probably do not affect the validity of the data (Marshall 1988).

Various groups have been investigated by DNA-DNA hybridization, ranging from insects (Hunt & Carson 1983), to rodents (Bromwell 1983, Catzeflis et al. 1987), and to primates (Sibley & Ahlquist 1984, Caccone & Powel 1989, Sibley et al. 1990), but most DNA hybridization studies have concerned the phylogeny of birds (see Sibley & Ahlquist 1983 for a review). Among echinoderms, DNA investigations deal especially with asteroids (Smith et al. 1982), irregular echinoids (Marshall & Swift 1992, for sand dollars) and mainly camarodonts whose classically grounded phylogeny appears quite uncertain because their fossil record is poor (Kier 1977) and which are the most common shallow-water regular echinoids (e.g. see Angerer et al. 1976, Hall et al. 1980, Poltaraus 1981, Yanagisawa 1988, Roberts et al. 1985). However, few attempts have been proposed to reconcile molecular with palaeontological and morphological data (Smith 1988a for camarodonts, Marshall 1991 for sand dollars).

4.2.4 *Mitochondrial DNA*

Mitochondrial DNA is a relatively small molecule which is convenient for analysis of gene sequences by mapping of restriction sites. It corresponds to an approach which provides reliable character data about the taxa compared, while DNA-DNA hybridization gives distance data. Such a method has both advantages and limitations. Mitochondrial DNA is independent of the nuclear genome and does not undergo genetic exhange. It is inherited clonally from the mother line. Consequently, mitochondrial DNA is free of recombination, but it can cross species boundaries after chance hybridization. Lateral transfer may exist in a hybridization area of two species: species of genotype 'A' may possess mitochondrial DNA of species'B' and vice versa (a process observed for the mice *Mus musculus* and *M. domesticus,* Ferris et al. 1983). The mitochondrial DNA exhibits two almost independent molecular 'clocks': 1) it displays a higher rate of silent nucleotide substitutions than in the case of the nuclear DNA; 2) it is a very stable molecule, 'with rearrangements in gene order occuring only with extreme infrequency' (Jacobs et al. 1988, p. 123). It thus appears well suited to establish relationships at the genus-level as well as between groups whose divergence is rather ancient, at least as long as 50 Myr as suggested by Jacobs (1988) for echinoids.

Most mitochondrial DNA studies deal with *Drosophila* and rodents, and examples involving echinoderms are few (Smith et al. 1982). Jacobs et al. (1988) compared an echinoid *(Strongylocentrotus purpuratus),* vertebrates (humans) and *Drosophila,* showing that only two rearrangements of genes have occured in the mitochondrial genome since echinoderm and vertebrate phyla separated. They provided another comparison at a smaller scale which involved four species of regular echinoids belonging to four families: Arbaciidae, Strongylocentrotidae, Toxopneustidae, and Echinidae, representing both superorders Stirodonta and Camarodonta whose approximate divergence time is over 190 Myr (Smith 1981). No differences were found in mapping the three analyzed genes, showing that their arrangement could be characteristic of the echinoid lineage. In addition, the amount of silent nucleotide substitutions between Echinidae and Strongylocentrotidae was estimated at 38%, indicating a divergence time of 32 Myr which is consistent with the data from the fossil record. Recently, mtDNA has been used for inter- and intraspecific comparisons. Restriction analyses lead to estimates of genotypic similarity and diversity between closely related species (Palumbi 1990, Palumbi & Wilson 1990, McMillan et al. pers. comm.) as well as between geographically separated populations (Palumbi & Kessing 1991 for trans-Arctic comparison of populations of *Strongylocentrotus pallidus)* or between morphotypes (Palumbi & Metz 1991 for a measure of the genetic distance separating the four varieties of *Echinometra mathei).* Such analyses illustrate how molecular methods may provide clues to scale evolutionary processes, namely speciation.

4.2.5 *Ribosomal RNA*

Sequence analysis of ribosomal RNA is a relatively recent method which involves different types of RNA molecules: 5S, 18S or 28S. The first attempts (Lu et al. 1980 in echinoids) correspond to the mapping of the small 5S ribosomal RNA molecule since the necessity of a prior end-labelling prevented the use of long molecules. However, the 5S ribosomal RNA molecules are very short, restricting the scope of investigations (Ohama et al. 1983). The mapping of larger ribosomal RNA molecules such as 18S and 28S (following a method developed by Qu et al. 1983) corresponds to the most promising technique. Sequence data from rRNA molecules seem well suited for phylogenetic analyses between distantly related taxa, as well as between closely related ones. Such a range of use is made possible by the particular structure of ribosomal RNA which displays both highly conserved and variable zones. An illustration of a 28S RNA sequence phylogeny in echinoderms is given by the study of the relationships of brooding Antarctic schizasterid echinoids (Féral & Derelle 1991). Analyses are performed with

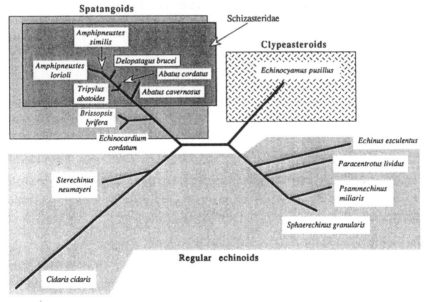

Figure 12. Unrooted phylogenetic tree for fifteen species of echinoids established from partial sequences of the D2 domain of 28S rRNA. The tree adequately separates some groups. It supports the molecular homogeneity of brooding schizasterids in comparison with other echinoids, and illustrates the possibility of separating species of the same genus. On the other hand, the tree does not recognize camarodonts as a natural group as shown by the position of *Sterechinus* close to *Cidaris* (modified from Féral & Derelle 1991). See text for complementary explanations.

the D2 variable domain of the molecule which allows exploration of short range phylogenetic distances. They are calibrated by comparison with non-brooding spatangoids and regular species. The results of the analysis are expressed as a tree showing the topology with the distances between the compared forms (Fig. 12). The tree shows: 1) the isolation of spatangoids; 2) the isolation of brooding schizasterids from other spatangoids; 3) a discrimination between brooding schizasterids both at the genus and species-level; 4) a likely relationship between *Delopatagus brucei* (previously belonging to the polyphyletic family Asterostomatidea) and *Amphipneustes*. These four results are consistent with morphological data and further studies. Taking into account other brooding species, they should allow the question of the origin of brooding in Antarctic spatangoids to be resolved. But some difficulties remain in the topology of the tree: 1) there is no clear differentiation of cidaroids from camarodonts; 2) clypeasteroids appear as the sister group of the set including *Echinus* and related forms. Addition of supplementary taxa should solve these difficulties.

4.3 *Relationships between the five extant classes*

Although echinoderms constitute a very well defined and homogeneous phylum, and although numerous proposals of relationships grounded on stratigraphy (Smith 1988b), morphology (Fell 1963a, Paul & Smith 1984), and embryology (Smiley 1988) have been done, the relationships between the five extant classes are still disputed (Fig. 13A-B). After all these attempts failed to reach a clear consensus, molecular methods have opened a new field of investigation, enhancing rather than resolving the debate.

Comparative analyses of the amino acid composition of collagens of the five extant echinoderm classes show similarities between asteroids and holothuroids, and between ophiuroids and echinoids, while crinoids remain separated (Matsumura et al. 1979). The resulting tree differs thus from former trees based on embryological or cladistic data (Fig. 13C). Trees based on 18S ribosomal RNA confirm the separation of crinoids from other classes (Field et al. 1988, Raff et al. 1988). They indicate a common ancestry for holothuroids and echinoids, as do trees founded on adult morphologies, while ophiuroids and asteroids occupy intermediate positions (Fig. 13D). Analyses of 28S RNA (Hindenach & Stafford 1984, Ratto & Christen 1990) produce an arrangement which differs from that obtained by analysis of 18S RNA by a switch between ophiuroids and asteroids, but which agrees with classical stratigraphic and morphological data (Fig. 13E). In summary, molecular phylogenies provide no common answer to the puzzling problem of the relationships among the five extant classes (except perhaps for the crinoids). It appears there have been too many convergence or reversion events since their divergence time. Moreover Smith (1989), in testing the

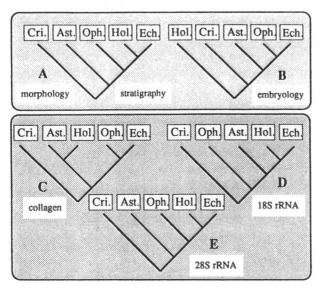

Figure 13. Comparison between different hypotheses of relationships among the five extant classes of echinoderms. Cladogram A illustrates the consensus tree between adult morphological characters (Paul & Smith 1984) and stratigraphical data (Smith 1988b); cladogram B is grounded on embryological characters (Smiley 1988); cladogram C is grounded on collagen C analysis (Matsumura & Shigei 1988); cladogram D is grounded on 18S RNA analysis (Raff et al. 1988); cladogram E is grounded on 28S RNA analysis (Ratto & Christen 1990).

limits of the use of 18S RNA, demonstrated that the question of the origin of the five extant classes could not be answered by this method probably because 'echinoderm classes diverged from one another over such a relatively brief interval of time that few point changes within chain regions or transversions occurred between successive divergences and this weak signal has since been masked through saturation effects' (p. 341).

4.4 *Limitations and prospects*

In conclusion with respect to these molecular approaches to evolution, I would like to emphasize that the comparison of molecular sequences actually involves several hundreds of combinations of the four bases. Molecules are information rich (studies still consider a small part of this great potential) and they should provide in the future more direct evidence of relationships than morphology. Obviously, the comparison of molecular features represents a consistent tool by which to gain phylogenetic information. However, this tool still needs to become technically more accurate in order to improve the

30

reliability of the inferred phylogenies, and it is particularly important to draw attention to the way molecular distances are used to derive trees. Indeed, the appropriateness of the metric used for measuring phylogenetic distances must be discussed (Sarich et al. 1989). In this respect, Farris (1985) has postulated that numerous published results have been grounded on insufficiently reliable calculations which compute trees from a distance matrix whereas parsimony analyses are deemed to be a more effective technique to achieve optimum goodness of fit. Molecular approaches still need to be calibrated accurately by joint studies combining molecular, stratigraphic, and morphological data to obtain critical comparisons of the methods. However, they probably represent an highly promising bet for the future, and will contribute to the knowledge of evolution in echinoderms, from the level of the intraspecific diversity to that of the phylogenetic patterns at high taxonomic rank.

5 THE ONTOGENETIC APPROACH TO EVOLUTION

All the methods dealing with molecules or dealing with morphology (i.e. cladistics) discussed above provide powerful and reliable ways of establishing the relationships between taxa and of reconsidering evolutionary patterns, but they contribute almost nothing to our understanding of the underlying evolutionary processes. The most promising line of research into this aspect of evolution is probably genetics (as expressed by a large number of papers published throughout the issues of 'Evolution') but this is beyond the scope of this paper. Another interesting field of research, allowing evolutionary processes to be addressed at the scale of organisms, concerns ontogeny.

5.1 *Ontogenetic requisites*

In the last few years, there has been an increase in the interest in the study of evolution through its relationships with ontogeny. Progress has come in the understanding of genomic structure (Britten & Davidson 1969, McClintock 1984) and of the genetics of development (Bonner 1982, Raff & Kaufman 1983). These advances have led to the idea that ontogeny has an important role in structuring evolution. This tendency might be regarded as a renewal of concepts rooted as far back as the early nineteenth century, notably recapitulation as illustrated by Louis Agassiz's (1849) explanation of the diversity of the life (he advocated that the divine plan of creation resulted in a threefold parallelism between adults of extant taxa, ontogenetic stages, and series of fossils).

Nevertheless, the recent tendency is on no account a simple reversion to the nineteenth century evolutionary philosophies (e.g. von Baer's laws of

development 1828, or Haeckel's biogenetic law, 1866). The renewal constitutes the incorporation of the concepts of a complex and hierarchically
organized ontogeny into the paradigm of the modern synthesis (including
adjustments required by theories such as punctuated equilibria). Gould
(1977, 1982) and subsequent authors (see Alberch et al. 1979, McNamara
1982a, Dommergues et al. 1986, McKinney 1988 for basic ideas) showed
how ontogeny can be a reliable and powerful key to understanding the
diversity and evolution of groups, and specifically how changes in developmental timing – which is termed heterochrony – can be involved in evolutionary processes. Such an approach is rooted in the idea that morphogenetic
processes are highly interactive, and that interactions create developmental
constraints. These constraints result in a discontinuous phenotypic space:
possible phenotypes tend to cluster around 'steady states' corresponding to
bounded domains of morphological variation (Alberch 1982, 1989, Jacob
1981). Developmental pathways are generally canalized. Consequently,
some phylogenetic changes appear channeled by inherent properties of the
evolving system itself, more than by the external selection. 'Epigenetic
interactions (...) impose directionality in morphological transformations
through phylogeny.' (Alberch 1980, p. 654). In this perspective, the evolving
organisms are not only passively submitted to natural selection, but partly
control their own evolution as ontogenetic constraints limit the emergence of
novelties which are then submitted to natural selection (Devillers 1985). All
this implies that evolutionary changes are channeled both by constructional
necessities (i.e. internal constraints, Laurin & David 1988) and by ecological
stresses (i.e. external constraints, Grime 1989). Evolution follows, at least
partly, an underlying logic. The deciphering of heterochronies amounts to
discovering this logic (i.e. ontogeny provides important clues to evolutionary
paths).

Such a conceptual background has determined the concomitant development of practical rules allowing the inference of evolutionary processes from
the analysis of morphological variation. The basic principle is to establish
and then to compare ontogenetic trajectories. The successive states crossed
by an organism throughout its development are plotted in a three dimensional
space involving size, shape and age, to generate an ontogenetic trajectory.
Heterochronies are modifications of trajectories registered when comparing
two forms, one ancestral and the other descendent. Schematically, six major
heterochronic processes are possible, corresponding to pure alterations of
each standard (Fig. 14). In the first three heterochronies, the ancestral
juvenile feature is retained by the descendent adult: these heterochronies
express paedomorphosis, and correspond to neoteny (slowing of change in
shape), progenesis (early attainment of sexual maturity), or post-
displacement (retardation of the onset age of the character). In the other three
heterochronies, the descendant exhibits new hyperadult features: these hete-

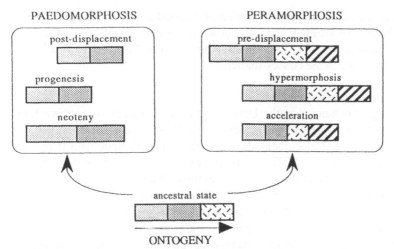

PAEDOMORPHOSIS   PERAMORPHOSIS

post-displacement   pre-displacement

progenesis   hypermorphosis

neoteny   acceleration

ancestral state

ONTOGENY

Figure 14. Main ontogenetic heterochronies. The horizontal rectangles represent the ontogenetic pathways of a supposed ancestral group (bottom line) and of evolved groups (boxed); the length of the rectangles is proportional to adult size and the different stipplings illustrate successive ontogenetic stages. The descendant morphologies are grouped into those resulting from paedomorphosis (the descendant retains juvenile characters of the ancestor) and those resulting from peramorphosis (the descendant attains new features) respectively.

rochronies express peramorphosis (meaning 'beyond the shape') and corre-spond to acceleration (acceleration of change in shape), hypermorphosis (delayed sexual maturity), or pre-displacement (earlier onset age).

5.2 *Ontogenetic analyses in echinoderms*

Providing that growth stages are available, an ontogenetic approach to evolution can be adopted for virtually all echinoderm groups, extant or fossil, and numerous examples of heterochrony have been described for echino-derms. But probably because of the capricious interest of researchers, studies are not evenly distributed among the classes. Some groups have been neglected (carpoids, holothuroids), or scarcely analyzed such as regular echinoids (Smith 1984a), ophiuroids (Hotchkiss 1980, Vadon 1990), aste-roids (Breton 1990), and edrioasteroids (Bell 1976, Sprinkle & Bell 1978). Conversely, most of the studies are focussed on irregular echinoids due to their relative abundance and widespread distribution (McNamara & Philip 1980, McNamara 1982b, 1985, 1989, McKinney 1984, Beadle 1989, David 1990, Mooi 1990b), and to a lesser extant on blastoids (Brett et al. 1983, Waters et al. 1985) and on crinoids (Roux 1977, 1978, Eckert 1987, Brower 1990, Maples et al. 1990). These examples reveal that different heterochronic

33

processes are involved in the evolution of echinoderms, (McNamara 1988). Nevertheless the available data do not allow an assessment of whether the relative frequencies of peramorphosis and paedomorphosis might be a function of the nature of the classes studied.

5.2.1 A peramorphic trend

The strangest echinoids of the Upper Cretaceous chalk from northwestern Europe belong to the genera *Infulaster* and *Hagenowia*. The evolution of these groups is well documented by accurate stratophenetic data (Ernst & Schultz 1971, Schmid 1972). The main lineage illustrates a gradual evolutionary transformation throughout several intermediate species ranging from the Turonian to the Campanian (Fig. 15). The Turonian *Infulaster excentricus* is the oldest member of the lineage. It possesses a conical test, anteriorly accuminate, with a sharp and deep anterior groove which leads to the mouth. All of these features are increasingly developed in the next species, *I. tuberculatus,* which is also conspicuously smaller. During the subsequent evolution of the lineage, the rostrum becomes more and more slender and

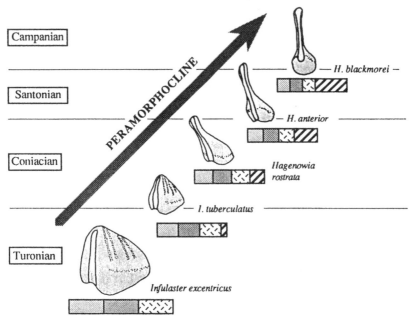

Figure 15. Example of a peramorphic trend: evolution of the lineage *Infulaster – Hagenowia* during the Upper Cretaceous. The progressive elongation and straightening of the apical rostrum along the lineage results from an acceleration of development and the trend corresponds to a peramorphocline (modified from Dommergues et al. 1986). See Figure 14 for explanation of the symbols.

34

vertical, the apical system is separated into a bivium (posterior) and a trivium (anterior), the mouth shifts forwards on the anterior face, and the groove becomes narrower and enclosed by spines. All of these morphological changes can be related to an allometry involving the meridian growth of plates, particularly evident after the apex has been separated into two parts. The architectural repatterning of the trivium is greatly involved in this process. Indeed the extension of the rostrum is the result of a great elongation of the plates combined with loss of other plates and with exclusion of entire plate rows from this part of the test (Gale & Smith 1982). The increasing allometry throughout the lineage and the related plate repatterning lead to hyper-adult morphologies which resulted primarily from an acceleration or a pre-displacement process. Such a directional morphological evolution can be considered as a peramorphocline (sensu McNamara 1982a).

But all these morphological changes can also be related to adaptive changes as suggested by Gale and Smith (1982), who interpreted the trend as an adaptation to a new feeding technique and to a new style of burrowing. Both *Infulaster* and *Hagenowia* were infaunal dwellers, burrowing just beneath the surface layer. But *Infulaster* collected the bulk of its food from the sediment close to the peristome, while *Hagenowia* probably fed only on sediment from the surface layer, using its anterior groove as a food collector. Moreover, the outline of *Hagenowia* suggests that the track left by the moving individual on the sea floor was less conspicuous than that of *Infulaster,* thus allowing it to escape predation. These observations imply an adaptive interpretation of the peramorphocline, but cannot be clearly related to changes in the nature of the sediment, the chalk facies apparently remaining uniform during the period involved.

5.2.2 *A paedomorphic trend*
A reverse example of a paedomorphic trend can be documented in spatangoids from the Tertiary of Australia (McNamara 1987). Five species of *Hemiaster (Bolbaster)* succeed one another, without overlapping stratigraphic ranges, from the Late Eocene to the Middle Miocene. They are regarded as members of a single lineage by McNamara. Thirteen directional morphological trends along the lineage, involving the general shape of the test, the petals, the fasciole, and the plastronal area, have been documented. Comparison with the ontogenetic changes recorded in the Palaeocene species *H. targari,* recognized as ancestral, allows the type of heterochronic transformation of the trends to be assessed (Fig. 16). Of the analyzed morphological trends, the greater number display increasingly juvenile characters in the descendent adult, and could be achieved through a neotenic process (e.g. the lateral profile of the test as illustrated on Figure 16). The lineage is thus considered as a paedomorphocline.

Analysis of the sediment in which the five *H. (Bolbaster)* species lived

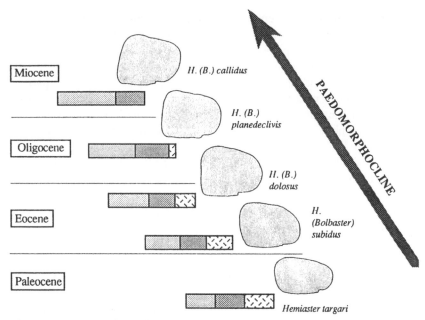

Figure 16. Example of a paedomorphic trend: evolution of the lineage *Hemiaster targari* – *H. (Bolbaster) callidus* in the Tertiary of Australia. The changes in lateral test profile result from neoteny and the trend corresponds to a paedomorphocline (see Figure 14 for explanation of the symbols) probably driven by a reduction in sediment grain size (modified from McNamara 1987).

reveals that the paedomorphocline is concomitant with a progressive decrease in the sediment grain size. This environmental gradient can be correlated to spatangoid morphofunctional data (Smith 1980, 1984a), and consequently led McNamara to interpret the morphological changes recorded along the lineage as reflecting adaptation for life in finer grained sediment.

5.2.3 *Analysis of a diversity: The deep-sea holasteroids (part 1)*
The two former examples illustrate how heterochronies can help in understanding the dynamics of morphological change along a lineage more or less considered as a gradual trend. They give information about an evolutionary mechanism, but they do not provide a critical appraisal of the phyletic link between the successive species, which remains basically grounded on stratigraphy. The contribution of ontogenetic changes to the study of evolution can be expanded to the search for diversity by using heterochrony to reveal relationships within groups. This will be illustrated by a third example dealing with the diversity of an extant group, and will also provide the

36

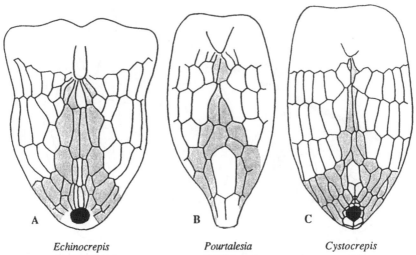

Echinocrepis   *Pourtalesia*   *Cystocrepis*

Figure 17. Plate pattern and shape of the adoral side of *Echinocrepis cuneata* (A), *Pourtalesia alcocki* (B), and *Cystocrepis setigera* (C) illustrating different degrees of disjunction of interambulacrum 5 (posterior ambulacra are stippled).

opportunity to develop a methodological account of how to conduct this type of study.

Holasteroids are an order of irregular echinoids whose earliest representatives were epicontinental forms from the Lower Cretaceous, and whose recent forms are successful deep-sea inhabitants. They constitute a monophyletic group which offers a very large diversity of forms, and which includes 55 genera belonging to five families (adapted from Wagner & Durham 1966, and Foster & Philip 1978). Among these families, the Pourtalesiidae constitute a small monophyletic grouping (David 1985), including seven genera which exhibit highly transformed architectures and shapes (Fig. 17A-C). Such morphologically bizarre deep-sea animals are ideal for retracing ontogenetic trajectories, and for observing how morphological diversification proceeded within the group. Numerous morphological features are involved in the differentiation of the family, but for sake of clarity, I shall focus on a single feature chosen as typical (for detailed analyses see David 1987, 1990). This feature is the architecture of the central part of the ventral side of the test, i.e. the plate array of the plastronal area. Plastronal architecture offers the advantage of displaying major transformations giving an almost complete scope of growth, from the newly metamorphosed juvenile to the adult.

5.2.3.1 *An ontogenetic model.* The first step in this example is devoted to the analysis of one species, *Pourtalesia miranda,* with the aim of establishing a

37

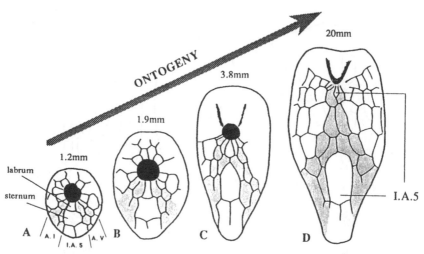

Figure 18. Main ontogenetic stages of *Pourtalesia miranda* (plate pattern and shape of the adoral side): A) post-metamorphic juvenile with a plate pattern typical of holasterids; B) juvenile with a reduced, narrow boundary between labrum and sternum C) juvenile showing a single architectural separation of interambulacrum 5 (I.A. 5); D) adult with a three-plate separation of interambulacrum 5 (posterior ambulacra are stippled).

model of ontogeny. It is possible to observe the plate array of the plastronal area of this species at a body size of 1.2 mm. At this size, the plate array is almost regular (Fig. 18A). The odd interambulacrum (5, according to Lovén's rule) is continuous, and the posterior ambulacra (I and V) exhibit a regular, alternating pattern. The next growth stage illustrates the same architectural state, but the boundary between the two first plates of interambulacrum 5 (the so-called labrum and sternum) is slightly narrower (Fig. 18B). Allometry of growth rates between the plates leads, at a body size of 4 mm, to a separation of the labrum and sternum by the adjacent ambulacral plates which grow in the direction of the axis of symmetry until they meet (Fig. 18C). A single architectural break exists in interambulacrum 5. During the following stage, another pair of ambulacral plates meets along the axis of symmetry. At a body size of 8 mm, the test achieves an architectural pattern with a two-plate separation, corresponding to the basic adult condition. However in adults (at a body size of ca. 25 mm), supplementary disjunctions can be induced by the insertion of other ambulacral or also laterointerambulacral plates (Fig. 18D) leading to three-plate or four-plate separations.

5.2.3.2 *Heterochronies in Pourtalesiidae.* The second step of the example is devoted to the description of heterochronies which have led to the diver-

38

sification of the whole family Pourtalesiidae. In doing this, it is necessary to establish an inventory of all the states taken by the architecture of the plastronal area in adults of other genera within the family. They all share the same basic architectural plan, but some genera go further than others in the transformation described above. For instance, in *Echinocrepis* the architectural break of interambulacrum 5 is limited to the junction of one pair of ambulacral plates (Fig. 17A), while in *Cystocrepis* the break expands to three pairs of ambulacral plates (Fig. 17C). From such observations, it is possible to arrange sequentially the six configurations of this feature seen in the different genera of the family. The logic of this arrangement is achieved by reference to the ontogenetic results given by the study of *P. miranda* (Fig. 19). This succession defines a morphocline which ranges from a continuous interambulacrum 5 (state A on Fig.19) to an interambulacrum 5 interrupted by three (states D and E) or four series of plates (state D + E). The different genera are distributed along the morphocline according to a precise order. It is possible to polarize this order, and the morphocline, by comparison with other holasteroids showing the ancestral condition of the feature (i.e. by using an outgroup criterion). In doing this, the trend appears peramorphic, and from

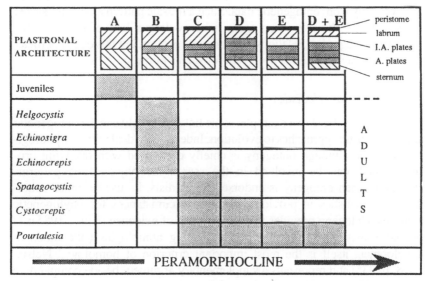

Figure 19. Morphocline of plastronal architecture in Pourtalesiidae. Distinction among six discrete states (upper row) sequentially arranged and lettered from A to D + E. The usual states recorded in adults are denoted for each genus by the shaded areas (excepted for *Ceratophysa* which is insufficiently known). The morphocline is considered peramorphic by outgroup comparison (state A is characteristic of the outgroup condition) (redrawn from David 1990).

the relative size of the genera, it is possible to suggest that the heterochonic process involved in this peramorphic trend is acceleration. So, the different genera can be classified according to their evolutionary level, and an evolutionary process can be proposed to explain their diversity. On such bases, it may be theoretically possible to construct a cladogram by using the rule that the more advanced the state of the plastron architecture, the greater the degree of apomorphy. *Helgocystis, Echinosigra* and *Echinocrepis* appear thus as the most primitive, while *Pourtalesia* appears to be the most derived genus of the family. However, when the analysis is expanded to features other than the architecture of the plastron, the order of the genera along morphoclines is observed to vary greatly according to the feature examined, even though all of the analyzed morphoclines are peramorphic. For instance *Pourtalesia,* which occupies the most derived position within the cline of architecture (see Fig. 19), occupies a rather conservative position within the clines of the ambitus and back profile of the test. This reflects a mosaic pattern of heterochronies within the family, and each genus can be viewed as an heterogeneous combination of more or less contradictory trends. The general evolution of the family is the result of a rather intricate history implementing a differential distribution of acceleration, that depends both on the genera and on the characters. There is much homoplasy (reversion and convergence), and no consensus phylogeny of Pourtalesiidae can be fully achieved only from ontogenetic data. Such a mosaic pattern also emphasizes a limitation of the use of ontogeny in providing a line of evidence for assumptions regarding relationships for some groups, but this does not mean that the approach in itself is irrelevant for such a purpose.

5.3 *Ontogeny and cladistics*

As indicated by the above example, some links exist between the ontogenetic and the cladistic approaches to evolution. Indeed, both deal with morphological data, and although ontogeny is chiefly concerned with resolving processes, both can be involved in the search for phylogenetic patterns. However, when ontogeny is endorsed by cladists, its use becomes subordinate to cladistics in providing facts or circumstances for phylogenetic analyses. Ontogeny provides three basic lines of evidence (Kluge 1985): 1) ontogeny may be a criterion for establishing homologies ('the mode of development itself is the most important criterion of homology,' Nelson 1978, p. 335) and represents a supplementary argument beside comparative anatomy; 2) 'ontogenetic transformation series may serve as a source of raw character-state data, in addition to those traditionally recorded from the adult stage' (Kluge 1985, p. 14); 3) ontogeny may be used to order the states taken by a character (i.e. to define a morphocline) and to polarize character transformations. This third use of ontogeny is the most salient in cladistics

40

and it is the only one that involves ontogeny as a truly dynamic process. But it is also intimately entangled with the problem of an unambiguous recognition of the underlying heterochronic processes responsible for the transformation. Indeed, the diversity of heterochronies 'is claimed to provide too many exceptions to argument linking ontogenetic pathways with decision on polarity' (Novacek 1987, p. 188). As a result the legitimacy of this third use is controversial (debating exchanges of views are stated in Nelson 1978, 1985, Beatty 1982, Fink 1982, Voorzanger & van der Steen 1982, Brooks & Wiley 1985, and Kluge 1985).

The expansion of the example of the Pourtalesiidae to include other families of the order Holasteroida provides an illustration of that third use of ontogenetic criteria in a cladistic context.

5.3.1 *An analysis of diversity: The deep-sea holasteroids (part 2)*
When pursuing the analysis of the architecture of the plastronal area, and considering the states taken by this feature throughout all the holasteroid genera, it becomes possible to distinguish a basic plate pattern from other, more particular patterns (David 1988). The basic plastron arrangement is called meridosternous and corresponds to the plesiomorphic condition (Fig. 20A). From the posterior border of the peristome it displays successively: the labrum, two sternal plates of which only one meets the labrum, and a series of plates that alternate as usual. Among the other known plate patterns, several can be organized into morphoclines grounded on ontogenetic trajectories, corresponding to different apomorphic conditions. For instance, the structure of the plastron can be modified towards a uniserial plate array (metasternous condition, Fig. 20B). The plates extend and occupy the whole width of interambulacrum 5, the sutures becoming transversal. The process begins adorally, and extends backwards including more and more plates. In another modification of the plastron a large unpaired sternum (Fig. 20C) becomes characteristic of the so-called orthosternous condition. The basic orthosternous condition can itself be subdivided into two more derived patterns. One corresponds to the addition of a new, small, central plate: the rostral plate (Fig. 20C-I); the other forms a trend leading to an increasingly disjunct interambulacrum 5 and corresponds to the morphocline described above in Pourtalesiidae (Figs 19, 20C-II). All of the recorded morphoclines can be polarized as peramorphic by using out-group criteria, and they gain the status of progressive apomorphies for cladistic analysis. A consensus cladogram can then be obtained when other features are added and analyzed in the same way (Fig. 21). Despite its lack of congruence within the Pourtalesiidae (see section 5.2.3.2), the plastron architecture is now congruent enough with other features and fits well with the consensus cladogram at the scale of the entire order Holasteroida, even if two small discrepancies remain: first, the Urechinidae are split into two subgroups according to this feature; and second, the

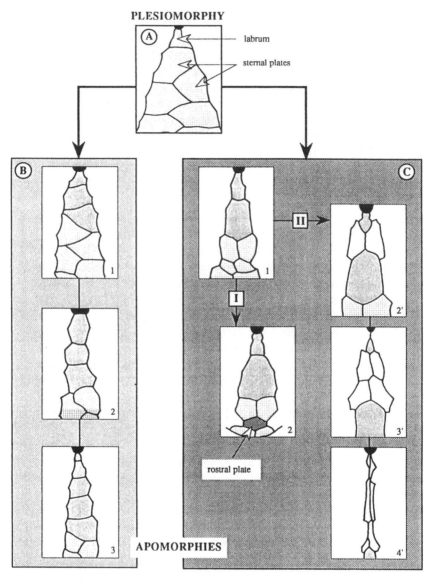

PLESIOMORPHY

labrum

sternal plates

APOMORPHIES

Figure 20. Diversity of plastronal architecture in the order Holasteroida. Plesiomorphic (A) and apomorphic (B-C) conditions: A) basic meridosternous plastron of *Holaster intermedius* (from Lambert 1893); B) metasternous plate patterns of *Sternotaxis icaunensis* (B1) (from Raabe 1964), *Pseudananchys credneriana* (B2) (from Elbert 1901), and *Cardiotaxis heberti* (B3) (from Ernst 1972); C) orthosternous plate patterns of *Urechinus naresianus* (C1), *Corystus disasteroides* (C2) (from Foster & Philip 1978), *Plexechinus spectabilis* (C2′) (from Mortensen 1950), *Pourtalesia miranda* (C3′), and *Cystocrepis setigera* (C4′) (from Agassiz 1904). Peristome in solid black, interambulacrum 5 shaded; in B and C the darkest plates exhibit apomorphic patterns (modified from David 1988).

42

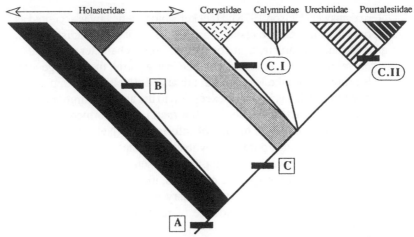

Figure 21. Consensus cladogram of relationships of the main groups of Holasteroida. Apomorphic groups are represented by triangles, and the plesiomorphic sets (not fully resolved) by trapezoidal areas. The cladogram is simplified from David (1988) and the taxonomic units refer to classical family names taken from Smith (1984a), amended by Foster & Philip (1978); the Stenonasteridae are not included in the cladogram because they are insufficiently known, and the Somaliasteridae are not considered members of the Holasteroida, so they are omitted. The architectural apomorphics of the plastronal area are indicated (lettering refers to Figure 20).

Calymnidae, although considered as an independent group on the basis of other features, remain unresolved by plastronal architecture. This example indicates that it is possible to propose hypotheses of relationship using ontogeny as the basic guideline for the interpretation of character polarities. The previous sentence is rather comforting, and I am rather confident in such a use of ontogeny. However in the previous example, things were made easier because the trends involved were all peramorphic and thus in accordance with the law of differentiation of characters on which the cladistics is grounded. Subsequently, this raises the question of how paedomorphosis can be related to cladistics.

5.3.2 *Paedomorphosis in cladistics*
A basic contradiction exists in the definition of paedomorphosis and its recognition in a cladistic way, and it is generally considered that paedomorphosis blurs the knowledge of relationships between taxa (Stevens 1980). Indeed, from an ontogenetic point of view, paedomorphosis should correspond to a derived state, while in a pure cladistic sense paedomorphosis should result in morphologies which may be interpreted as being primitive (Fig. 22). This conflict may be resolved in two different approaches: internalist and externalist.

43

An internal answer (Nelson 1978): Haeckel's biogenetic law is accepted as a general rule. This law can be restated as follows to fit with cladistic aims: 'given an ontogenetic character transformation, from a character observed to be more general to a character observed to be less general, the more general character is primitive and the less general advanced' (Nelson 1978, p. 327). In this case paedomorphosis is just an exception to the above definition, since it is assumed that characters are not lost but merely transformed, 'perhaps neoteny is (...) only a reflection of lack of information' (Nelson & Platnick 1981, p. 353). Consequently, ontogeny by itself becomes sufficient to polarize character states, but always in a recapitulatory sense ! On the contrary if we accept the diversity of ontogenetic pathways, including paedomorphosis as a real and not particularly rare evolutionary process, ontogeny in itself cannot provide an unambiguous evidence of relationships, and additional data are required.

An external answer (Eldredge & Novacek 1985, Kluge 1985): the generalization of the biogenetic law is not accepted and reference is made to non-ontogenetic arguments to offset the uncertainty of the ontogenetic criterion. The answer may be given by comparative anatomy (Eldredge & Cracraft 1980). Indeed, outgroup comparisons can be used to polarize morphoclines, and thus to detect paedomorphosis (in a way similar to that used to demonstrate peramorphosis in the example of Pourtalesiidae described above). Multiple character congruence (grounded on parsimony

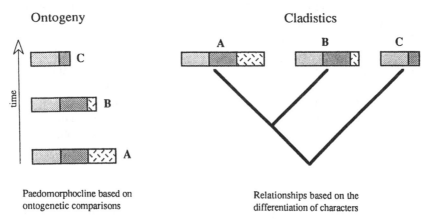

Figure 22. Antagonism between paedomorphosis and primitiveness. A, B and C are three taxa which can be sequentially arranged according to the degree of completeness of their ontogeny (explanation of the symbols is given in Figure 14). Two opposite hypotheses of relationship could be proposed depending whether priority is given to ontogeny (implemented by outgroup comparisons and stratigraphy) or to adult morphology. The ontogenetic criterion allows C to be considered as a derived paedomorphic form, while pure cladistics supports C as being primitive.

analyses of other supportive characters) can advocate hypotheses of partial paedomorphosis involving some traits (Fink 1982). But comparative anatomy can only work as far as paedomorphic transformations are not complete and do not concern all of the somatic characters (Kluge 1985). In fact, in the absence of a mosaic pattern of heterochronies, no decision will be made on cladistic grounds and no morphological detection of paedomorphosis is possible. Therefore, the answer should be founded in data taken from the frame of evolution. Palaeontology could provide stratigraphic evidence to propose an hypothesis of general paedomorphosis (progenesis) when the presumed paedomorphic taxon is the most recent of the lineage (see the *Hemiaster* lineage described above). General paedomorphosis of one taxon could also be deduced from an incongruency between its spatial distribution and the biogeographic patterns recorded in other independant groups (Janvier 1989).

Fortunately, real organisms are generally more complex than the rec-

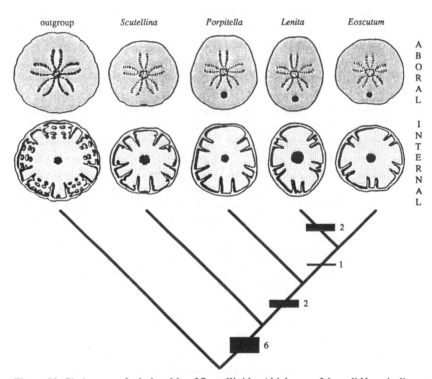

Figure 23. Cladogram of relationship of Scutellinidae (thickness of the solid bars indicate the number of paedomorphic features mapped onto the suggested topology). Comparison with the outgroup allows the observed pattern of relationship to be interpreted as part of a paedomorphocline (modified from Mooi 1987).

tangles of Figure 22, and possess a number of features that conveniently resolve the problem of paedomorphosis. The analysis of the irregular echinoid family Scutellinidae, comprising *Scutellina, Porpitella, Lenita,* and *Eoscutum* (Mooi 1987) provides an example of a preliminary cladistic analysis of such a problem. If the topology of the cladogram can be corroborated by future analyses, the morphological comparison of the four scutellinid genera with an outgroup allows interpretation of a paedomorphocline (Fig. 23). Several increasingly paedomorphic features can be mapped onto the tree. Test size generally decreases, internal buttresses become more simple, petaloid areas become smaller, and the periproct becomes more centrally located on the aboral surface. Cladistic analyses using characters independent of the heterochronic trend to be studied should permit accurate identification and placement of events in both pera- or paedomorphoclines.

5.4 *Conclusions about ontogeny and prospects*

Ontogeny can be used in two ways: as a key to understanding evolutionary processes and as a source of information when searching for patterns of relationships. The former use resorts to the description of morphological results to detect heterochrony. Obviously this research into heterochronic processes is fundamentally based on a prior recognition of an ancestor to which changes in the developmental timing of the descendant could be compared. Even though an ancestral development program may be sometimes detected by stratigraphic data, a phylogenetic hypothesis of the relevant taxa is a prerequisite for further attempts to detect heterochrony (Fink 1988). Heterochronies should reasonably be sought after cladistic analyses have established clear patterns of relationships. The latter use considers heterochronies as basic arguments to define apomorphies (see above the example of Pourtalesiidae). In this case ontogeny comes uppermost in the study, and it must be referred to outgroup comparisons. Reference to the fossil record can also be used to detect heterochronies before phylogenetic analysis is undertaken. Such an approach can be criticized from a strictly cladistic point of view (the pattern cladists consider heterochronies as ad hoc hypotheses and reject the practice of *a priori* recognition). This is part of the debate between Nelson (1973, 1978) and Gould (1973, 1977). Nevertheless, with the caveat for careful use of the methods, I would like to defend these two implementations of ontogeny as necessary for a complete understanding of evolutionary patterns and processes.

In addressing new perspectives in the field of ontogenetic studies, I shall call attention to a practical problem. As stated in the introduction of this section, the detection of heterochrony implies a formal description of three parameters: age, size, and shape. When available, age and size data can be derived from raw observations without complicated quantification. On the other hand, shape is always a complex parameter that requires elaborate techniques when more than one feature has to be quantified. Classically, information about shape is derived from statistical analyses performed on various measurements of specimens (i.e. factor analyses). More sophisticated descriptions of shape are now available. They primarily comprise outline analysis by Fourier series (Waters 1977 for blastoids), and representations based on the use of landmarks (characteristic points digitized on the specimens). The latter leads to the use of powerful methods which allow quantification of shape in terms of truss networks (Fink 1988) or the visualization of shape changes by vector fields, maps, and grids (David & Laurin 1989). Such technical improvements provide very precise spatial views of allometries and thus represent a promising tool for future analyses of the relationship between ontogeny and phylogeny. The landmarks technique seems particularly relevant in echinoderms whose test architecture exhibits numerous characteristic points that are easy to define (see Laurin & David 1990 for an example using echinoids). Other promising developments for ontogenetic approaches to evolution should probably emerge from studies at the scale of cellular biology: studies of cell lineages (Davidson 1988 for echinoids) including molecular and cellular analysis of heterochronies (Parks et al. 1988 for the direct developing echinoids *Heliocidaris erythrogramma),* or research into the biochemical controls of heterochronic processes (see Raff & Kaufman 1983 for a short review).

6 THE BIOGEOGRAPHIC APPROACH TO EVOLUTION

6.1 *Methods and principles*

Besides stratigraphy, other extrinsic data are available to deal with evolutionary problems. They can be found in the spatial distribution of organisms. At first sight, it seems unlikely that space (biogeography) can provide some direct evidence of phylogenetic relationships. This feeling is partly due to the fact that biogeography has generally been little involved as a primary argument in evolutionary studies despite the extensive data available.

The patterns of spatial distribution of organisms are complex phenomena resulting from many factors which are strongly subordinate to scale effects in time and space (Myers & Giller 1988a). Although quite a continuum exists, they can be conveniently subdivided into two main sets: large scale historical factors which are closely related to the geological history of the area concerned, and ecological factors which depend on the fitness between organisms and their environment. This twofold aspect of the distribution of organisms allows the definition of two categories of biological communities: biotas which correspond to communities sharing a common geological

history, and biomes which include taxa sharing similar ecological requirements. For instance, the Recent inter-tropical marine fauna constitutes an
ecological unit which could be considered as a biome, while the South
American fauna represents a historical unit comprising a biota. The problem
of the spatial distribution of groups throughout these two types of biogeographical units has given rise to various approaches which can be conveniently
explained in two main ways (see Dommergues & Marchand 1988 for a brief
review of the approaches). The first way is perceived as 'narrative,' and refers
to case by case studies generally grounded on empirical observations. The
second way is recognized as 'analytical' and refers to formal methods for
suggesting hypotheses about how patterns of distribution are achieved.

Due to their abundance and wide distribution in all the seas of the world,
and because they occur in almost all the marine environments from the
deep-sea to hypo- or hypersaline shore waters (Stickle & Diehl 1987),
echinoderms have been frequently involved in biogeographic works.
Examples include both Recent and fossil forms, but most deal with the
narrative approach.

6.2 Narrative biogeography

In this type of traditional approach, causal explanations of the distribution of
taxa are deduced from the raw data without the assistance of an unambiguous
research method, even if quantitative components may be concerned. Within
this field of research, a continuum exists which links ecological to historical
biogeography. It is thus possible to propose a diagrammatic array varying
between very large scale attempts that are purely descriptive with no explana-
tory aims (Eckman 1953), equally large scale studies performed on historical
bases, and smaller scale ecological studies.

6.2.1 Historical narrative biogeography in echinoderms
The aim of historical narrative biogeography is to explain how large scale
events have shaped the spatial distribution of groups using scenarios
grounded on hypotheses concerning centres of origin and dispersal patterns.
These hypotheses are usually established from maps of taxa in space and time
which are plotted in relation to the arrangement of oceanic areas. In echinoderms, such studies concern extant forms (Mironov 1983, 1985), but more
frequently they deal with palaeontological data which offer the advantage of
allowing a historical dimension to be introduced (Breton 1990 for Mesozoic
asteroids, Oji 1990 for Cenozoic crinoids, Waters 1990 for blastoids). But
echinoids, as they are abundant in Mesozoic and Cenozoic deposits (two eras
for which the geomorphological changes of the oceans are well known), have
been the most often involved: regulars (Durkin 1980) as well as irregulars
(Devriès 1960, Rey 1972, Foster & Philip 1978, Seilacher 1979, Ali 1983,

Roman et al. 1989). The reconstruction of dispersal centres and pathways of migration are generally viewed on the geological scale in comparison with large-scale events such as continental rifting and collision, opening and isolation of oceanic basins or eustatic sea level variations. Nevertheless the results may be reliable especially when accurate plate tectonic data are included. A representative example of such attempts can be illustrated from the Cenozoic story of the genus *Echinolampas* (Roman 1977). From a centre of origin in north-east Africa (Paleocene of Egypt), the genus spread throughout the Tethys sea during the Lower Eocene, before reaching Asia and the tropical part of America during the Middle Eocene (Fig. 24). After a short recess, caused by a general cooling of the sea water (Upper Eocene), *Echinolampas* dispersed again in the Oligocene with some species reaching

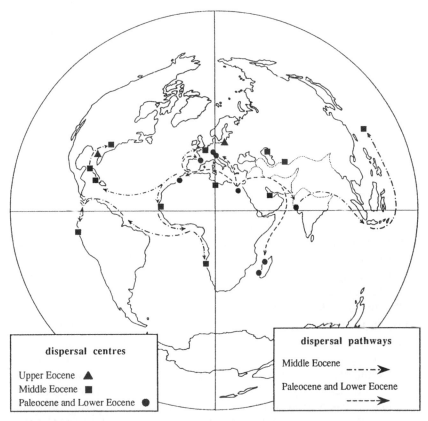

Figure 24. Historical narrative biogeography in echinoids. Map showing the main areas of distribution (symbols) and dispersal pathways (arrows) for the genus *Echinolampas* from the Paleocene to the Upper Eocene (modified from Roman 1977).

Australia and New Zealand. The end of the story (Miocene to Recent) corresponds to a progressive decrease in the area occupied by the genus. Roman explained the migratory processes which occurred during the first part of the history by reference to the position of European, American and Australian plates. He suggested that larvae may have been transported from Europe to America by equatorial currents across the Atlantic (not as wide then as it is now), and that the late colonization of Australia from the Indo-Malayan area may have come about through the continental drift of Australia and a supposed oceanic current. This example shows how tectonic events can be used to establish scenarios interpreting biogeographical distributions of groups.

However in this type of study, dispersal hypotheses are untestable because they address individual case explanations (Rosen 1978). Any strict method precludes the establishment of dispersal pathways which at best represent a particular hypothesis, among a wide range of other equally plausible hypotheses (for instance, on Figure 24, consider the migratory path from Europe to South America). Moreover, the historical aspect is never directly inferred from the biogeographic data, but is deduced from the stratigraphic framework. Biogeography only provides complementary data to phylogenies primarily established by classical (e.g. stratophenetic) studies.

6.2.2 Ecological narrative biogeography in echinoderms

When the scale of observation is reduced, the effects of ecological factors, involving biotic processes, become more and more evident and can decrease the importance of the historical dimension. Such studies are particularly frequent for Recent forms since ecological parameters are potentially available from direct field records. It thus becomes possible to compare the spatial distribution of echinoderm groups with changes in environmental characteristics such as salinity (Pagett 1980, Himmelman et al. 1984), hydrodynamics (Ferber & Lawrence 1976, Harold & Telford 1982), depth (Nichols 1980), or the organic content of the sediment (Sibuet et al. 1984, Sibuet 1985). Studies involving purely biotic factors (e.g. life-history characteristics) also contribute to the explanation of the distribution of organisms over the world (Guille & Albuquerque 1990 for ophiuroids, Lawrence 1990 for a review in echinoderms). In addition, comparisons between faunistic assemblages from different areas may be performed in order to establish patterns of relationships between geographic provinces. Sibuet (1979), for example, has used diversity and similarity indexes derived from asteroid faunas to identify the main abyssal basins across the Atlantic. Rowe (1985) has discussed the problem of the origin of the Australian tropical echinoderms by comparing different provinces. He suggested that the Indo-Malay region cannot be considered as a centre of origin, but that a Pacific origin could be more easily envisaged.

Adding a historical dimension, palaeontological data allow echinoderm distribution to be considered simultaneously in time and space. On the other hand, they offer less precise ecological data because the environmental parameters are not directly available, but have to be deciphered from the associated lithofacies. The evolutionary history and distribution of the groups are then analyzed with respect to changes in geological and sedimentological conditions through successive time planes. Thierry (1985) thus considered the evolution of collyritid echinoids in the epicontinental sea covering the Paris Basin during the Middle and Upper Jurassic. Studies have also been performed for numerous other groups and times: Carboniferous blastoids (Waters & Sevastopulo 1985), Cretaceous echinoids (Rat et al. 1987, Néraudeau & Moreau 1989), Paleogene clypeasteroids (Mooi 1990a), as well as Middle Jurassic (Meyer 1990) and Lower Cenozoic stalked crinoids (Roux & Plaziat 1978). An example, covering all aspects of the relationship between different sedimentary environments and the distribution of species, is given by the analysis of the spatial distribution of the spatangoid echinoid *Hemiaster* in northern Spain during the Upper Cretaceous (Néraudeau 1990). Various species of *Hemiaster* are found in the Cenomanian, Turonian and Coniacian deposits on the north Castillan platform from onshore deposits located on the southern margin of the platform to the offshore slope and basin extending into the Pyrenean trough (Fig. 25). A strong relationship exists between the morphology of the different species of *Hemiaster* and their bathymetric range. Species with a depressed, undulating aboral surface inhabit shallow environments while globose species with a rounded aboral surface are found in the mudstone deposits of the distal end of the platform. This distribution pattern can be observed in the successive *Hemiaster* assemblages which accompany the transgression and regression events all through the Upper Cretaceous. Such a bathymetric influence on *Hemiaster* has also been recorded in Tunisia where the most globose forms come from bathyal deposits (Zaghbib-Turki 1989, 1990). More generally this distribution, conspicuously evident in Spain, can be viewed as the local consequence of a large scale phenomenon, at first stated by Zoeke (1951), which separates the genus *Hemiaster* into two geographic groups: a north European group comprising round-shaped species with short petals and relatively few ambulacral pores, and a north African group comprising species with long petals and an undulating test. Both groups occur in intermediate regions. Because the length of the petals, and correlatively the number of ambulacral pores, are closely involved with respiration, the separation between these two groups (and concomitantly shapes of the test) can therefore be related to oxygen consumption (and concomitantly water temperature). All these considerations, along with morphological data concerning the ancestry of *Hemiaster,* have led Néraudeau (1990) to question the monophyly of the genus, and to propose two distinct origins for the *Hemiaster* group. According to this hypothesis, it becomes necessary to reconsider the meaning of some lineages

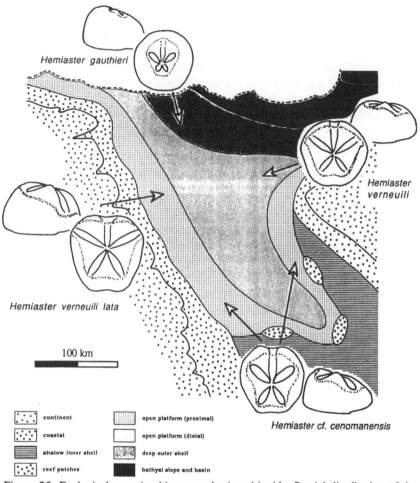

Figure 25. Ecological narrative biogeography in echinoids. Spatial distribution of the echinoid genus *Hemiaster* on the north Castillan platform (Spain) during the Late Cenomanian – Early Turonian . The different species are distributed according to several palaeogeographic sub-units reflecting the topography of the platform (modified from Néraudeau 1990).

which appear as heterogeneous ecological groupings. For instance, a transgressive trend occuring in an intermediary region where both forms are potentially present (e.g. Spain) will induce a cooling of the bottom sea water and results in the replacement of undulating species by globose ones. The stratigraphic order of species will represent an ecological succession, not a phylogenetic lineage. At each time of the *Hemiaster* story, horizontal distri-

52

butions correspond to a balance between the adaptive ranges of the two groups, and several of the observed vertical distributions along geological sections could solely reflect changes in this balance.

This type of study illustrates how ecological biogeography can be profitably expanded to a phyletic frame, and can help to improve the evolutionary interpretation of groups.

6.2.3 *A synthetic approach*

As stated above, ecological and historical approaches are part of a continuum and it is possible to propose a comprehensive explanation of the biogeographical patterns that include both ecological causes and geological events (Waters 1988 for blastoids, Hoggett & Rowe 1988 for echinoderms from coral reefs). A synthetic view of biogeography of modern stalked crinoids has been thoroughly discussed by Roux (1979, 1982, and 1987 for a complete discussion). The evolution of stalked crinoids since the Triassic is explained in relation to certain keyfactors involved in the vertical and horizontal distribution of the oceanic benthos. The plate tectonic history (e.g. suturing and expansion events) provides a frame in which to explain provincialism and migratory patterns, particularly propagating dispersal along mid-oceanic ridges. Moreover, the distribution of the energy sources (i.e. primary carbon production including the oceanic surface phyto-production, chemoautotroph production from thermal vents, and continental detritism) determines the characteristics of environments (chiefly stability and predictability). These factors result in different adaptive strategies (r or K) which are used to explain the bathymetrical distribution of stalked crinoids. By discussing the effects and interactions of these different factors, Roux (1987) proposed a pattern of distribution for the modern bathyal and abyssal stalked crinoids. The origin of this fauna is then related to ocean floor spreading since the Upper Cretaceous, primarily by involving the migration of opportunistic species along passive margins and oceanic ridges.

The interest of such a study is to propose a link between ecological data, biogeography and evolution, and to demonstrate that underlying processes are probably highly interactive. But the approach remains narrative: it provides an *a posteriori* explanation of evolution. To obtain *a priori* arguments, it becomes necessary to address other methods, such as vicariance biogeography.

6.3 *Analytical biogeography*

In response to narrative biogeography, which results in more or less idiosyncratic and unfalsifiable scenarios of dispersal, other approaches have been developed. All are grounded on unambiguous methods that search for common distribution patterns among different geographic areas. But they do not share the same basic principles and for some years now, there has been a fierce debate among biogeographers about the appropriateness of different types of data and methods.

6.3.1 *About methods*

The simplest methods correspond to phenetic approaches to biogeography which treat raw distributional data and often avoid historical aims, but which are useful for depicting areas of endemism. Similarities between areas are deduced from hierarchical classifications, frequently based on the use of similarity indices (e.g. Jaccard's coefficient). Some examples dealing with echinoderms exist (Maluf 1988, Perez-Ruzafa & Lopez-Ibor 1988, Ghiold & Hoffman 1989). However because the clustering technique provides straight-forward results, the completeness of the data set takes on utmost importance. This is particularly evident for deep sea groups where sampling may be poor. Only partial views of distribution are found, which may be quite misleading (e.g. inaccurate and incomplete data [Ghiold 1988] have led to unreliable results about the biogeography of holasteroid echinoids).

Other methods are less technical and include explicit historical aims. They state that if several unrelated taxa have similar patterns of distribution, their biogeography probably results from disjunction events (i.e. vicariance). These methods include Croizat's panbiogeography and attempts to combine cladistics with biogeography (i.e. expansion of Croizat's method within a cladistic approach). These methods have produced numerous theoretical developments (e.g. Croizat 1964, Craw 1982, 1984, 1985, Craw & Weston 1984, Craw 1988 for panbiogeography; Platnick & Nelson 1978, Wiley 1980, Nelson and Platnick 1981, Cracraft 1982 for cladistic biogeography; Humphries & Parenti 1986 for a historical and methodological review). Proponents of cladistic biogeography (or vicariance biogeography sensu stricto) advocate analytical biogeography which assumes that large-scale spatial distributions of organisms are in majority due to vicariant events, i.e. factors other than dispersal and ecology. In practical terms the purpose of the analyses can be summarized as follows: 'are areas of endemism interrelated among themselves in a way analogous to the interrelationships of the species of a certain group of organisms?' (Humphries & Parenti 1986, p. 1). To answer this question, cladistic biogeography searches for congruent patterns of areas (i.e. spatial patterns which are common to several unrelated taxa) and uses them to construct consensus cladograms where taxa are employed instead of characters and where the tips of the branches are areas. It allows the common history of groups sharing the same area (biotas) to be traced back in time. Biogeography thus involves a direct evolutionary dimension. In return cladistic biogeography introduces a methodological gap with ecology and neglects the ecological components of distribution; biomes are out of the scope of the study (Rosen 1978).

In spite of a large number of theoretical examples, few applied examples exist, except for extant terrestrial groups (e.g. Rosen 1979, Cracraft 1980, Legendre 1986), shallow-water dwellers (e.g. Knox 1980), or epicontinental groups (e.g. Nelson & Platnick 1981 for fishes and crustaceans). This situation is due, to some degree, to the relatively small amount of information available (Nelson & Platnick 1981), but it is also due to extrinsic limitations of the method, when applied to marine taxa. Indeed, the post-Triassic history of the earth does not allow one to place both the continental and the marine realms in the same vicariant frame. Vicariance methods are very well fitted to land taxa, because plate tectonics since the Triassic have mainly produced disjunction of the continental areas (Sengör et al. 1988). These methods are not always appropriate to marine biotas, except for strictly onshore groups, since the history of the oceans is mostly a history of progressive openings (e.g. the palaeo-Tethys and the Atlantic Ocean which suggest dispersal patterns), even if some closings have also occurred (e.g. the suturing of the Tethys leading to the formation of the Mediterranean sea). Moreover all the great oceanic domains have always been more or less in connection one with the other. For these reasons examples dealing with echinoderms (Knox 1980) are few. But in fact they have led to some very interesting compromise solutions combining both vicariance and convergence, fossil and Recent data.

6.3.2 A model

Rosen (1985) proposed a very elegant approach (parsimony analysis of endemism, fully developed in Rosen & Smith 1988) which considers contemporaneous data from successive stratigraphic time planes. Indeed, a single time plane is not enough to infer an unequivocal history when the vicariance hypothesis is partly rejected. Besides, this allows old biogeographic patterns to be detected which could have been partly erased by later events. An area cladogram expressing relationships between biotas is drawn for each sampling time (time planes numbered t1 to t6 on Fig. 26). Each cladogram is inferred from a cladistic analysis using a parsimony method, in which shared endemism is analogous to synapomorphies. On the cladograms, vicariance events correspond to dichotomies. They can be detected as the occurrence of a new biota in the subsequent cladogram (e.g. time planes t1 and t2 illustrate a vicariance event affecting biota A). From a historical point of view, 'a cladogram provides a sequence (...) in which dichotomies nearer to the origin of the cladogram represent older divergences than those further from the origin' (Rosen & Smith 1988, p. 282). Convergence between biotas can be deduced from the transposition of sister-relationships between three biotas over a time sequence. An example of convergence is given by the composite biota A2BC1 at time planes t4 and t5. In Figure 26, biota A2B is related to the biota A1 at time t4, and becomes

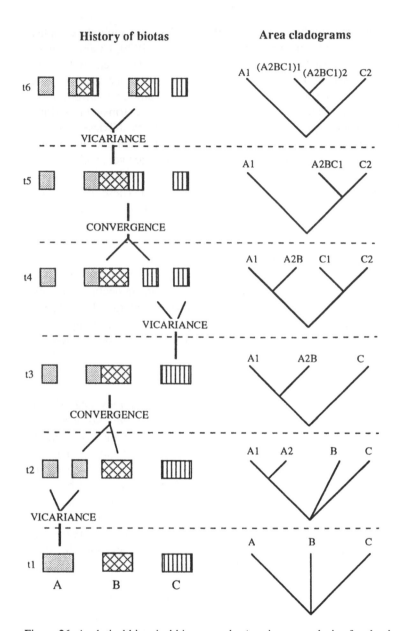

History of biotas   Area cladograms

Figure 26. Analytical historical biogeography (parsimony analysis of endemism), general principle. A hypothetical history of three initial biotas (A, B and C) is depicted throughout 6 successive time planes. Geographic sample localities are represented by rectangles and biotas are represented by different patterns (dots, diamonds and stripes). The corresponding area cladograms illustrate how the major biogeographic events can be read from the relationships between biotas. Biotas are represented by capital letters on the cladograms (modified from Rosen & Smith 1988).

linked to the biota C1 at time t5. All the divergent or convergent events occurring at a given time are recorded in the series of subsequent cladograms where they appear closer to the basal node as geological time progresses. From the analysis of successive cladograms, the history of the geographic areas and biotas taken into account can be recovered. It then becomes possible to test the reliability of this history by reference to geological events (tectonic as well as climatic) which have occurred during the surveyed span of time.

Rosen & Smith (1988) applied this method to a concrete study, and they proposed area cladograms for corals and echinoids at different geological periods (Fig. 27). Their study depicts the consensus area cladograms for Recent and Lower Miocene time periods. The Recent cladogram distinguishes two main sets: 1) an Indo-Pacific set which gathers together the

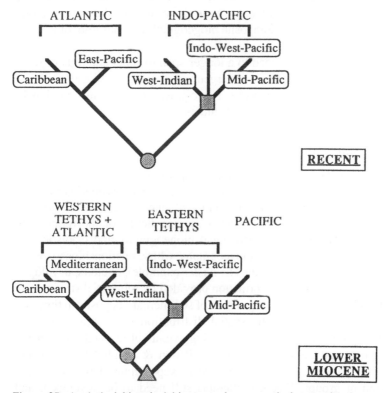

Figure 27. Analytical historical biogeography, a practical example. Area consensus cladograms for Recent and Lower Miocene distributions of clypeasteroid echinoids and reef corals. The different symbols illustrate some of the main vicariance events (highly modified from Rosen & Smith 1988).

West Indian Ocean, Indo-West Pacific, and Mid-Pacific biotas; 2) an Atlantic set which gathers together the Caribbean and the East Pacific biotas. The Lower Miocene cladogram separates an Eastern Tethys fauna from an Atlantic and Western Tethys fauna, while the Pacific Ocean biota appears very distinct. A historical interpretation of these patterns suggests that some convergence or vicariance events have occurred, among them it is possible to quote.

1. The oldest vicariance event (shown by a triangle on Figure 27) separated the Pacific Ocean fauna from the whole Tethyan fauna. This event is too old to have left traces on the Recent cladogram.

2. The separation of the Eastern Tethys (including the Indian Ocean) from the Western Tethys (including the Atlantic Ocean) (circle on Figure 27) appears to be older than the Miocene, but it has still not been erased from the Recent cladogram.

3. The modern Indo-Pacific grouping results from a convergence event which can be deduced from the switch in sister-area relationships of the Pacific Ocean fauna between the two time periods. Thus the present Indo-Pacific fauna is a young and composite biogeographical entity, not older than the Lower Miocene.

Such a study can be extended to a longer span of time, and thus provide some clues about the history of divergent or convergent biogeographical events. But at first, it can also test the phylogenetic relationships previously inferred from classical morphological studies.

6.4 *Biogeographic prospects*

What could be the future of biogeography for echinoderms? Narrative historical biogeography will probably continue to propose scenarios based on future improvements in our understanding of plate tectonics, but without major conceptual changes. The most promising fields for the future may came from the two extremes: ecological biogeography and cladistic biogeography. Ecological narrative biogeography, whose relationships to evolution appear crucial (see above, the *Hemiaster* example), could be developed by incorporating additional information provided by sequential stratigraphy (i.e. for fossils, the geological regulation of sedimentary deposits in time and space could pre-determine the spatial distribution of groups). Cladistic biogeography represents another promising avenue of research (see above, the reef-corals and echinoids example of Rosen & Smith 1988) allowing evolutionary patterns to be unambiguously superimposed on the spatial distribution of groups. Practical attempts still need to be developed, and numerous questions have to be addressed. For instance, the history of the Mediterranean biota since the Messinian salinity crisis could provide a model to estimate the influence of ecological events (e.g. Pleistocene glaciations) versus geological events (e.g. the suturing of the Tethys Sea or the closure of the Atlantic-Mediterranean gateway) on a benthic marine fauna (e.g. the echinoderms). Finally, because biogeography appears strongly dependent on basic data, new primary sources will also largely determine the prospect for further progress.

7 CONCLUSIONS

7.1 *Conceptual framework for evolution*

Evolution is a historical process. As such it is closely related to the notion of time. The notion of deep geological time emerged at the end of the eighteenth century, but until rather recently, time was viewed as permanent or cyclic with 'no vestige of a beginning, no prospect for an end' (Hutton 1795), a position which has greatly favoured the introduction and development of the determinist thought. With such an attitude life's history became a boundless series of steps following one after another and whose progression is governed by specific laws (e.g. the 'biogenetic law' which assumes that phylogeny is recorded in ontogeny, or Cope's rule which supports the idea that there exists

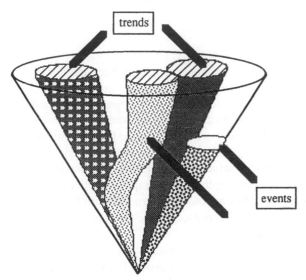

Figure 28. Theoretical design depicting the two main aspects of evolution. Evolution combines trends (internal shadowed cones) which are bounded by structural constraints appropriate to the studied taxon (external conical envelope), and random events which can disrupt the course of the lineages. The previous trends are either interrupted or their pathway is diverted by an external factor (modified from Laurin & David 1988).

a general tendancy towards phyletic increase in size). For some years, science in general has deserted this purely determinist point of view (Popper 1984) and geology in particular has once again promoted the part played by time (Simpson 1950, Allègre 1985). In this reassertion of the value of history, evolution gains a status of unequivocal sequence which includes both obviously oriented trends and random, undetermined, drastic changes (Simpson 1953). The former channeled trends correspond to what can be conveniently called the serial component of evolution. Conversely, random changes are unpredictable events occurring more or less sporadically during the course of lineages. They correspond to what can be called the extrinsic component of evolution (Tintant 1986).

This twofold aspect means that evolution is not a well ordered sequence of successive interrelated stages following predictable pathways, but neither does it correspond to a random juxtaposition of independent facts. Evolution gathers these two aspects in proportions which vary according to the groups considered, the level of observation, or the equilibrium of the system (Laurin & David 1988). In broad outline, the course of evolution can be viewed as a succession of channeled trends disrupted by random events (Fig. 28). At each step there exists a choice. This choice is mainly under the control of selective external factors. In contrast, evolution within trends is channeled and 'controlled by the interaction between developmental dynamics (...), stabilizing selection, and ecological parameters' (Alberch 1980, p. 664).

7.2 Discussing the approaches

The two principal sets of methods available for tracing back the history of groups reflect this duality of evolution. 'Morphological' methods (i.e. cladistics, molecular methods, ontogeny) favour the serial component of evolution. Indeed, they are based on rules that assume a hierarchical link between descendants and ancestors. But they are also bounded by events which disrupt the rules upon which they are grounded: 1) reversion and convergence; 2) ontogenetic trajectories which do not follow the heterochronic model. 'Historical' methods (stratigraphy, biogeography) implicitly support an extrinsic view of evolution. They are basically grounded on facts viewed as independent one from the other, but consequently, they are limited by the incompleteness of their respective records.

Because evolution is a composite process, each of these two sets of data and related methods give only a more or less probable picture of the real history. Gould accurately underlined the dichotomy: 'palaeontology (...) resides in the middle of a continuum stretching from idiographic to nomothetic disciplines' (1980a, p. 116). Consequently, evolution cannot be encompassed by purely deterministic approaches, but the purely idiographic approaches neglect a part of the information. It is thus better to combine the two

60

types of investigation to increase the degree of corroboration for the proposed phylogenies. But it also remains necessary to accord leadership to one of them, to define a preference for a main approach leading to the establishment of evolutionary models. Other approaches can then act as tests of these models. This choice of priority in phylogeny reconstruction reflects underlying conceptual controversies, and determines the orientation of research. But I do not believe that the choice might be unequivocally made, because it also will depend on the nature of the taxa analysed.

Because echinoderms exhibit a rather complex morphology and because of the ease with which the diverse events of their ontogeny can be explored, they can provide an accurate picture of the serial part of evolution. Therefore, it may be preferable that 'morphological' methods should be used first to reconstruct phylogenies. This could lead to the construction of cladograms enhanced by precise ontogenetic studies or complemented by molecular data. 'Historical' data could then be integrated secondarily to corroborate the hypotheses of relationships previously inferred, as well as to convert a cladogram into an evolutionary tree. In fact, these different methodological approaches may be used selectively or concomitantly depending on whether emphasis has to be put on the serial or on the extrinsic component of evolution.

Major progress will probably come from multidisciplinary approaches. A convincing example of such an attempt was proposed recently during the 1990's International Echinoderm Conference in Japan. The diversity expressed by the regular sea urchin *Echinometra mathaei* was documented by gathering statements from ecology (Tsuchiya & Nishihira 1985, 1986, Nishihira et al. 1991), morphology (Uehara & Shingaki 1985, Uehara et al. 1986, Uehara 1991), karyotypes (Uehara et al. 1991), gamete fertilization (Palumbi & Metz 1991), and both mitochondrial and nuclear DNA (Metz et al. 1991). All participated to demonstrate that *E. mathaei* may be a complex undergoing speciation. Such a combined study would be a propitious model for future studies.

Much of this paper has been devoted to questions of method, of how different ways of analysis can be used to reconstruct echinoderm phylogenies. I have explored the relative value and efficacy of these methods in different contexts. In this way, this paper has been a review article. But it was also not free of advocacy, since discussions about methodology involve generally philosophical questions, and because the choice of facts itself is never purely objective (Gould 1980a). To conclude, I would advocate a pluralistic approach in order to enhance the probability of discovering the real course of evolutionary pathways. To come back to the analogy with a theatrical play, I shall point out that both good actors and beautiful scenery are necessary to provide an ideal program.

ACKNOWLEDGEMENTS

I am grateful to J. M. Lawrence and M. Jangoux for providing the opportunity to write this review and for their helpful editorial comments. I also thank J. P. Féral who shared his intellectual insight into molecular evolution, B. Laurin, D. Néraudeau, H. Tintant, the whole palaeontological staff of the University of Burgundy for debating evolutionary issues, and C. R. Marshall and S. R. Palumbi for bibliographic information. I would like to especially thank two anonymous reviewers for their copious and helpful comments and R. Mooi for one final bash at the MS.

REFERENCES

Agassiz, A. 1904. Reports on an exploration off the West Coasts of Mexico, Central and South America, and off the Galapagos Islands, by the U. S. Fish Commission Steamer 'Albatros' during 1891. The Panamic Deep Sea Echini. *Mem. Mus. compar. Zool.* 31: 1-243.

Agassiz, L. 1836. Prodrome d'une monographie des Radiaires ou échinodermes. *Mém. Soc. Sci. nat. Neuchâtel* 1: 168-199.

Agassiz, L. 1839. Description des échinodermes fossiles de la Suisse; première partie: Spatangoïdes et Clypeastroïdes. *Mém. Soc. helvét. Sci. nat.* 3: 1-101.

Agassiz, L. 1840. *Catalogus systematicus ectyporum echinodermatum fossilium Musei neocomiensis.* Neuchâtel: O. Petitpierre.

Agassiz, L. 1849. *Twelve lectures on comparative embryology.* Boston: Henry Flanders.

Agassiz, L. & E. Desor 1846-1847. Catalogue raisonné des Echinides. *Ann. Sc. nat. Zool.* Sér.3, 6(1846): 305-374, 7(1847): 129-168 & 8(1847): 5-35, 355-380.

Alberch, P. 1980. Ontogenesis and morphological diversification. *Amer. Zool.* 20: 653-667.

Alberch, P. 1982. Developmental constraints in evolutionary processes. In J.T. Bonner (ed.), *Evolution and development, Dahlem Konferenzen*: 313-332. Berlin: Springer Verlag.

Alberch, P. 1989. Orderly monsters: Evidence for internal constraint in development and evolution. In B. David, J.L. Dommergues, J. Chaline & B. Laurin (eds), *Ontogenèse et évolution.*Géobios Mém. Spéc.12: 21-58.

Alberch, P., S.J. Gould, G.F. Oster & D.B. Wake 1979. Size and shape in ontogeny and phylogeny. *Paleobiology* 5: 296-317.

Ali M.S. 1983. The paleogeographic distribution of *Clypeaster* (Echinoidea) during the Cenozoic era. *N. Jb. Geol. Paläont. Mh.* 8: 449-464.

Allègre, C. 1985. *De la pierre à l'étoile.* Paris: Fayard.

Angerer, R.C., E.H. Davidson & R.J. Britten 1976. Single copy DNA and structural gene sequence relationships among four sea urchin species. *Chromosoma* 56: 213-226.

Ausich, W.I. & N.G. Lane 1982. Evolution of arm structure in the early Mississippian crinoid *Cydrocrinus*. In J. Lawrence (ed.), *Echinoderms, Proceed. 4th internat. Echinoderm Conf., Tampa Bay, 1981*: 139-143. Rotterdam: Balkema.

Ayala, F.J. & J.W. Valentine 1974. Genetic variability in the cosmopolitan deep-water ophiuran. *Ophiomusium lymani. Mar. Biol.* 27: 51-57.

Ayala, F.J., J.W. Valentine, D. Hedgecock & L.G. Barr 1975. Deep-sea asteroids: high genetic variability in a stable environment. *Evolution* 25: 203-212.

Baer, K.E. von 1828. *Entwicklungsgeschichte der Thiere: Beobachtung und Reflexion.* Königsberg: Bornträger.

Beadle, S.C. 1989. Ontogenetic regulatory mechanisms, heterochrony, and eccentricity in dendrasterid sand dollars. *Paleobiology* 15: 205-222.

Beatty, J. 1982. Classes and cladists. *Syst. Zool.* 31: 25-34.

Bell, B.M. 1976. Phylogenetic implications of ontogenetic development in the class Edrioasteroidea (Echinodermata). *J. Paleont.* 50: 1001-1019.

Benveniste, R.E. 1985. The contributions of retro-viruses to the study of mammalian evolution. In R.J. MacIntyre (ed.), *Molecular evolutionary genetics*: 359-417. London: Plenum Press.

Bisol, P.M., R. Costa & M. Sibuet 1984. Ecological and genetical survey on two deep-sea holothurians: *Benthogone rosea* and *Benthodytes typica*. *Mar. Ecol. Prog. Ser.* 15: 275-281.

Blake, D.B. 1987. A classification and phylogeny of post-Paleozoic sea stars (Asteroidea: Echinodermata). *J. Nat. Hist.* 21: 481-528.

Bock, W.J. 1977. Foundations and methods of evolutionary classification. In M.K. Hecht & P.C. Goody (eds), *Major pattern in vertebrates evolution*: 851-895. New York: Plenum Press.

Bockelie, J.F. 1981. Functional morphology and evolution of the cystoid *Echinosphaerites*. *Lethaia* 14: 189-202.

Bonis De, L. 1983. L'évolution des primates supérieurs: un exemple d'équilibres ponctués? In J. Chaline (ed.), *Modalités rythmes et mécanismes de l'évolution biologique* 330: 67-72. CNRS: Coll. internat.

Bonner, J.T. (ed.) 1982. *Evolution and Development, Dahlem Konferenzen.* Berlin: Springer Verlag.

Breimer, A. & D.E. Macurda 1972. The phylogeny of fissiculate blastoids. *Verhandel. Koninklijke Nederlandse Akad. Wetenschap.* 26: 1-390.

Breton, G. 1990. Les Goniasteridae (Asteroidea, Echinodermata) jurassiques et crétacés de France: taphonomie, systématique, biostratigraphie, paleobiogéographie, évolution. Thèse Etat Univ. Caen: 1-492, 48 pl. (unpublished).

Brett, C.E., T.J. Frest, J. Sprinkle & C.R. Clement 1983. Coronoidea: a new class of blastozoan echinoderms based on taxonomic reevaluations of *Strephanocrinus*. *J. Paleont.* 57: 627-651.

Breynius, J.P. 1732. *De Echinis et Echinitis sive methodica Echinorum distributione schediasma*. Danzig:edani.

Britten, R.J. 1986. Rates of DNA sequence evolution differ between taxonomic groups. *Science* 231: 1393-1398.

Britten, R.J., A. Cetta & E.H. Davidson 1978. The single copy DNA sequence polymorphism of the sea urchin *Strongylocentrotus purpuratus*. *Cell* 15: 1175-1186.

Britten, R.J. & E.H. Davidson 1969. Gene regulation for higher cells: a theory. *Science* 165: 349-357.

Britten, R.J. & D.E. Kohne 1968. Repeated sequences in DNA. *Science* 161(3841): 529-540.

Bromwell, E. 1983. DNA/DNA hybridization studies of muroid rodents: symmetry and rates of molecular evolution. *Evolution* 37: 1034-1051.

Brooks, D.R. & E.O. Wiley 1985. Theories and methods in different approaches to phylogenetic systematics. *Cladistics* 1: 1-11.

63

Brower, J.C. 1982. Phylogeny of primitive calceocrinids. In J.Sprinkle (ed.), *Echinoderm faunas from the Bromide formation (Middle Ordovician) of Oklahoma.* Univ. Kansas Paleont. Contr. Monograph 1: 90-110.

Brower, J.C. 1990. Ontogeny and phylogeny of the dorsal cup in calceocrinid crinoids. *J. Paleont.* 64: 300-318.

Bruguieres, J.G. 1791. *Histoire naturelle des vers échinodermes.* Encyclopédie Méthodique, Paris: atlas.

Bucaille, E. 1883. Etude sur les échinides fossiles du département de la Seine Inférieure. *Bull. Soc. géol. Normandie* 8: 16-39.

Bullimore, B. & R.G. Crump 1982. Enzyme electrophoresis and taxonomy of two species of *Asterina* (Asteroidea). In J. Lawrence (ed.), *Echinoderms, Proceed. 4th internat. Echinoderm Conf., Tampa Bay, 1981*: 185-188. Rotterdam: Balkema.

Caccone, A. & J.R. Powell 1989. DNA divergence among hominoids. *Evolution* 43: 925-942.

Catzeflis, F.M., F.H. Sheldon, J.E. Ahlquist & C.G. Sibley 1987. DNA-DNA hybridization evidence of the rapid rate of Muroid Rodent DNA evolution. *Mol. Biol. Evol.* 4: 242-253.

Chauvel, J. 1941. Recherches sur les cystoïdes et les carpoïdes armoricains. *Soc. géol. minéral. Bretagne Mém.* 5: 1-286.

Costa, R., P.M. Bisol & M. Sibuet 1982. Genetic variability in deep-sea holothurians. In J. Lawrence (ed.), *Echinoderms, Proceed. 4th internat. Echinoderm Conf., Tampa Bay, 1981*: 189-191. Rotterdam: Balkema.

Cotteau, G. 1883. Note sur les échinides jurassiques, crétacés, éocènes du sud-ouest de la France. *Bull. Soc. géol. Fr. Sér.* 3, 12: 179-188.

Cracraft, J. 1967. Comments on homology and analogy. *Syst. Zool.* 16: 355-359.

Cracraft, J. 1980. Biogeographic patterns of terrestrial vertebrates in the southwest Pacific. *Palaeogeogr. Palaeoclim. Palaeoecol.* 31: 353-369 .

Cracraft, J. 1982. Geographic differentiation, cladistics, and vicariance biogeography: reconstructing the tempo and mode of evolution. *Amer. Zool.* 22: 411-424.

Craw, R.C. 1982. Phylogenetics, areas, geology and the biogeography of Croizat: a radical view. *Syst. Zool.* 31: 304-316.

Craw, R.C. 1984. Biogeography and biogeographical principles. *New Zealand Entomol.* 8: 49-52.

Craw, R.C. 1985. Classic problems of southern hemisphere biogeography re-examined. *Z. zool. Syst. Evol.* 23: 1-10.

Craw, R. 1988. Panbiogeography: method and synthesis in biogeography. In A.A. Myers & P.S. Giller (eds), *Analytical biogeography*: 405-435. London: Chapman & Hall.

Craw, R.C. & P. Weston 1984. Panbiogeography: a progressive research program? *Syst. Zool.* 33: 1-13.

Croizat, L. 1964. *Space, time and form, the biological synthesis.* Caracas: Publish. by the author.

David, B. 1985. Significance of architectural patterns in the deep-sea echinoids Pourtalesiidae. In B.F. Keegan & B.D.S. O'Connor (eds), *Echinodermata, Proceed. 5th internat Echinoderm Conf., Galway, 1984*: 237-243. Rotterdam: Balkema.

David, B. 1987. Dynamics of plate growth in the deep-sea echinoid *Pourtalesia miranda* Agassiz: a new architectural interpretation. *Bull. mar. Sci.* 40: 29-47.

David, B. 1988. Origins of the deep-sea holasteroid fauna. In C.R.C. Paul & A.B. Smith (eds), *Echinoderm phylogeny and evolutionary biology*: 331-346. Oxford: Clarendon Press.

David, B. 1990. Mosaic pattern of heterochronies: variation and diversity in Pourtalesii-dae (deep-sea echinoids). *Evol. Biol.* 24: 297-327.

David, B. & M. Fouray 1984. Variation et disjonction évolutive des caractères dans les populations de *Micraster* (Echinoidea, Spatangoida) du Crétacé supérieur de Picardie. *Géobios* 17: 447-476.

David, B. & B. Laurin 1989. Déformations ontogénétiques et évolutives des organismes: l'approche par la méthode des points homologues. *C.R. Acad. Sci. Paris* 309, *Sér.*2: 1271-1276.

Davidson, E.H. 1988. What molecular biology tells us about the genomic programme for development. In C.R.C. Paul & A.B. Smith (eds), *Echinoderm phylogeny and evolutionary biology*: 139-146. Oxford: Clarendon Press.

Des Moulins, C. 1835-1837. Etudes sur les échinides. *Act. Soc. linn. Bordeaux* 7 & 9: 1-520.

Desor, E. 1855-1858. *Synopsis des échinides fossiles*; 1 atlas, 44 pl. Paris: Reinwald.

Devillers, C. 1985. Quelques remises en cause de la théorie synthétique de l'évolution *Ann. Biol.* 24: 153-177.

Devriès, A. 1960. Contribution à l'étude de quelques groupes d'échinides fossiles d'Algérie. *Publ. Serv. Carte géol. Alg.* (nlle. Série) (Paléont.) 3: 1-278.

Dommergues, J.L., B. David & D. Marchand 1986. Les relations ontogenèse-phylogenèse: Applications paléontologiques. *Géobios* 19: 335-356.

Dommergues, J.L. & D. Marchand 1988. Paleobiogeographie historique et ecologique, application aux Ammonites du Jurassique. In J. Wiedmann & J.Kullmann (eds), *Cephalopods present and past*: 351-364. Stuttgart: Schweizer Verlagsbuchhandl.

Donovan, S.K. 1988. The early evolution of the Crinoidea. In C.R.C. Paul & A.B. Smith eds), *Echinoderm phylogeny and evolutionary biology*: 235-244. Oxford: Clarendon Press.

Donovan, S.K. & Cr.C. Paul 1985. Coronate echinoderms from the Lower Palaeozoic of Britain. *Palaeontology* 28: 527-543, pl. 62-63.

Doyle, R.W. 1972. Genetic variation in Ophiomusium lymani populations in the deep-sea. *Deep-Sea Res.* 19: 662-664.

Dupuis, C. 1984. Willi Hennig's impact on taxonomic thought. *Ann. Rev. Ecol. Syst.* 15: 1-24.

Durham, J.W. 1966. Evolution among the Echinoidea. *Biol. Rev.* 41: 368-391.

Durham, J.W., H.E. Fell, A.G. Fisher, R.V. Melville, D.L. Pawson & C.D. Wagner 1966. Echinoids. In R.C. Moore (ed.), *Treatise on invertebrate palaeontology, Part U: Echinodermata* 3, 1,2: 211-640. Lawrence: Geol. Soc. America & Univ. Kansas Press.

Durkin, M.K. 1980. The Saleniidae in time and space. In M. Jangoux (ed.), *Echinoderms: present and past, Proceed. 1st European Echinoderm Conf., Bruxelles, 1979*: 3-14. Rotterdam: Balkema.

Eckert, J.D. 1987. Heterochrony (progenesis) in the Silurian crinoid *Homocrinus* Hall. *Can. J. Earth Sci.* 24: 2568-2571.

Eckman, S. 1953. *Zoogeography of the sea*.London: Sidgwick & Jackson.

Elbert, J. 1901. Das untere Angoumien in der Osningbergketten des Tentoburger Waldes. *Verh. Natur. Verein. Preuss. Rheinl.* 58: 77-167.

Eldredge, N. 1971. The allopatric model and phylogeny in Paleozoic invertebrates. *Evolution* 25: 156-167.

Eldredge, N. 1979. Cladism and common sense. In J.Cracraft & N. Eldredge (eds), *Phylogenetic analysis and paleontology*: 165-198. New York: Columbia Univ. Press.

Eldredge, N. & J. Cracraft 1980. *Phylogenetic patterns and the evolutionary process: method and theory in comparative biology.* New York: Columbia Univ. Press.

Eldredge, N. & S.J. Gould 1972. Punctuated equilibria: an alternative to phyletic gradualism. In T.J.M. Schopf (ed.), *Models in paleobiology*: 82-115. San Francisco: Freeman, Cooper.

Eldredge, N. & M.J. Novacek 1985. Systematics and paleobiology. *Paleobiology* 11: 65-74.

Ernst, G. 1970. Zur Stammesgeschichte und stratigraphischen Bedeutung der Echiniden-Gattung *Micraster* in der nordwestdeutschen Oberkreide. *Mitt. Geol.-Paläont. Inst. Univ. Hamburg* 39: 117-135.

Ernst, G. 1972. Grundfragen der Stammesgeschichte bei irregulären Echiniden der nordwesteuropäischen Oberkreide. *Geol. Jb.* A(4): 63-175.

Ernst, G. & M.G. Schultz 1971. Die Entwicklungsgeschichte der hochspezialisierten Echiniden-Reihe *Infulaster – Hagenowia* in der borealen Oberkreide. *Paläont. Zeit.* 45: 120-143.

Farris, J.S. 1980. The information content of the phylogenetic system. *Syst. Zool.* 28: 483-519.

Farris, J.S. 1982. Outgroups and parsimony. *Syst. Zool.* 31: 328-334.

Farris, J.S. 1985. Distance data revisited. *Cladistics* 1: 67-85.

Fell, H.B. 1963a. The evolution of the Echinoderms. *Smithson. Report* 1962: 457-490.

Fell, H.B. 1963b. The phylogeny of sea-stars. *Phil. Trans. Roy. Soc. London* (B)246: 381-435.

Féral, J.P. & E. Derelle 1991. Partial sequence of the 28S ribosomal RNA and the echinid taxonomy and phylogeny. Application to the Antarctic brooding schizasterids. In T. Yanagisawa et al. (eds), *Proceed. 7th internat. Echinoderm Conf., Atami, 1990*: 331-337. Rotterdam: Balkema.

Féral, J.P., E. Derelle, P. Schatt & J.L. Toffart 1989. Polymorphisme enzymatique observé chez l'oursin eurybathe *Brissopsis lyrifera* (Forbes, 1841). In M.B. Régis (ed.), *Echinodermes: actuels et fossiles*. Vie Marine H.S. 10: 79-81.

Féral, J.P., J.G. Ferrand & A. Guille 1990. Macrobenthic physiological responses to environmental fluctuations: the reproductive cycle and enzymatic polymorphism of a eurybathic sea-urchin on the northwestern Mediterranean continental shelf and slope. *Continental Shelf Res.* (in press).

Ferber, I. & J.M. Lawrence 1976. Distribution, substratum preference and burrowing behaviour of *Lovenia elongata* (Gray) (Echinoidea: Spatangoida) in the Gulf of Elat (Aquaba), Red Sea. *J. exp. mar. Biol. Ecol.* 22: 207-225.

Ferris, S.D., R.D. Sage, R.C. Huang, J.T. Nielsen, U. Ritte & A.C. Wilson 1983. Flow of mitochondrial DNA across a species boundary. *Proc. Natl. Acad. Sci. USA* 80: 2290-2294.

Field, K.G., G.J. Olsen, D.J. Lane, S.J. Giovannoni, M.T. Ghiselin et al. 1988. Molecular phylogeny of the animal kingdom. *Science* 239: 748-753.

Fink, W.L. 1982. The conceptual relationship between ontogeny and phylogeny. *Paleobiology* 8: 254-264.

Fink, W.L. 1988. Phylogenetic analysis and the detection of ontogenetic patterns. In M.L. McKinney (ed.), *Heterochrony in Evolution. A Multidisciplinary approach*: 71-91. New York: Plenum Press.

Fitch, W.M. & E. Margoliash 1967. Construction of phylogenetic trees: a method based on mutation distances as estimated from cytochrome C sequences is of general applicability. *Science* 155: 279-284.

Forbes, E. 1841. *A history of British starfishes and other animals of the Class Echinodermata*. London.

Forbes, E. 1850. Figures and descriptions illustrative of British organic remains. *Mem. geol. Surv. U.K.* Decade 4: 10 pls.

Forey, P. 1982. Neontological analysis versus palaeontological stories. In K.A. Joysey & A.E. Friday (eds), *Problems of phylogenetic reconstruction*, Syst. Assoc. Spec. Vol. 21: 119-157. London: Academic Press.

Foster, R.J. & G.M. Philip 1978. Tertiary holasteroid echinoids from Australia and New-Zealand. *Palaeontology* 21: 791-822.

Fouray, M. 1981. L'évolution des *Micraster* (Echinides, Spatangoïdes) dans le Turonien-Coniacien de Picardie occidentale (Somme). Intérêt biostratigraphique. *Ann. Paléont. (Invert.)* 67: 81-134.

Fouray, M. & B. Pomerol 1985. Les *Micraster* (Echinoidea, Spatangoida) de la limite Turonien-Sénonien dans la région stratotypique du Sénonien (Sens, Yonne) implications stratigraphiques. *Ann. Paléont.* 71: 59-82.

Frizzell, D.L., H. Exline & D.L. Pawson (1966) Holothurians. In R.C. Moore (ed.), *Treatise on invertebrate palaeontology, Part U: Echinodermata* 3, 2: 641-672. Lawrence: Geol. Soc. America & Univ. Kansas Press.

Gale, A.S. 1987. Phylogeny and classification of the Asteroidea (Echinodermata). *Zool. J. linn. Soc.* 89: 107-132.

Gale, A.S. & A.B. Smith 1982. The palaeobiology of the Cretaceous irregular echinoids *Infulaster* and *Hagenowia*. *Palaeontology* 25: 11-42.

Gauthier, V. 1886. Description de trois échinides nouveaux recueillis dans la craie de l'Aube et de l'Yonne. *Ass. française Avanc. Sci.*, Notes Mém. C.R.14ème Sess.: 356-362.

Ghiold, J. 1988. Echinoid biogeography: Cassiduloida, Holasteroida, Holectypoida, Neolampadoida. In R.D. Burke, P.V. Mladenov, P.Lambert & R.L. Parsley (eds), *Echinoderm Biology. Proceed. 6th internat. Echinoderm Conf., Victoria, 1987*: 349-354. Rotterdam: Balkema.

Ghiold, J. & A. Hoffman 1989. Biogeography of spatangoid echinoids. *N. Jb. Geol. Paläont. Abh.* 178: 59-83.

Gingerich, P.D. 1979. The stratophenetic approach to phylogeny reconstitution in vertebrate paleontology. In J. Cracraft & N. Eldredge (eds), *Proceed. of phylogenetic Models symposium, Lawrence*. Kansas: 41-77. New York.

Gingerich, P.D. 1985. Species in the fossil record: concepts, trends, and transitions. *Paleobiology* 11: 27-41.

Gingerich, P.D. 1987. Evolution and the fossil record: patterns, rates, and processes. *Can. J. Zool.* 65: 1053-1060.

Goldfuss, A. 1826. *Petrefacta Germaniae*. Atlas 199 pl. Leipzig: Von List & Francke.

Gooch, J.L. & T.J.M. Schopf 1973. Genetic variability in the deep sea: relation to environmental variability. *Evolution* 26: 545-552.

Goodman, M. 1981. Globin evolution was apparently very rapid in early vertebrates: a reasonable case against the rate constancy hypothesis. *J. Mol. Evol.* 17: 114-120.

Gould, S.J. 1973. Systematic pluralism and the uses of history. *Syst. Zool.* 22: 322-324.

Gould, S.J. 1977. *Ontogeny and phylogeny*. Cambridge (Mass.): Belknap press.

Gould, S.J. 1980a. The promise of paleobiology as a nomothetic, evolutionary discipline. *Paleobiology* 6: 96-118.

Gould, S.J. 1980b. Is a new and general theory of evolution emerging? *Paleobiology* 6: 119-130.

67

Gould, S.J. 1982. Change in developmental timing as a mechanism of macroevolution. In J.T. Bonner (ed.), *Evolution and development, Dahlem Konferenzen*: 333-346. Berlin: Springer Verlag.

Gould, S.J. & N. Eldredge 1977. Punctuated equilibria: the tempo and mode of evolution reconsidered. *Paleobiology* 3: 115-151.

Grime, J.P. 1989. The stress debate: symptom of impending synthesis? *Biol. J. Linnean Soc.* 37: 3-17.

Grula, J.W., T.J. Hall, J.A. Hunt, T.D. Guigni, G.J. Graham, E.H. Davidson & R.J. Britten 1982. Sea urchin DNA sequence variation and reduced interspecies differences of the less variable DNA sequences. *Evolution* 36: 665-676.

Guille, A. & M.N. Albuquerque 1990. Stratégies de dispersion et insularité: les ophiures littorales de la chaîne des seamounts Vitoria-Trindade (Brésil); résultats préliminaires. In C. De Ridder, P. Dubois, M.C. Lahaye & M. Jangoux (eds), *Echinoderm research. Proceed. 2nd European Echinoderm Conf., Bruxelles, 1989*: 125-129. Rotterdam: Balkema.

Haeckel, E. 1866. *Generelle Morphologie der Organismen.* Berlin: Georg Reimer.

Hall, T.J., J.W. Grula, E.H. Davidson & R.J. Britten 1980. Evolution of sea urchin non-repetitive DNA. *J. Mol. Evol.* 16: 95-110.

Harold, A.S. & M. Telford 1982. Substrate preference and distribution of the northern sand dollar, *Echinarachnius parma* (Lamarck). In J. Lawrence (ed.), *Echinoderms. Proceed. 4th internat. Echinoderm Conf., Tampa Bay, 1981*: 243-249. Rotterdam: Balkema.

Harold, A.S. & M. Telford 1990. Systematics, phylogeny and biogeography of the genus *Mellita* (Echinoidea: Clypeasteroida). *J. Nat. History* 24: 987-1026.

Hecht, M.K. & J.L. Edwards 1977. The methodology of phylogenetic inference above the species level. In M.K. Hecht, P.C. Goody & B.M. Hecht (eds), *Major patterns in vertebrate evolution*: 3-51. New York: Plenum Press.

Hennig W. 1950. *Grundzüge einer theorie der phylogenetischen Systematik.* Berlin: Deutscher Zentralverlag.

Hennig, W. 1965. Phylogenetic systematics. *Ann. Rev. Entom.* 10: 97-116.

Hill, R.L., J. Buettner-Janusch & V. Buettner-Janusch 1963. Evolution of hemoglobin in primates. *Proc. Natl. Acad. Sci. USA* 50: 885-893.

Himmelman, J.H., H. Guderley, G. Vigneault, G. Drouin & P.G. Wells 1984. Response of the sea urchin *Strongylocentrotus droebachiensis* to reduced salinities, importance of size, acclimatation and interpopulation differences. *Can. J. Zool.* 62: 1015-1021.

Hindenach, B.R. & D. Stafford 1984. Nucleotide sequence of the 18-28S rRNA intergene region of the sea urchin. *Nucleic Acid Res.* 12: 1737-1747.

Hochachka, P.W. 1971. Enzyme mechanisms in temperature and pressure adaptation of offshore benthic organisms: the basic problem. *Amer. Zool.* 11: 425-435.

Hoggett, A.K. & F.W.E. Rowe 1988. Zoogeography of echinoderms on the world's most southern coral reefs. In R.D. Burke, P.V. Mladenov, P. Lambert. & R.L. Parsley (eds), *chinoderm Biology. Proceed. 6th internat. Echinoderm Conf., Victoria, 1987*: 379-387. Rotterdam: Balkema.

Horowitz, A.S., D.B. Macurda Jr & J.A. Waters 1986. Polyphyly in the Pentremitidae (Blastoidea, Echinodermata). *Geol. Soc. Amer. Bull.* 97: 156-161.

Hotchkiss, F.H.C. 1980. The early growth state of a Devonian ophiuroid and its bearing on echinoderm phylogeny. *J. nat. Hist.* 14: 91-96.

Hoyer, B.H., B.J. McCarthy & E.T. Bolton 1964. A molecular approach in the systematics of higher organisms. *Science* 144: 959-967.

Humphries, C.J. & L.R. Parenti 1986).Cladistic biogeography. Oxford: Clarendon Press.

Hunt, J.A. & H.L. Carson 1983. Evolutionary relationships of four species of Hawaiian *Drosophila* as measured by DNA reassociation. *Genetics* 104: 353-364.

Hutton, J. 1795. Theory of the earth. *Trans. Roy. Soc. Edinburgh* 1: 209-305.

Hyman, L.H. 1955. *The Invertebrates (vol.4): Echinodermata*. New York: McGraw-Hill.

Jacob, F. 1981. *Le jeu des possibles*. Paris: Fayard.

Jacobs, H.T. 1988. Rates of molecular evolution of nuclear and mitochondrial DNA in sea urchins. In R.D. Burke, P.V. Mladenov, P. Lambert & R.L. Parsley (eds), *Echinoderm Biology. Proceed. 6th internat. Echinoderm Conf., Victoria, 1987*: 287-295. Rotterdam: Balkema.

Jacobs, H.T., P. Balfe, B.L. Cohen, A. Farquharson & L. Comito 1988. Phylogenetic implications of genome rearrangement and sequence evolution in echinoderm mitochondrial DNA. In C.R.C. Paul & A.B. Smith (eds), *Echinoderm phylogeny and evolutionary biology*: 121-137. Oxford: Clarendon Press.

Jacobs, H.T, J.W. Posakony, J.W. Grula, J.W. Roberts, J.H. Xin, R.J. Britten & E.H. Davidson 1983. Mitochondrial DNA sequences in the nuclear genome of *Strongylocentrotus purpuratus. J. Mol. Biol.* 165: 609-632.

Janvier, P. 1984. Cladistics: theory, purpose, and evolutionary implications. In J.W. Pollard (ed.), *Evolutionary theory: paths into the future*: 39-75. John Wiley & Sons.

Janvier, P. 1989. Ontogénie, phylogénie et homologie: les tests de l'hétérochronie. In B. David, J.L. Dommergues, J. Chaline & B. Laurin (eds), *Ontogenèse et évolution Géobios Mém. spéc.* 12: 245-255.

Jefferies, R.P.S. 1979. The origin of chordates – a methodological essay. In M.R. House (ed.), *The origin of major invertebrates groups*: 443-477. London & New York: Academic Press.

Jensen, M. 1981. Morphology and classification of Euechinoidea Bronn, 1860. A cladistic analysis. *Vidensk. Meddr dansk naturh. Foren.* 143: 7-99.

Jukes, T.H. & R. Holmquist 1972. Evolutionary clock: nonconstancy of rate in different species. *Science* 177: 530-532.

Kellogg, D.E. 1983. Phenology of morphologic change in radiolarian lineages from deep-sea cores: implications for macroevolution. *Paleobiology* 9: 355-362.

Kermack, K.A. 1954. A biometrical study of *Micraster coranguinum* and *M. (Isomicraster) senonensis. Phil. Trans. Roy. Soc. Lond.* (B)237: 375-428 .

Kier, P.M. 1962. Revision of the cassiduloid echinoids. *Smithson. misc. Coll.* 144: 1-262.

Kier, P.M. 1965. Evolutionary trends in Paleozoic echinoids. *J. Paleont.* 39: 436-465.

Kier, P.M. 1974. Evolutionary trends and their functional significance in the Post-Paleozoic echinoids. (The paleontological society, memoir n°5). *J. Paleont.* 48 (suppl. n°3): 1-95.

Kier, P.M. 1977. The poor fossil record of the regular echinoid. *Paleobiology* 3: 168-174.

Kier, P.M. 1982. Rapid evolution in echinoids. *Palaeontology* 25: 1-9.

Kier, P.M. 1984. Echinoids from the Triassic (St. Cassian) of Italy, their lantern supports, and a revised phylogeny of Triassic echinoids. *Smithson Contr. Paleobiology* 56: 1-41.

Kimura, M. 1983. *The neutral theory of molecular evolution*. Cambridge, England: Cambridge Univ. Press

Kimura, M. 1987. Molecular evolutionary clock and the neutral theory. *J. Mol. Evol.* 26: 24-33.

Klein, J.T. 1734. *Naturalis dispositio Echinodermatum. Accessit lucubratiuncula de aculeis Echinorum marinorum cum spicilegio de Belemnitis.* T.J. Schreiber.

Kluge, A.G. 1985. Ontogeny and phylogenetic systematics. *Cladistics* 1: 13-27.

Knox, G.A. 1980. Plate tectonics and the evolution of intertidal and shallow-water benthic biotic distribution patterns of the southwest Pacific. *Palaeogeogr., Palaeoclimat., Palaeoecol.* 31: 267-297.

Lamarck, J.B. 1816. *Histoire naturelle des animaux sans vertèbres.* Echinides. Paris: Baillière.

Lambert, J. 1893. Etudes morphologiques sur le plastron des spatangidés. *Bull. Soc. Sc. hist. nat. Yonne* 46(1892): 55-98.

Lambert, J. 1895. Essai d'une monographie du genre *Micraster* et notes sur quelques échinides. In de Grossouvre (ed.), *Recherches sur la craie supérieure. Mém. Carte géol. Fr.* Part 1 (Strati.): 149-267.

Lambert, J. 1898. *Micraster coranguinum* Park. (Sub. *Spatangus*), 1811. In Fortin *Notes de géologie normande V. Bull. Soc. Am. Sc. nat. Rouen* Sér. 4, 34: 358-361.

Lambert, J. 1902-1907. Description des Echinides de la province de Barcelone. *Mém. Soc. Géol. Fr. (Paléonto.)* 9,14: 1-128.

Lambert, J. 1920. Etude sur quelques formes primitives de Spatangidés. *Bull. Soc. Sc. hist. nat. Yonne* Sér.5, 73: 1-41.

Lambert, J. 1931-1932. Etude sur les échinides fossiles du Nord de l'Afrique. *Mém. Soc. géol. Fr.* Nlle.Sér. 7(2,4): 1-228.

Lambert, J. & P. Thiery 1909-1925. *Essai de nomenclature raisonnée des Echinides.* Chaumont: Ferrière L.

Laurin, B. & B. David 1988. L'évolution morphologique: un compromis entre contraintes du développement et ajustements adaptatifs. *C.R. Acad. Sci. Paris* 307, Sér. 2: 843-849.

Laurin, B. & B. David 1990. Mapping morphological changes in the spatangoid Echinocardium; applications to ontogeny and interspecific comparisons. In C. De Ridder, P. Dubois, M.C. Lahaye & M. Jangoux (eds), *Echinoderm research. Proceed. 2nd European Echinoderm Conf., Bruxelles, 1989*: 131-136. Rotterdam: Balkema.

Lawrence, J.M. 1990. The effects of stress and disturbance on echinoderms. *Zool. Sci.* 7: 17-28.

Legendre, P. 1986. Reconstructing biogeographic history using phylogenetic tree analysis of community structure. *Syst. Zool.* 35: 68-80.

Lessios, H.A. 1981. Divergence in allopatry: molecular and morphological differentiation between sea urchins separated by the isthmus of Panama. *Evolution* 35: 618-634.

Lewin, R. 1988. DNA clock conflict continues. *Science* 241: 1756-1759.

Lewis, D.N. & R.P.S. Jefferies 1980. *Salenia trisuranalis* sp. nov. (Echinoidea) from the Eocene (London Clay) of Essex, and notes on its phylogeny. *Bull. Br. Mus. nat. Hist. Geol.* 33: 115-121.

Liddell, W.D. & S.L. Ohlhorst 1982. Morphological and electrophoretic analyses of Caribbean comatulid crinoid populations. In J. Lawrence (ed.), *Echinoderms. Proceed. 4th internat. Echinoderm Conf., Tampa Bay, 1981*: 173-182. Rotterdam: Balkema.

Lu, A.L., D.A. Steege & D.W. Stafford 1980. Nucleotide sequence of a 5S ribosomal RNA gene in the sea urchin *Lytechinus variegatus. Nucleic Acids Res.* 8: 1839-1853.

Macurda, D.B. & D.L. Meyer 1977. Adaptative radiation of the comatulid crinoids. *Paleobiology* 3: 74-82.

Maluf, L.Y. 1988. Biogeography of the central eastern Pacific shelf echinoderms. In R.D. Burke, P.V. Mladenov, P. Lambert & R.L. Parsley (eds), *Echinoderm Biology. Proceed. 6th internat. Echinoderm Conf., Victoria, 1987*: 389-398. Rotterdam: Balkema.

Manwell, C. & C.M.A. Baker 1963. A sibling species of sea cucumber discovered by starch gel electrophoresis. *Comp. Biochem. Physiol.* 10: 39-53.

Maples, C.G., J.A. Waters & H.T. Walsh 1990. *Harmostocrinus jonesi* n.sp. (Crinoidea): an evolutionary intermediate between *Abrotocrinus* and *Harmostocrinus*. *J. Paleont.* 64: 141-146.

Marcus, N.H. 1977. Genetic variation within and between geographically separated populations of the sea urchin *Arbacia punctulata*. *Biol. Bull.* 153: 560-576.

Marcus, N.H. 1980. Genetics of morphological variation in geographically distant populations of the sea urchin *Arbacia punctulata*. *J. exp. mar. Biol. Ecol.* 43: 121-130.

Margoliash, E. 1963. Primary structure and evolution of cytochrome c. *Proc. Natl. Acad. Sci. USA* 50: 672-679.

Marshall, C.R. 1988. DNA-DNA hybridization, the fossil record, phylogenetic reconstruction, and the evolution of the clypeasteroid echinoids. In C.R.C. Paul & A.B. Smith (eds), *Echinoderm phylogeny and evolutionary biology*: 107-119. Oxford: Clarendon Press.

Marshall, C.R. 1990. The fossil record and estimating divergence times between lineages: maximum divergence times and the importance of reliable phylogenies. *J. Mol. Evol.* 30: 400-408.

Marshall, C.R. 1991. DNA-DNA hybridization phylogeny of sand dollars: lack of concordance with morphological phylogenies. In T. Yanagisawa et al. (eds), *Proceed. 7th internat. Echinoderm Conf., Atami, 1990*: 347.

Marshall, C.R. 1992. Character analysis and the intergration of molecular and morphological data in an understanding of sand dollar phylogeny. *Mol. Biol. Evol.* 9: 309-322.

Marshall, C.R. & H. Swift 1992. DNA-DNA hybridization phylogeny of sand dollars and highly reproductible extent of hybridization values. *J. Mol. Evol.* 34: 31-44.

Matsumura, T., M. Hasegawa & M. Shigei 1979. Collagen biochemistry and phylogeny of echinoderms. *Comp. Biochem. Physiol.* 62B: 101-105.

Matsumura, T. & M. Shigei 1988. Collagen biochemistry and the phylogeny of echinoderms. In C.R.C. Paul & A.B. Smith (eds), *Echinoderm phylogeny and evolutionary biology*: 43-52. Oxford: Clarendon Press.

Matsuoka, N. 1984. Electrophoretic evaluation of the taxonomic relationship of two species of the sea-urchin, *Temnopleurus toreumaticus* (Leske) and *T. hardwickii* (Gray). *Proc. Jap. Soc. syst. Zool.* 29: 30-36 .

Matsuoka, N. 1985. Biochemical phylogeny of the sea-urchins of the family Toxopneustidae. *Comp. Biochem. Physiol.* 80B: 767-771.

Matsuoka, N. 1987. Biochemical study on the taxonomic situation of the sea-urchin, *Pseudocentrotus depressus. Zool. Sci.* 4: 339-347.

Matsuoka, N. 1989. Biochemical systematics of four sea-urchin species of the family Diadematidae from japanese waters. *Biochem. Syst. Ecol.* 17: 423-429.

Matsuoka, N. & H. Suzuki 1987. Electrophoretic study on the taxonomic relationship of the two morphologically very similar sea-urchins, *Echinostrephus aciculatus* and *E. molaris. Comp. Biochem. Physiol.* 88B: 637-641.

Matsuoka, N. & H. Suzuki 1989a. Electrophoretic study on the phylogenetic relationships among six species of sea-urchins of the family Echinometridae found in the japanese waters. *Zool. Sci.* 6: 589-598.

Matsuoka, N. & H. Suzuki 1989b. Genetic variation and differentiation in six local japanese populations of the sea-urchin, *Anthocidaris crassispina:* electrophoretic analysis of allozymes. *Comp. Biochem. Physiol.* 92B: 1-7.

McClintock, B. 1984. The significance of responses of the genome to challenges. *Science* 226: 792-801.

McKinney, M.L. 1984. Allometry and heterochrony in an Eocene lineage: morphological change as a by-product of size selection. *Paleobiology* 10: 207-219.

McKinney, M.L. 1988. Classifying heterochrony. Allometry, size, and time. In M.L. McKinney (ed.), *Heterochrony in Evolution. A multidisciplinary approach, Topics in Geobiology* 7: 17-34.

McMillan, W.O., R.A. Raff & S.R. Palumbi (pers. comm.) Population genetic consequences of developmental evolution and reduced dispersal in sea urchins (genus *Heliocidaris*).

McNamara, K.J. 1982a. Heterochrony and phylogenetic trends. *Paleobiology* 8: 130-142.

McNamara, K.J. 1982b. Taxonomy and evolution of living species of *Breynia* (Echinoidea: Spatangoida) from Australia. *Rec. West. Aust. Mus.* 10: 167-197.

McNamara, K.J. 1985. Taxonomy and evolution of the Cainozoic spatangoid echinoid *Protenaster. Palaeontology* 28: 311-330.

McNamara, K.J. 1987. Taxonomy, evolution, and functional morphology of southern Australian Tertiary hemiasterid echinoids. *Palaeontology* 30: 319-352.

McNamara, K.J. 1988. The abundance of heterochrony in the fossil record. In M. McKinney (ed.), *Heterochrony and evolution*: 287-325. New York: Plenum Press.

McNamara, K.J. 1989. The role of heterochrony in the evolution of spatangoid echinoids. In B. David, J.L. Dommergues, J. Chaline & B. Laurin (eds), *Ontogenèse et évolution. Géobios Mém. sp.* 12: 283-295.

McNamara, K.J. & G.M. Philip 1980. Australian Tertiary schizasterid echinoids. *Alcheringa* 4: 47-65.

Metz, E.C., H. Yanagimachi & S.R. Palumbi 1991. Genetic differentiation and reproductive isolation of indo-pacific sea urchins, genus *Echinometra*. In T. Yanagisawa et al. (eds), *Proceed. 7th internat. Echinoderm Conf., Atami, 1990*: 131-137. Rotterdam: Balkema.

Meyer, C.A. 1990. Depositional environment and palaeoecology of crinoid- communities from the Middle Jurassic Burgundy-Platform of Western Europe. In C. De Ridder, P. Dubois, M.C. Lahaye & M. Jangoux (eds), *Echinoderm research. Proceed. 2nd European Echinoderm Conf., Bruxelles, 1989*: 25-31. Rotterdam: Balkema.

Mironov, A.N. 1983. Accumulation effects in distribution of sea urchins. *Zool. Zh.* 62: 1202-1208 (in russian).

Mironov, A.N. 1985. Role of dispersion in the development of the present-day echinoid faunal complexes of the tropical zone. *Oceanology* 25: 593-595.

Mooi, R. 1987. A cladistic analysis of the sand dollars (Clypeasteroida: Scutellina) and the interpretation of heterochronic phenomena. PhD Thesis, University of Toronto (unpublished).

Mooi, R. 1990a. Paedomorphosis, Aristotle's lantern, and the origin of the sand dollars (Echinodermata: Clypeasteroida). *Paleobiology* 16: 25-48.

Mooi, R. 1990b. Progenetic miniaturization in the sand dollar *Sinaechinocyamus:*

implications for clypeasteroid phylogeny. In C. De Ridder, P. Dubois, M.C. Lahaye & M. Jangoux (eds), *Echinoderm research. Proceed. 2nd European Echinoderm Conf., Bruxelles, 1989*: 137-143. Rotterdam: Balkema.

Moore, R.C. & L.R. Laudon 1943. Evolution and classification of Paleozoic crinoids. *Geol. Soc. Amer. Spec. Pap.* 46: 1-167.

Moore, R.C. & C. Teichert (eds) 1978. *Treatise of invertebrate paleontology. Part T, Echinodermata* 2. Lawrence: Geol. Soc. America & Univ. Kansas Press.

Mortensen, T. 1950. *A monograph of the Echinoidea* . T. 5 Spatangoida, vol. 1. Reitzel, Copenhagen.

Murphy, L.S., G.T. Rowe & R.L. Hardwick 1976. Genetic variability in deep-sea echinoderms. *Deep-Sea Res.* 23: 339-348.

Myers, A.A. & P.S. Giller 1988a. Process, pattern and scale in biogeography. In A.A. Myers & P.S. Giller (eds), *Analytical biogeography*: 3-21. London: Chapman & Hall.

Myers, A.A. & P.S. Giller (eds) 1988b *Analytical biogeography.* London: Chapman & Hall.

Nagaoki, S. 1985. Comparative analyses of free amino-acid composition in eggs and ovaries of sea urchins (the subclass Regularia). In B.F. Keegan & B.D.S. O'Connor (eds), *Echinodermata. Proceed. 5th internat Echinoderm Conf., Galway, 1984*: 392. Rotterdam: Balkema.

Nei, M. 1975. *Molecular population genetics and evolution.* North-Holland, Amsterdam.

Nelson, G. 1973. The higher-level phylogeny of vertebrates. *Syst. Zool.* 22: 87-91.

Nelson, G. 1978. Ontogeny, phylogeny, paleontology and the biogenetic law. *Syst. Zool.* 27: 323-345.

Nelson, G. 1985. Outgroups and ontogeny. *Cladistics* 1: 29-45.

Nelson, G. & N. Platnick 1981. *Systematics and biogeography. Cladistics and vicariance.* New York: Columbia Univ. Press.

Néraudeau, D. 1990. Ontogenèse, paléoécologie et histoire des *Hemiaster*, échinides irréguliers du Crétacé. Thèse Doctorat, Univ. Bourgogne (unpublished).

Néraudeau, D. & P. Moreau 1989. Paléoécologie et paléogéographie des faunes d'échinides du Cénomanien nord-aquitain (Charente-Maritime, France). *Géobios* 22: 293-324.

Nichols, D. 1959a. Changes in the Chalk heart-urchin *Micraster* interpreted in relation to living forms. *Phil. Trans. roy. Soc. London* (B)242: 347-437 .

Nichols, D. 1959b. Mode of life and taxonomy in irregular sea-urchins. *Spec. Publ. Syst. Assoc.* (Function and taxonomic importance) 3: 61-80.

Nichols, D. 1980. A biometrical study of the British sea-urchin *Echinus esculentus* from diving surveys of three areas over successive years. In Jangoux (ed.). *Echinoderms: present and past. Proceed. 1st European Echinoderm Conf., Bruxelles, 1979*: 209-218. Rotterdam: Balkema.

Nishihira, M., Y. Sato, Y. Arakaki & M. Tsuchiya 1991. Ecological distribution and habitat preference of four types of the sea urchin *Echinometra mathaei* on the Okinawan coral reefs. In T. Yanagisawa et al. (eds), *Proceed. 7th internat. Echinoderm Conf., Atami, 1990*: 91-104. Rotterdam: Balkema.

Nisiyama, S. 1966. The echinoid fauna from Japan and adjacent regions. Part. I. *Palaeont. Soc. Japan Spec. Pap.* 11: 1-277.

Novacek, M.J. 1987. Characters and cladograms: examples from zoological systematics. In H.M. Hœnigswald & L.F. Wiener (eds), *Biological metaphor and cladistic classification*: 181-192. Philadelphia: Univ. Pennsylvania Press.

Ohama, T., H. Hori & S. Osawa 1983. The nucleotide sequence of 5S RNAs from a sea cucumber, a starfish, and a sea urchin. *Nucleic Acids Res.* 11: 5181-5184.

Oji, T. 1990. Miocene Isocrinidae (stalked crinoids) from Japan and their biogeographic implication. *Trans. Proc. Palaeont. Soc. Japan* N.S., 157: 412-429.

Orbigny, A. D' 1850. *Prodrome de paléontologie et stratigraphie universelle des animaux mollusques et rayonnés.* 1: 1-394; 2: 1-427; 3: 1-188. Paris: Masson.

Orbigny, A. D' 1853-1855. *Paléontologie française, terrains crétacés. Echinodermes.* 6: 1-596. Paris: Masson.

Pagett, R.M. 1980. Tolerance to brackish water by ophiuroids with special reference to a Scottish sea loch, Loch Etive. In M. Jangoux (ed.), *Echinoderms: present and past. Proceed. 1st European Echinoderm Conf., Bruxelles, 1979*: 223-229. Rotterdam: Balkema.

Pajaud, D., J. Roman & M. Collignon 1976. La longue vie de Jules Lambert (1848-1940) et son legs scientifique. *Bull. Soc. géol. Fr.* Sér.7, 18: 661-674.

Palumbi, S.R. 1990. Mitochondrial DNA diversity in the sea urchins *Strongylocentrotus purpuratus* and *S. droebachiensis. Evolution* 44: 403-415.

Palumbi, S.R. & B.D. Kessing 1991. Population biology of the trans-Arctic exchange: MtDNA sequence similarity between Pacific and Atlantic sea urchins. *Evolution*: 45(8): 1790-1805.

Palumbi, S.R. & E.C. Metz 1991. Strong reproductive isolation between closely related tropical sea urchins (genus *Echinometra*). *Mol. Biol. Evol.* 8: 227-239.

Palumbi, S.R. & C.A. Wilson 1990. Mitochondrial DNA diversity in the sea urchins *Strongylocentrotus purpuratus* and *S. droebachiensis. Evolution* 44: 403-415.

Parks, A.L., B.A. Parr, J.E. Chin, D.S. Leaf & R.A. Raff 1988. Molecular analysis of heterochronic changes in the evolution of direct developing sea urchins. *J. evol. Biol.* 1: 27-44.

Patterson, C. 1981. Significance of fossils in determining evolutionary relationships. *Ann. Rev. Ecol. Syst.* 12: 195-223.

Patterson, C. 1982. Morphological characters and homology. In K.A. Joysey & A.E. Friday (eds), *Problems of phylogenetic reconstruction* Syst. Assoc. Spec. 21: 21-74. London: Academic Press.

Paul, C.R.C. 1988. The phylogeny of the cystoids. In C.R.C. Paul & A.B. Smith (eds), *Echinoderm phylogeny and evolutionary biology*: 199-213. Oxford: Clarendon Press.

Paul, C.R.C. & A.B. Smith 1984. The early radiation and phylogeny of echinoderms. *Biol. Rev.* 59: 443-481.

Penniston, J.T. 1971. High hydrostatic pressure and enzymatic activity: inhibition of multimeric enzymes by dissociation. *Arch. Biochem. Biophys.* 142: 322-332.

Perez-Ruzafa, A. & A. Lopez-Ibor 1988. Echinoderm fauna from the south-western Mediterranean – Biogeographic relationships. In R.D. Burke, P.V. Mladenov, P. Lambert & R.L. Parsley (eds), *Echinoderm Biology. Proceed. 6th internat. Echinoderm Conf., Victoria, 1987*: 355-362. Rotterdam: Balkema.

Peron, A. 1895. Notice biographique de Gustave Cotteau. *Bull. Soc. géol. Fr.* Sér.3, 23: 231-270.

Platnick, N. & G. Nelson 1978. A method of analysis for historical biogeography. *Syst. Zool.* 27: 1-16.

Plott, R. 1676. *Natural History of Oxfordshire.*

Poltaraus, A.B. 1981. The estimation of relative connections between nine species of echinodermata by molecular hybridization of their DNA. *Zh. Obsh. Biol.* 42: 55-59.

Popper, C. 1984. *L'univers irrésolu. Plaidoyer pour l'indéterminisme.* Paris: Hermann.

Qu, L. H., B. Michot & J. P. Bachellerie 1983. Improved methods for structure probing in large RNAs: a rapid 'heterologous' sequencing approach is coupled to the direct mapping of nuclease accessible sites. Application to the 5' terminal domain of eukaryotic 28S rRNA. *Nucleic Acids Res.* 11: 5903-5920.

Raabe, H. 1964. Untersuchungen am plastron von *'Holaster'. N. Jb. Geol. Paläont.* 5: 306-311.

Raff, R. A., K. G. Field, M. T. Ghiselin, D. J. Lane, G. J. Oslen, N. R. Pace, A. L. Parks, B. A. Parr & E. C. Raff 1988. Molecular analysis of distant phylogenetic relationships in echinoderms. In C. R. C. Paul & A. B. Smith (eds), *Echinoderm phylogeny and evolutionary biology*: 29-41. Oxford: Clarendon Press.

Raff, R. A. & T. C. Kaufman 1983. *Embryos, genes and evolution.* New York: Macmillan Publ. Co.

Rat, P., B. David, F. Magniez & O. Pernet 1987. Le golfe du Crétacé inférieur sur le Sud-Est du Bassin parisien: milieux (échinides, foraminifères) et évolution de la transgression. In J. Salomon (ed.), *Transgressions et régressions au Crétacé (France et régions voisines).* Mém. géol. Univ. Dijon 11: 15-29.

Ratto, A. & R. Christen 1990. Phylogénie moléculaire des échinodermes déduite de séquences partielles des ARN ribosomiques 28S. *C.R. Acad. Sci. Paris* Sér. 3, 310: 169-174.

Rey, J. 1972. Recherches géologiques sur le Crétacé inférieur de l'Estramadura (Portugal). Thèse Sci. Univ. Paul Sabatier 465 (unpublished).

Riedl, R. 1979. *Order in living organisms.* Chichester: Wiley.

Roberts, J.W., S.A. Johnson, P.M. Kier, T.J. Hall, E.H. Davidson & R.J. Britten 1985. Evolutionary conservation of DNA sequences expressed in sea urchin eggs and early embryos. *J. Mol. Evol.* 22: 99-107.

Roman, J. 1965. Morphologie et évolution des *Echinolampas* (Echinides, Cassiduloïdes). *Mém. Mus. natn. Hist. nat. (Sciences Terre)* Sér. C, 15: 1-341.

Roman, J. 1977. Biogéographie d'un groupe d'échinides cénozoïques *(Echinolampas* et ses sous-genres *Conolampas* et *Hypsoclypus). Géobios* 10: 337-349.

Roman, J., J. Roger, J. P. Platel & C. Cavelier 1989. Les échinoïdes du Crétacé et du Paléogène du Dhofar (Sultanat d'Oman) et les relations entre les bassins de l'océan Indien et de la Méditerranée. *Bull. Soc. géol. Fr.* Sér.8, 5: 279-286.

Rosen, B. R. 1985. Long-term geographical controls on regional diversity. *Open Univ. Geol. Soc. J.* 6: 25-30.

Rosen, B. R. & A. B. Smith 1988. Tectonics from fossils? Analysis of reef-coral and sea-urchin distributions from late Cretaceous to Recent, using a new method. In M.G. Audley-Charles & A. Hallam (eds), *Gondwana and Tethys. Geol. Soc.* (spec. Publ.) 37: 275-306.

Rosen, D. E. 1978. Vicariant patterns and historical explanation in biogeography. *Syst. Zool.* 27: 159-188.

Rosen, D. E. 1979. Fishes from the uplands and intermontane basins of Guatemala: revisionary studies and comparative geography. *Bull. Amer. Mus. nat. Hist.* 162: 271-375.

Rosenberg, V.A & R.P. Wain 1982. Isozyme variation and genetic differentiation in the decorator sea urchin *Lytechinus variegatus* (Lamarck, 1816). In J. Lawrence (ed.), *Echinoderms. Proceed. 4th internat. Echinoderm Conf., Tampa Bay, 1981*: 193-197. Rotterdam: Balkema.

Roux, M. 1977. Les Bourgueticrinina (Crinoidea) recueillis par la 'Thalassa' dans le

Golfe de Gascogne: anatomie comparée des pédoncules et systématique. *Bull. Mus. natn. Hist. nat. (Zool.)* Sér.3, 426: 25-83.

Roux, M. 1978. Ontogenèse et évolution des crinoïdes pédonculés depuis le Trias. Implications océanographiques. Thèse Univ. Paris Sud (Orsay) (unpublished).

Roux, M. 1979. Un exemple de relation étroite entre la géodynamique des océans et l'évolution des faunes benthiques bathyales et abyssales: l'histoire des crinoïdes pédonculés du Mésozoïque à l'Actuel. *Bull. Soc. géol. Fr.* Sér.7, 21: 613-618.

Roux, M. 1982. De la biogéographie historique des océans aux reconstitutions paléobio-géographiques: tendances et problèmes illustrés par des exemples pris chez les échinodermes bathyaux et abyssaux. *Bull. Soc. géol. Fr.* Sér. 7, 24: 907-916.

Roux, M. 1987. Evolutionary ecology and biogeography of recent stalked crinoids as a model for the fossil record. In M. Jangoux & J.M. Lawrence (eds), *Echinoderm Studies* 1: 1-53. Rotterdam: Balkema.

Roux, M. & J.C. Plaziat 1978. Inventaire des Crinoïdes et interprétation paléobathymétrique de gisements du Paléogène pyrénéen franco-espagnol. *Bull. Soc. géol. Fr.* Sér. 7, 20: 299-308.

Rowe, A.W. 1899. An analysis of the genus *Micraster,* as determined by rigid zonal collecting from the zone of *Rhynchonella cuvieri* to that of *Micraster coranguinum. Quart. J. Geol. Soc. London* 55: 494-547.

Rowe, F.W.E. 1985. Preliminary analysis of distribution patterns of Australia's non-endemic, tropical echinoderms. In B.F. Keegan & B.D.S. O'Connor (eds), *Echinodermata. Proceed. 5th internat Echinoderm Conf., Galway, 1984*: 91-98. Rotterdam: Balkema.

Sadler, P.M. 1981. Sediment accumulation rates and the completeness of stratigraphic sections. *J. Geol.* 89: 569-584.

Sarasin, C. 1909. Perceval de Loriol. *Verhandl. schweiz. naturf. Gesell.*: 1-13.

Sarich, W.M., C.W. Schmidt & J. Marks 1989. DNA hybridization as a guide to phylogenies: a critical analysis. *Cladistics* 5: 3-32.

Schindel, D.E. 1982. Resolution analysis: a new approach to gaps in the fossil record. *Paleobiology* 8: 340-353.

Schmid, F. 1972. *Hagenowia elongata* (Nielsen), ein hochspezialisierter Echinide aus dem höheren Untermaastricht NW-Deutschlands. *Geol. Jb.* A4: 177-195.

Schoch, R.M. 1986. *Phylogenetic reconstruction in palaeontology.* New York: Van Nostrand Reinhold.

Schopf, T.J.M. & L.S. Murphy 1973. Protein polymorphism of the hybridizing seastars *Asterias forbesi* and *Asterias vulgaris* and implications for their evolution. *Biol. Bull.* 145: 589-597.

Schulz, M.G. & W. Weitschat 1975. Phylogenie und Stratigraphie der Asteroideen der nordwestdeutschen Scheibkreide. Teil 1: *Metopaster / Recurvaster* und *Calliderma / Chomataster* Gruppe. *Mitt. Geol. Paläont. Inst. Univ. Hamburg* 44: 249-284.

Schulz, M.G. & W. Weitschat 1981. Phylogenie und Stratigraphie der Asteroideen der nordwestdeutschen Scheibkreide. Teil 2: *Crateraster / Teichaster* Gruppe und Gattung *Ophryraster. Mitt. Geol. Paläont. Inst. Univ. Hamburg* 51: 27-42.

Seilacher, A. 1979. Constructional morphology of sand dollars. *Paleobiology* 5: 191-221.

Sengör, A.M.C., D. Altiner, A. Cin, T. Ustaomer & K.J. Hsu 1988. Origin and assembly of the Tethyside orogenic collage at the expense of Gondwana-land. In M.G. Audley-Charles & A. Hallam (eds), *Gondwana and Tethys. Geol. Soc. Spec. Publ.* 37: 119-181.

Sibley, G.C. & J.E. Ahlquist 1983. Phylogeny and classification of birds based on the data of DNA-DNA hybridization. In R.F. Johnston (ed.), *Current ornithology*: 245-292. New York: Plenum Press.

Sibley, G.C. & J.E. Ahlquist 1984. The phylogeny of the hominoid primates, as indicated by DNA-DNA hybridization. *J. Mol. Evol.* 20: 2-15.

Sibley, G.C., J.A. Comstock & J.E. Ahlquist 1990. DNA hybridization evidence of hominoid phylogeny: a reanalysis of the data. *J. Mol. Evol.* 30: 202-236.

Sibuet, M. 1979. Distribution and diversity of Asteroids in Atlantic abyssal basins. *Sarsia* 64: 85-91.

Sibuet, M. 1985. Quantitative distribution of echinoderms (Holothuroidea, Asteroidea, Ophiuroidea, Echinoidea) in relation to organic matter in the sediment, in deep sea basins of the Atlantic Ocean. In B.F. Keegan & B.D.S. O'Connor (eds), *Echinodermata. Proceed. 5th internat Echinoderm Conf., Galway, 1984*: 99-108. Rotterdam: Balkema.

Sibuet, M., C. Monniot, D. Desbuyeres, A. Dinet, A. Khripounoff, G. Rowe & M. Segonzac 1984. Peuplements benthiques et caractéristiques trophiques du milieu dans la plaine abyssale de Demerara. *Oceanol. Acta* 7: 345-358 .

Simms, M.J. 1988. The phylogeny of post-Palaeozoic crinoids. In C.R.C. Paul & A.B. Smith (eds), *Echinoderm phylogeny and evolutionary biology*: 267-284. Oxford; Clarendon Press.

Simpson, G.G. 1950. Evolutionary determinism and the fossil record. *Scientif. Month.* 71: 262-627.

Simpson, G.G. 1953. *The major features of evolution.* Columbia Biol. Ser. 27.

Smiley, S. 1988. The phylogenetic relationship of holothurians: a cladistic analysis of the extant echinoderm classes. In C.R.C. Paul & A.B. Smith (eds), *Echinoderm phylogeny and evolutionary biology*: 69-84. Oxford; Clarendon Press.

Smith, A.B. 1980. The structure and arrangement of echinoid tubercules. *Phil. Trans. roy. Soc. London* (B)289: 1-54.

Smith, A.B. 1981. Implications of lantern morphology for the phylogeny of post-Paleozoic echinoids. *Palaeontology* 24: 779-801.

Smith, A.B. 1983. British Carboniferous Edrioasteroidea (Echinodermata). *Bull. Br. Mus. nat. Hist. (Geol)* 37: 113-138.

Smith, A.B. 1984a. *Echinoid Palaeobiology* (Special topics in palaeontology: 1). London: Allen & Unwin,

Smith, A.B. 1984b. Classification of the Echinodermata. *Palaeontology* 27: 431-459.

Smith, A.B. 1985. Cambrian eleutherozoan echinoderms and the early diversification of edrioasteroids. *Palaeontology* 28: 715-756.

Smith, A.B. 1988a. Phylogenetic relationship, divergence times, and rates of molecular evolution for camarodont sea urchins. *Mol. Biol. Evol.* 5: 345-365.

Smith, A.B. 1988b. Fossil evidence for the relationships of extant echinoderm classes and their times of divergence. In C.R.C. Paul & A.B. Smith (eds), *Echinoderm phylogeny and evolutionary biology*: 85-97. Oxford: Clarendon Press.

Smith, A.B. 1989. RNA sequence data in phylogenetic reconstruction: testing the limits of its resolution. *Cladistics* 5: 321-344.

Smith, A.B. & C.W. Wright 1989. British Cretaceous Echinoids. Part 1, General introduction and Cidaroida. *Monogr. Palaeontogr. Soc. London Publ.* 578: 1-101.

Smith, M.J., R. Nicholson, M. Stuerzul & A. Lui 1982. Single copy DNA homology in sea stars. *J. Mol. Evol.* 18: 92-101.

Spencer, W. K. 1913. The evolution of Cretaceous Asteroida. *Phil. Trans. Roy. Soc. London* (B)204: 99-177.

Sprinkle, J. 1973. Morphology and evolution of blastozoan echinoderms. *Spec. Publ., Mus. comp. Zool. Harvard Univ.*: 1-283.

Sprinkle, J. 1982. Cylindrical and globular rhombiferans. In J. Sprinkle (ed.), *Echinoderm faunas from the Bromide formation (Middle Ordovician) of Oklahoma.* Paleont. Contr. Monogr. 1: 231-273. Univ. Kansas.

Sprinkle, J. 1983. Patterns and problems in echinoderm evolution. In M. Jangoux & J. M. Lawrence (eds), *Echinoderm Studies* 1: 1-18. Rotterdam: Balkema.

Sprinkle, J. & M. Bell 1978. Paedomorphosis in edrioasteroid echinoderms. *Paleobiology* 4: 82-88.

Stanley, S. M. 1975. A theory of evolution above species level. *Proc. Nat. Acad. Sci. USA* 72: 646-650.

Stanley, S. M. 1979. *Macroevolution. Pattern and process.* San Francisco: Freeman and Co.

Stevens, P. F. 1980. Evolutionary polarity of character states. *Ann. Rev. Ecol. Syst.* 11: 333-358.

Stickle, W. B. & W. J. Diehl 1987. Effects of salinity on echinoderms. In M. Jangoux & J. M. Lawrence (eds), *Echinoderm Studies* 2: 235-285. Rotterdam: Balkema.

Stokes, R. B. 1975. Royaumes et provinces fauniques du Crétacé établis sur la base d'une étude systématique du genre *Micraster. Mém. Mus. natn. Hist. nat. Nlle.* Sér.(C), 31: 1-94.

Thierry, J. 1985. Evolutionary history of Jurassic Collyritidae (Echinoidea, Disasteroidea) in the Paris basin (France). In B. F. Keegan & B. D. S. O'Connor (eds), *Echinodermata. Proceed. 5th internat Echinoderm Conf., Galway, 1984*: 125-133. Rotterdam: Balkema.

Tintant, H. 1963. Les Kosmocératidés du Callovien inférieur et moyen d'Europe occidentale. *Publ. Univ. Dijon* 29: 1-500; 1 atlas, 58 pl.

Tintant, H. 1986. La loi et l'événement. Deux aspects complémentaires des Sciences de la Terre. *Bull. Soc. géol. Fr.* Sér. 8, 2: 185-190.

Tsuchiya, M. & M. Nishihira 1985. Agonistic behavior and its effect on the dispersion pattern in two types of the sea urchin, *Echinometra mathaei* (Blainville). *Galaxea* 4: 37-48.

Tsuchiya, M. & M. Nishihira 1986. Re-colonization process of two types of the sea urchin, *Echinometra mathaei* (Blainville), on the Okinawa reef flat. *Galaxea* 5: 283-294.

Ubaghs, G. 1968. Stylophora. In R. C. Moore (ed.), *Treatise on invertebrate palaeontology, Part S: Echinodermata* 1: 496-565. Lawrence: Geol. Soc. America & Univ. Kansas Press.

Ubaghs, G. 1975. Early paleozoic echinoderms. *Rev. Earth planet. Sci.* 3: 79-98.

Uehara, T. 1991. Speciation of indo-pacific *Echinometra.* In T. Yanagisawa et al. (eds), *Proceed. 7th internat. Echinoderm Conf., Atami, 1990*: 139. Rotterdam: Balkema.

Uehara, T. & M. Shingaki 1985. Taxonomic studies in the four types of the sea urchin, *Echinometra mathaei,* from Okinawa, Japan. *Zool. Sci.* 2: 1009 .

Uehara, T., M. Shingaki & K. Taira 1986. Taxonomic studies in the sea urchin, genus *Echinometra,* from Okinawa and Hawaii. *Zool. Sci.* 3: 1114.

Uehara, T., M. Shingaki, K. Taira, V. Arakaki & H. Nakatomi 1991. Chromosome studies in eleven Okinawan species of sea urchins, with special reference to four species of

Indo-Pacific *Echinometra*. In T. Yanagisawa et al. (eds), *Proceed. 7th internat. Echinoderm Conf., Atami, 1990*: 119-129. Rotterdam: Balkema.

Vadon, C. 1990. *Ophiozonella novacaledoniae* n. sp. (Ophiuroidea, Echinodermata): description, ontogeny and phyletic position. *J. nat. Hist.* 24: 165-179.

Voorzanger, B. & W.J. van der Steen 1982. New perspectives on the Biogenetic Law? *Syst. Zool.* 31: 202-205.

Wagner, C.D. & J.W. Durham 1966. Holasteroids. In R.C. Moore (ed.), *Treatise on invertebrate palaeontology, Part U: Echinodermata* 3, 2: 523-543. Lawrence: Geol. Soc. America & Univ. Kansas Press.

Waters, J.A. 1977. Quantification of shape by use of Fourier analysis: the Mississipian blastoid genus *Pentremites*. *Paleobiology* 3: 288-299.

Waters, J.A. 1988. The evolutionary palaeoecology of the Blastoidea. In C.R.C. Paul & A.B. Smith (eds), *Echinoderm phylogeny and evolutionary biology*: 215-233. Oxford: Clarendon Press.

Waters, J.A. 1990. The paleobiogeography of the Blastoides (Echinodermata). In W.S. McKerrow & C.R. Scotese (eds), *Palaeozoic Palaeogeography and Biogeography*. *Geol. Soc. Mem.* 12: 339-352.

Waters, J.A., T.W. Broadhead & A.S. Horowitz 1982. The evolution of *Pentremites* (Blastoidea) and Carboniferous crinoid community succession. In J. Lawrence (ed.), *Echinoderms. Proceed. 4th internat. Echinoderm Conf., Tampa Bay, 1981*: 133-138. Rotterdam: Balkema.

Waters, J.A., A.S. Horowitz & D.B. Macurda Jr 1985. Ontogeny and phylogeny of the Carboniferous blastoid Pentremites. *J. Paleont.* 59: 701-712.

Waters, J.A. & G.D. Sevastopulo 1985. The paleobiogeography of Irish and British Lower Carboniferous blastoids. In B.F. Keegan & B.D.S. O'Connor (eds), *Echinodermata. Proceed. 5th internat Echinoderm Conf., Galway, 1984*: 141-147. Rotterdam: Balkema.

Wiley, E.O. 1975. Karl R. Popper, systematics, and classification: a reply to Walter Bock and other evolutionary taxonomists. *Syst. Zool.* 24: 233-243 .

Wiley, E.O. (1979) Ancestors, species, and cladograms. In J. Cracraft & N. Eldredge (eds), *Phylogenetic analysis and paleontology*: 211-225. New York: Columbia Univ. Press.

Wiley, E.O. 1980. Phylogenetic systematics and vicariance biogeography. *Syst. Botany* 5: 194-220.

Wiley, E.O. 1981. *Phylogenetics: the theory and practice of phylogenetic systematics*. New York: Wiley.

Wright, T. 1878. Monograph on the British fossil Echinodermata from the Cretaceous formations. Vol.1, part 8, Spatangidae and Echinocoridae. *Monogr. Palaeontogr. Soc.* 32: 265-300, pl. 62-69.

Wu, C.I. & W.H. Li 1985. Evidence for higher rates of nucleotide substitution in rodents than in man. *Proc. Natl. Acad. Sci. USA* 82: 1741-1745.

Yanagisawa, T. 1988. Base sequence complexity of sea urchin DNA. In R.D. Burke, P.V. Mladenov, P. Lambert & R.L. Parsley (eds), *Echinoderm Biology. Proceed. 6th internat. Echinoderm Conf., Victoria, 1987*: 297-298. Rotterdam: Balkema.

Zaghbib-Turki, D. 1989. Les échinides indicateurs des paléoenvironnements: un exemple dans le Cénomanien de Tunisie. *Ann. Paléont. (Vert.-Invert.)* 75: 63-81.

Zaghbib-Turki, D. 1990. Stratégie adaptative des *Hemiaster* du Crétacé supérieur (Cénomanien-Coniacien) de la plate-forme carbonatée de Tunisie. In C. De Ridder,

P. Dubois, M.C. Lahaye & M. Jangoux (eds), *Echinoderm research. Proceed. 2nd European Echinoderm Conf., Bruxelles, 1989*: 49-56. Rotterdam: Balkema.

Zoeke, E. 1951. Etude des plaques des *Hemiaster* (Echinides). *Bull. Mus. natn. Hist. nat.* 23: 696-705.

Zuckerkandl, E. & L. Pauling 1962. Molecular disease, evolution, and genetic heterogeneity. In M. Kasha & B. Pullman (eds), *Horizons in biochemistry*: 189-225. New York: Academic Press.

Comparative physiology of echinoderm muscle

ROBERT B. HILL

Department of Zoology, University of Rhode Island, Kingston, Rhode Island, USA

Final manuscript acceptance: September 1990

CONTENTS

1 PHYSIOLOGICAL CHEMISTRY

1.1 *Calcium regulation*

Echinoderms have been classified among organisms with single myosin-control, since myosin-control and actin-control do not operate simultaneously in their muscles as in the majority of invertebrates (Lehman & Szent-Györgyi 1975). Calcium may trigger contraction by regulating the interaction between actin and myosin, acting on the thick myosin filaments in myosin-linked regulation or on components of the thin actin filaments in actin-linked regulation (Lehman et al. 1973). Lehman et al. reported myosin-

81

linked regulation in *Sclerodactyla briareus* (then called *Thyone briareus*). Lehman & Szent-Györgyi (1975) found myosin-control in lantern retractor muscles (LRM) of *S. briareus* and *Cucumaria frondosa*. Myosin control is supported by the observation that longitudinal muscles of the body wall (LMBW) of *Stichopus japonicus* contains myosin with ATPase activity which is weak, but similar in several properties to myosin ATPase of rabbit striated muscle (Furukohri 1971a, b).

Kerrick & Bolles (1982) used the longitudinal muscles of the body wall (LMBW) of *Parastichopus californicus* to test the hypothesis that, early in metazoan evolution, Ca^{++}-regulation of contraction was accomplished through a myosin light-chain kinase/phosphatase system. For this purpose, Kerrick & Bolles placed strips of LMBW of *P. californicus* in a chemically relaxing solution, which was believed to solubilize the sarcolemma, so that control of the ionic environment of the strips would provide control of the ionic environment of the intracellular proteins. Incorporation of ^{32}P into the myosin light chains corresponded to muscle contraction at a pCa of 5.0, and dephosphorylation of the myosin light chains corresponded to relaxation in 10^{-8} M Ca^{++}. Thus, according to the criteria used by Kerrick & Bolles, LMBW of *P. californicus* is activated by a myosin light-chain kinase/phosphatase system, but the myosin light chain kinase is different from that of vertebrate smooth muscle since agents such as calmodulin from pork and bovine brain had no effect on Ca^{++}-activated force. An interesting point noted by Kerrick & Bolles was that fibers of LMBW could not sustain force; that is, 'skinned' fibers contract maximally at first in high Ca^{++}, but the force then gradually diminishes, under the stress of a 'tension transducer.'*

1.2 *Ionic distribution*

Steinbach (1937) first reported the relative concentrations of potassium and chloride in LMBW and extracellular space of *Sclerodactyla briareus*.** Potassium appeared to be concentrated in the cells at about 300 mM, while the total potassium concentration of the LMBW was 15-20× greater than that of extracellular space. Chloride diffused freely into the extracellular space and remained proportional to the chloride concentration of the medium.

---

*This was similar to the transducer used by Hallam & Podolsky (1969) and would thus have had a compliance of 0.18 μm/mg over a force range from 0.5 mg to 900 mg.

**At that time, Steinbach followed Hall (1927) in regarding the muscle cells of the LMBW as 'fibrils,' the bundles of cells as 'multinucleate' fibers, and the external lamina of each bundle as an outer cell membrane. Modern terminology is used above in reporting work which originally used the Hall terminology. Later, Steinbach (1940) followed Olson (1938) and made the same corrections in terminology.

Freeman & Simon (1964) found the extracellular space of LMBW of *Stichopus mollis* to be about 40%. Extracellular space values of 29.7% ± 0.9% obtained by Simon, Muller & Dewhurst (1963) led to an estimate of intracellular ionic levels at 241 ± 9 mM/kg Na^+, 100 ± mM/kg K^+, and 214 ± 26 mM/kg Cl^- (see comments by Robertson, 1980). Both Steinbach (1940) and Simon et al. (1963) found that isolated holothuroid LMBW exchange Na^{++} and Cl^- rapidly with the medium, but show fast and slow fractions in exchange of K^+ with the medium.

Robertson (1980) showed that inorganic ions and phosphates provide 60-70% of the osmoticity of echinoderm muscle. His analysis of 16 components of the coelomic fluid and muscles (using LMBW of *Parastichopus tremulus* and lantern jaw muscles of *Echinus esculentus* and *Strongylocentrotus droebachiensis*) showed that the inorganic ion content of coelomic fluid was nearly identical to that of seawater, although bicarbonate ion concentration was somewhat greater in coelomic fluid (3.5-5.6 mM/kg, vs 2.1-2.3 mM/kg for seawater). Extracellular inulin space of *P. tremulus* LMBW was 28.5 ± 3.4% of total muscle water, compared to 17.8% extracellular tissue space in *S. briareus* (Steinbach 1937). The LMBW of *P. tremulus* has a low $[Na]_i$ and high $[K]_i$ compared to coelomic fluid (Robertson, 1980). Robertson provided a valuable review of Na^+ and K^+ levels reported from LMBW of *Caudina chilensis* (Koizumi 1935), and *Isostichopus badionotus* (Zanders & Herrera 1974, Madrid et al. 1976). Overall, these investigations indicate that while the perivisceral fluids bathing echinoderm muscle have the inorganic ion distribution of seawater, the intracellular ionic balance is relatively low in sodium, chloride, and magnesium but high in potassium and phosphorus. Calcium ion has a corrected value of 6.6 mM in muscle cytoplasm, but 10.3 mM in perivisceral fluid. These measurements indicate a 'conventional' distribution of extracellular vs. intracellular inorganic ions.

1.3 *Metabolism*

Mattisson (1959) asked whether or not the normal vertebrate pattern for the terminal oxidative stages of respiration was to be found in lower invertebrates, which may live in extreme environments. He then included the holothuroid *Parastichopus tremulus* in a group to be studied in order to contrast metabolism in very slow animals with metabolism in highly active animals (such as the lobster *Nephrops norvegicus*). *P. tremulus* was also considered to live exposed to a very low ambient oxygen level. Mattisson found that *P. tremulus* had no detectable level of cytochrome c or cytochrome oxidase in the muscles studied and only the extremely low qO_2 of 31 in air and 54 in oxygen (which may be compared to values of 298 and 502 for the walking legs of *N. norvegicus*). This is considered to correspond to only intermittent and slow mobility in the life of *P. tremulus*. The body wall muscles (LMBW) of *P. tremulus* were later (Mattisson 1961) found to show a cyanide-sensitive increase in oxygen consumption from about 30 to about 250 µl/g/wet weight/hour in the presence of ascorbate as a substrate. Mattisson concluded that oxidation of p-phenylamine diamine and ascorbate by LMBW of *P. tremulus* was not localized in mitochondria and was carried out by a metal compound, but not cytochrome oxidase and the copper-containing enzymes.

When the LMBW of *Stichopus mollis* is caused to contract by application of 100 mM KCl, oxygen uptake increases (Gay & Simon 1964). Possibly oxygen uptake may be used as an index of excitation-contraction coupling in echinoderm muscle.

Herrera & Plaza (1981) showed that oxygen consumption is about the same in LMBW of the Venezuela holothuroid *Holothuria glaberrima* (62 µl 0_2/h g wet wt) as in Gay & Simon's Australian holothuroid *Stichopus mollis* (24 µl 0_2/h/g wet wt) and in Mattisson's Norwegian holothuroid *Parastichopus tremulus*. Some question arises as to whether the low respiratory rate of LMBW is due to low availability of oxygen. On the one hand, the muscles are thin and highly permeable. On the other hand, they lack a vascular system and are located about as far from the lumen of the respiratory trees as possible. This question has been attacked by Bianconcini et al. (1985) with respect to the lantern muscles of a Brazilian echinoid (*Echinometra lucunter*). Their findings may be interpreted to mean that, in fact, increased accessibility of 0_2 does not lead to increased respiration. The lantern muscles of *E. lucunter* have a mean respiratory rate of about 250 µl/O_2/g dry wt/h in air or O_2, which is actually less than the 310 µl/O_2/g dry wt/hr in air or 540 µl/O_2/g dry wt/h in O_2 which they report from Mattisson (1959) for LMBW of *Parastichopus tremulus*. Bianconcini et al. (1985) thought this is because the dry wt is low in *P. tremulus*. This leads to the following comparison of Q_{O2} values, setting aside *P. tremulus*:
  – Lantern muscles of *E. lucunter* ca. 249,
  – LMBW of *Holothuria grisea* ca. 182,
  – LMBW of *Holothuria glaberima* ca. 137.
The difference does not seem very great, but Bianconcini et al. (1985) suggested that the lantern muscles respire more actively because they are continuously active.

2 VISCOUS-ELASTIC PROPERTIES

A number of unusual physical properties of holothuroid muscle have long attracted attention. The LMBW of holothuroids is extremely extensible without physical damage; for instance, that of *Sclerodactyla briareus* may contract to 5 mm after being extended to a relaxed length of 50 or 60 mm

Figure 1. Application of a series of quick releases to a tetanized LMBW of *Holothuria nigra* leads to redevelopment of force in decreasing degrees (Hill 1926).

(Olson 1938). Isolated LMBW of *Isostichopus badionotus*, which has a relaxed length of 200 mm, may stretch to 350 mm and contract to about 50 mm (Galambos 1941). Even under tension, LMBW of *Holothuria nigra* may shorten from a relaxed length of more than 150 mm to less than 25% of that (Hill 1926). Hill (1926) showed that LMBW of *H. nigra* redevelops force after a quick release during a twitch, but the redeveloped force never rises to the level of the initial twitch. However, force redeveloped repeatedly, following repeated quick releases during a tetanus, with the same time course as at the initiation of the tetanus (Fig. 1). Similarly, force redeveloped repeatedly after quick releases during a contracture of LMBW of *H. nigra* in 0.76% KCl. Force also redevelops after a quick release during isometric tension induced by acetylcholine (ACh) in LMBW of *Isostichopus badionotus* (Tsuchiya 1985). Hill (1926) pointed out that the ability to redevelop force after release during a tetanus or contracture must be due to a molecular rearrangement in the muscle fiber.

As Hill pointed out (1926), it had already been deduced that contraction indicated the formation of a molecular interaction through the muscle substance, whether at the beginning of a tetanus or later, when the tetanized muscle was still capable of redeveloping force after a sudden release. At that time, the 'viscous-elastic' properties of contractile tissue had been described on the basis of observations of fast amphibian voluntary muscle. Hill (1926), looking for a slow contractile tissue, visited the Citadel Hill laboratory of the Marine Biological Association of the United Kingdom, at Plymouth, where C. F. A. Pantin and G. P. Wells introduced him to the extremely slow longitudinal muscles of the body wall of *Holothuria nigra* (= *Holothuria forskali*). Using these muscles, Hill found that it was strikingly clear that stretch or release of activated muscle gave evidence of the same process involved in the 'original development of the response'. This could not be 'due to simple physical 'viscosity' but to some molecular rearrangement' or 'formation of some molecular pattern in the muscle fibre' (Hill 1926). Thus echinoderm muscle played an important role in the development of muscle theory, which

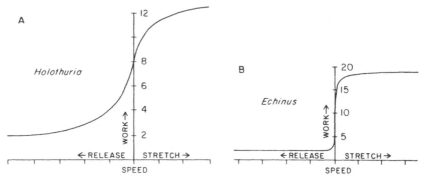

Figure 2. Work-speed curves from echinoderm muscles (Levin & Wyman 1927). A. LMBW of *Holothuria nigra*. B. Lantern muscle of *Echinus esculentus*. In these curves, arbitrary units proportional to the work done by the muscle appear on the vertical axis, while arbitrary units proportional to the speed of movement appear on the horizontal axis. Speed in stretches is plotted as positive (lengthening) while speed in releases is plotted as negative (shortening). The resulting work-speed curves show upper and lower limits for work, at high speeds of lengthening or shortening.

first led to the application of X-ray crystallography and finally to the 'creeping filament' theory.

Levin & Wyman (1927) obtained work-speed curves from both LMBW of *H. nigra* and lantern muscle of *Echinus* (Fig. 2); curves which were obtained by plotting work done by the muscle (during quick stretches or releases) against speed of movement. These curves show upper and lower asymptotic limits for work done, at higher speeds of stretch or release. In this way, the muscles behaved according to the well-known viscous-elastic model of Levin & Wyman (1927) in which muscle is modelled by a mechanical system containing damped and undamped springs in series. Later, Galambos restudied holothuroid muscle, with the aim of clarifying the opposing views of Bozler and Jordan (reviewed by Galambos, 1941).

Whereas Hill (1926) and Levin & Wyman (1927) had used slow holothuroid muscle to resolve questions raised in studies of rapid vertebrate voluntary striated muscle, Galambos (1941) wished to resolve questions raised in studies of vertebrate smooth muscle. Galambos reviewed the opposing points of view of Emil Bozler and Hermann Jordan. Essentially, Bozler considered that the viscous properties of the contractile elements of smooth muscle determined both the time course of physiological relaxation after a contraction and the time course of 'release of tension' in a maintained stretch. Jordan considered that plastic changes took place during 'release of tension,' while physiological relaxation was quite a different phenomen. Galambos (1941) visited the Bermuda Biological Station for Research, where he used a device suggested by A.C. Redfield to record 'release of tension' and physiological relaxation simultaneously from longitudinal muscles of the body wall of *Isostichopus badionotus* (then *Stichopus moebii*). Following Emil Bozler, Galambos himself used the term 'relaxation' to signify (in the usual sense) the loss of active force at the end of a physiological contraction, in contradistinction to 'release of tension' in a muscle stretched to a new length, at which there is passive tension, which gradually decays. (Following Hermann Jordan, release of tension may be due to plastic changes different from the physiological changes during relaxation). Galambos (1941) showed that release of tension from a stretched length, in LMBW of *Isostichopus badionotus*, was a slow steady process which proceeded on a curve unaffected by superimposed contractions and rapid relaxations. His results proved that release of tension is not due to the same fundamental process as relaxation.

Van Weel (1955) tested the hypotheses of H. Jordan and his pupils (see Galambos 1941) by means of experiments with deformation under maintained stretch, carried out with LMBW of *Holothuria atra*, *Holothuria monacaria*, *Holothuria edulis*, *Stichopus tropicalis*, *Opheodesoma spectabilis* and *Actinopyga mauritiana*. The results all dealt with passive behavior under stretch, which may follow either a plastic time course, of gradual continuing extension under a small load, or an elastic time course, in which a muscle may show resistance in a limited extension under a large load, and then elastic rebound when the load is removed. Following gradual extension, rebound is never complete (22% to 40%).

Application of a quick stretch to isolated LMBW of *Holothuria nigra* induced a quick rise in passive tension, sometimes followed by an active response consisting of a slowly developing rise and fall of force (Hill 1926). Prosser & Mackie (1980) asked whether applied stretch stimulated isolated holothuroid muscle by direct activation or by an action on nerves contained in the preparation. To determine this, they conducted experiments with LMBW of *Stichopus parvimensis*, *Parastichopus californicus*, *Cucumaria minata*, *Eupentacta pseudoquinquisemita*, and *Leptosynapta clarki*; and with pharynx retractors of *Cucumara minata*. These muscles all responded to quick stretches with slow active contractions, much as in *Holothuria nigra* (Hill 1926). Mechanical wave summation of active responses was elicited when quick stretches were applied in rapid succession. The active responses were slow but indicate maintained active state, since tension redeveloped following quick releases (Fig. 3). No electrical responses could be detected in association with the active responses to stretch-stimuli. These are probably neurally mediated, since agents such as lidocaine reduce the active responses to stretch-stimuli, without blocking responses to electrical stimulation. The system involved is probably cholinergic, since active responses to stretch-stimuli are blocked by tubocurarine and enhanced by physostigmine (Fig. 4).

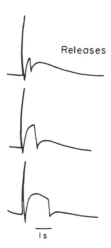

Releases

I s

Figure 3. Quick releases followed by redevelopment of tension in stretch responses of the LMBW of *Cucumaria minata* (Prosser & Mackie 1980). In each example, the first peak is a stretch-stimulus, the second peak is an initial muscle contraction and the third peak is redevelopment of tension.

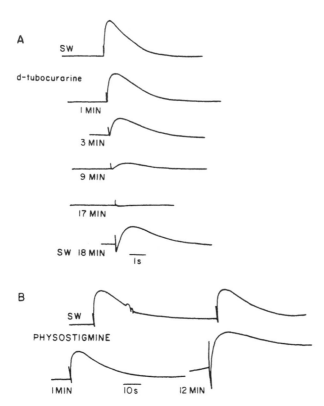

Figure 4A. Stretch responses of the pharynx retractor of *Cucumaria minata* are blocked by 0.5×10^{-7} M d-tubocurarine. B. Stretch responses are enhanced by physostigmine (Prosser & Mackie 1980).

88

3 ULTRASTRUCTURE

3.1 *Subcellular contractile apparatus*

The general nature of the subcellular organization of the myofilaments of echinoderm muscle began to be apparent in the first electron micrographs (e.g., Kawaguti & Kamishima 1964, Shida 1971). Muscle fibers of the LMBW of *Stichopus japonicus* contain thick filaments (15 to 35 nm diameter), thin filaments (5 nm diameter) and dense bodies (Shida 1971). The relaxed fibers averaged 2.5 μm in diameter, close to the diameter of 3 to 5 μm reported for cells of the LMBW of *Stichopus mollis*, which are 350 μm long (Freeman & Simon 1964). The smooth muscle cells of the tube feet of several echinoids are also quite similar. *Strongylocentrotus franciscanus, Arbacia lixula*, and *Echinus esculentus* have smooth muscle cells 1 μm to 3 μm in diameter and over 90 μm long, with thick filaments (35-40 nm diameter) surrounded by more numerous thin filaments (8-10 nm diameter) and associated with dense bodies (Florey & Cahill 1977). In general, muscle fibers of holothuroids, echinoids, and asteroids contain a rather simple organization of thick and thin filaments (reviewed by Candia Carnevali & Saita 1985).

The contractile apparatus of the LMBW cells of *Isostichopus badionotus* also has thick filaments and has thin filaments attached to dense bodies, which in turn may be attached to the inner surface of the plasma membrane (Hill et al. 1978). However, dense bodies appear to be scattered at random among the thick and thin filaments of fibers of the LMBW of *Sclerodactyla briareus* (Chen 1983). In this muscle, the tapered thick filaments range up to 3 μm long and 40 nm in diameter, with a 14.5 nm periodicity, suggesting paramyosin. There is a very high ratio of thin (8 nm diameter) filaments to thick filaments (Plate 1), with some fields of thin filaments lacking thick filaments. In a stretched muscle, there may be 12 thin filaments in orbit around one thick filament (calculated by counting thin filaments in a semicircular orbit; Plate 1).

Crinoid arm muscles have several fiber types and a more complex organization of myofilaments (Candia Carnevali & Saita 1985). Flexor muscle bundles of the arms contain central 'A' fibers, with an array of thick filaments (20 to 40 nm in diameter) that appears as a precise hexagonal pattern in cross sections. Thick filaments are surrounded by thin filaments. Peripheral 'B' fibers of the flexor muscles have also extremely thick filaments (more than 100 nm in diameter) which appear to be oversized fused myosin filaments, and which are also surrounded by thin actin filaments. Candia Carnevali & Saita (1985) consider the 'A' and 'B' fibers to be obliquely striated, but the flexor muscles also contain 'C' fibers, which are typical smooth muscle, without any particular organization of thick and thin filaments.

3.2 Subcellular vesicles

The smooth muscle fibers of the LMBW of *Stichopus japonicus* contain small vesicles, all at the periphery of the cells (Shida 1981). In *Isostichopus badionotus*, similar vesicles are found in the cytoplasm near the cell surface (Hill et al. 1978). The visibility of these vesicles in electron micrographs varies with the osmolarity of the fixative employed (Hill et al. 1982). When the fixative is prepared in filtered sea water, the vesicles are hard to see, since they are flattened and closely apposed to the sarcolemma. The vesicles become most visible in osmotically swollen preparations, when they may be 0.15 µm by 0.8 µm. Aside from such vesicles, the cytoplasm of LMBW myofibers lacks SR. However, cisternae of SR form an anastomosing network 8.5 nm from the inner surface of the sarcolemma in podial retractor muscles of the asteroid *Sylasterias forresi* (Cavey & Wood 1979). These cisternae are at places associated with the sarcolemma by a dyadic plaque and in freeze-fracture replicas arrays of particles are associated with the dyadic couplings. Fibers of the flexor muscle bundles of the arm of the comatulid *Antedon mediterranea* have SR represented by only a few small subsarcolemmal cisternae, which often form dyadic couplings with the sarcolemma (Candia Carnevali & Saita 1985). Chen (1983) prefers the term 'sacs,' for the subsarcolemmal cisternae or vesicles of *Sclerodactyla briareus* (Plate 2a), since these are long, flattened structures which run parallel to the adjacent sarcolemma for considerable distances. The sacs are generally about 25 nm thick, but up to 0.6 µm in width, and as long as 1.4 µm. The 15 nm gap between a sac and the sarcolemma is filled with 'regularly patterned material of medium electron density.' Since holothuroid muscle fibers have no t-tubules, this material may be the equivalent of SR feet, but arrayed between SR sacs and plasma membrane. The sacs may be half as long as a thick filament (Plate 2b), which is the equivalant of about half a sarcomere of vertebrate striated skeletal muscle. Chen proposed that spread of excitation along the plasma membrane may induce spread of release of calcium along the sacs, much as the spread of excitation into a t-tubule induces Ca-release from SR. In these small slow fibers, the long flattened sacs may provide Ca-release along a significant segment of fiber length.

3.3 Close apposition of cells in a bundle

Early electron micrographs of spine muscles of the echinoid *Anthocidaris crassispina* showed interlocking areas between adjacent muscle fibers, which in cross-section resemble interlocking pieces in a jigsaw puzzle (Kawaguti & Kamishima 1964). Frequently, some of the many small cellular extensions were prolonged, running through the intercellular space to provide contacts with other cells or with the extensions of other cells. At the

90

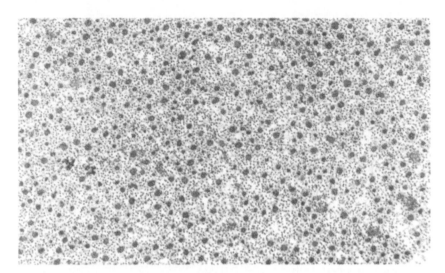

Plate 1. Thin filaments surround thick filaments in a cross-section of a muscle cell from a LMBW of *Sclerodactyla briareus*, stretched to twice resting length. × 57 600 (Chen 1983).

→

Plate 2a. Sr indicates adjacent subsarcolemmal sacs in adjoining muscle fibers of a LMBW of *Sclerodactyla briareus* pre-fixed with 2% formaldehyde, (which fixed the sacs relatively well although filaments were not well preserved). Transverse section. × 21 600 (Chen 1983).

Plate 2b. Sr indicates a subsarcolemmal sac, which can be seen to be about half the length of a thick filament. Electron dense material appears between the sac and the plasma membrane. Pre-fixed in 4% glutaraldehyde in filtered sea water and postfixed in 1% osmium tetroxide in sodium cacodylate buffer (Chen 1983).

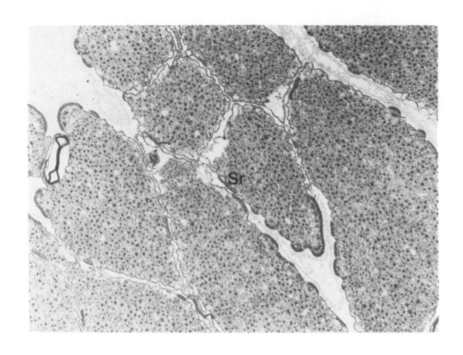

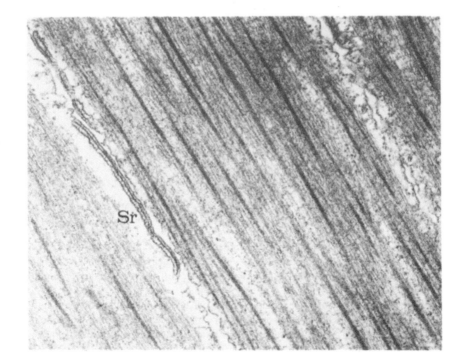

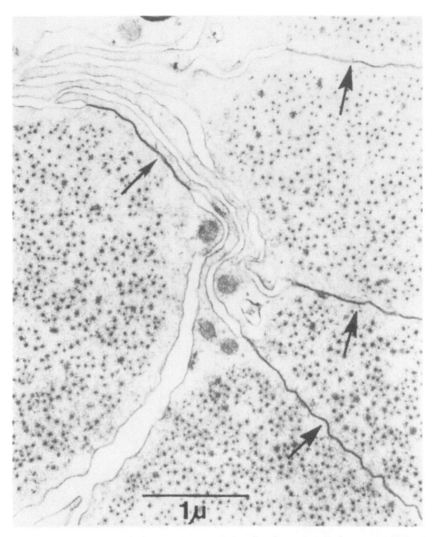

Plate 3. A cross-section through the lumen of a bundle of muscle cells from an LMBW of *Isostichopus badionotus* (Hill et al. 1978). Many projections stretch across the intercellular lumen and extension make close surface contacts (arrow) with adjacent cells. 1 μm calibration. Fixed with 4% glutaraldehyde in sea water, followed by 1% osmium tetroxide in cacodylate buffer. × 22 400.

regions where membranes approached each other most closely, there were noticeable adjacent thickenings of the sarcolemmata. Similarly, adjacent muscle cells of the lantern retractor muscles of *Echinus esculentus* have interdigitating projections, separated by a cleft of 0.1 to 0.4 µm, containing granular debris but not intercellular cement (Cobb & Laverack 1966c). However, the cellular extensions and areas of apposition are much more extensive in LMBW of *Isostichopus badionotus* (Hill et al. 1978). In each muscle bundle, the 3 to 7 small cells have an average of 35% of the surface area of each cell in close surface association with other cells in the bundle. In addition, frill-like cellular extensions arise from the main bodies of the cells and run side-by-side to the center of the bundle (Plate 3). Since the frills run longitudinally along the muscle fibers, they greatly increase the surface area of each cell, and delimit secluded intercellular space, as a lumen to each muscle bundle. Tight surface contacts are formed between adjacent extensions and between extensions and other muscle cells, as well as between main muscle cells. The double leaflets of the sarcolemmata of adjacent cells may fuse to form pentalaminar junctions (Hill et al. 1982), in which the central dark line (delimited by clear lines) is twice as thick as the outer dark lines (facing the cytoplasm of each cell). These fused membranes might be expected to provide mechanical cohesiveness in the bundle, and perhaps also provide a pathway for electrical current flow between cells in a bundle. At present, not enough is known about these fused membranes to compare them to tight junctions or gap junctions of other tissues.

Parallel extensions curling into the center of a muscle bundle are also seen in the central longitudinal muscle of the arm of *Antedon mediterranea* (Candia Carnevali & Saita 1985), where the thin extensions form complex coils along the central axis of each small bundle, and in tube feet of echinoids (Florey & Cahill 1977). Myo-muscle junctions are seen between muscle cells of the tube feet of *Echinus esculentus*, where local thickenings of adjacent sarcolemmata are apposed, but may not be in direct contact (Florey & Cahill 1977).

3.4 *Intramuscular nervous plexus*

Spines of the echinoid *Anthocidaris crassispina* contain a nerve plexus between the epidermis and the outer sheath of smooth muscle cells, which expands into a ring nerve at the base of each articulation (Kawaguti & Kamishima 1964). Kawaguti (1964) described a neural plexus in tube feet of an echinoid, but it was not clear whether or not nerve fibers form neuromuscular junctions with muscle fibers. Florey & Cahill (1977) found no axons entering the muscle layer in the tube feet of three species of echinoids, *Strongylocentrotus franciscanus, Arbacia lixula*, and *Echinus esculentus*. In those tube feet, the muscle layer is separated from the subepithelial nervous plexus by a dense connective tissue layer. The basal lamina of this underlying connective tissue layer is invested with nerve terminals, provided by axons in the plexus. Thus no neuromuscular junctions are seen, and yet the nervous plexus contains ACh and the muscle fibers are cholinoceptive. This appears to be a cholinergic system without synapses, functioning by diffusion of ACh across the connective tissue layer.

The innervation of the LMBW of *Sclerodactyla briareus* is much more conventional (Chen 1983). Small trunks containing a half-dozen or so axons can be seen running between muscle fibers, and nerve terminals contain several types of vesicles: Clear vesicles of 60-80 nm diameter, vesicles of 70-160 nm diameter almost filled with a dense core, and vesicles of 80-200 nm diameter with a small dense core. Other nerve terminals contain large electron-opaque granules.

3.5 *Gross organization*

Muscle cells of the LMBW of *Stichopus mollis* are organized into small bundles of 2 to 10 cells, separated from other bundles by a network of connective tissue (Freeman & Simon 1964). The individual LMBW cells of *S. mollis* are from 2.5 μm to 6 μm in diameter and from 240 μm to 555 μm long. The bundles of the LMBW of *Sclerodactyla briareus*, containing 1 to 12 cells, are wrapped together in a layer of external lamina about 35 nm thick but the individual cells of a muscle bundle are not separated by external lamina (Chen 1983). The external lamina surrounding the whole bundle is separated from the plasma membranes of abutting muscle fibers contained in the bundle, by a narrow electron-lucent space, which becomes filled with a dense substance wherever there are dense patches on the plasma membrane. The external lamina surrounding a bundle of muscle fibers thus seems to be firmly attached at the dense patches. In these areas there is a rather complex reinforcement. The basal lamina is joined by dense material to the normal trilaminar plasma membrane, which is reinforced on its sarcoplasmic side by a dense patch 0.2 μm to 0.6 μm long and 55 nm thick (a 'dense band'). The dense band in turn, may be reinforced by a dense pad on its sarcoplasmic side. These dense pads may be modified dense bodies, since thin filaments insert into them. The small muscle bundles of the LMBW of *S. briareus* are grouped into blocks of up to 5 bundles, delimited by connective tissue septa which radiate from the mid-line of the muscle, proximal to the body wall.

4 EXCITATION-CONTRACTION COUPLING

Contraction in LMBW fibers of *Isostichopus badionotus* involves both influx of calcium ions across the sarcolemma and release of calcium from intracellular storage sites, such as the subsarcolemmal vesicles (Suzuki 1982). Contractions induced by 10^{-3}M acetylcholine (ACh) were abolished by 30 minutes soaking in a 5 mM EGTA, Ca-free solution, or blocked by 20 mM Mn or 20 mM La, which blocked influx of Ca. This indicates that calcium for E-C coupling in ACh contractions enters through the cell membranes. However, the LMBW of *Isostichopus badionotus* has only a few small subsarcolemmal vesicles as apparent Ca-storage sites and yet the muscle fibers are not dependent on $[Ca^{++}]_o$ for contraction, unless pretreated in a Ca-free solution containing a chelating agent (Hill et al. 1978). (The pretreatment does not chemically 'skin' the fibers, which still have a resting potential). After 12 hours in a Ca-free solution of chelating agent, LMBW of *I. badionotus* had lost all contractile response to 50 mM KCl, but in 9 mM Ca^{++} (artificial sea water) the LMBW seems to quickly reload Ca^{++} (Hill et al. 1978). The reloaded muscle then reliably, reproducibly and reversibly lost contractility in a graded fashion, in Ca-free solution, with a 70% loss after one hour. Thus, even without SR, except for the subsarcolemmic vesicles, these results (Hill et al. 1978) suggest that intracellular stores of calcium were emptied into Ca-free solution, and reloaded from 9 mM Ca^{++}.

Treatment of the isolated LMBW of *Isostichopus badionotus* with repeated doses of 10 mM caffeine elicited a series of contractures which diminished progressively (Hill 1980). These contractures were not accompanied by any depolarization in LMBW of *I. badionotus* (Hill et al. 1978) or of *Holothuria cinerascens* (Hill 1987), and occurred even in muscles which did not contract in response to depolarization (Hill et al. 1978). Thus the progressively diminishing caffeine-contractures may be due to Ca-release into the cytoplasm from relatively quickly loading and unloading intracellular storage sites, at the level of the sarcolemma and the subsarcolemmic vesicles. In experiments using isotonic recording from isolated LMBW of *I. badionotus* in perfused muscle baths, contractions in 10 mM caffeine diminished progressively on repetition, from 29% of resting length to 2.5% of resting length (Hill 1980). Once reduced, the amplitude of responses was not restored by 30 minutes of incubation in sea water (9 mM Ca^{++}), or in 27 mM Ca^{++} solution, or in a solution with 27 mM Ca^{++} and 2 mM caffeine. However, the amplitude of responses was restored to about 50% of the initial value by 60 minutes of incubation in 100 mM Ca^{++} solution, or to 100% of the initial value by 3 minutes depolarization by 10^{-7} M ACh or 50 mM KCl (Hill 1980). Thus the quickly loading calcium stores seem to depend on extracellular Ca^{++}, made available either by an enhanced transmembrane calcium gradient or by membrane depolarization. However, the situation is transformed by treatment with the calcium ionophore X-537A (Hill 1980). The ionophore (X-537A) was used to test for the hypothesized caffeine-depleted storage sites, since once the ionophore has partitioned into the cell membrane, the submembrane storage sites should reload more readily. In the presence of X-537A, a few minutes incubation in sea water was sufficient to restore contractility lost in a series of caffeine contractures. After 60 minutes treatment with X-537A, caffeine contractures remained reproducible, even after the ionophore had been washed out of the bath. (The LMBW does not become chemically 'skinned' in ionophore solution, since it still contracts in response to treatment with elevated $[K^+]_o$, ACh, or caffeine but does not contract in response to elevated bath calcium). These results suggested that molecules of X-537A had partitioned into the cell membrane, where they acted as calcium carriers and enhanced reloading of the stores from which caffeine releases calcium (Hill 1980).

Five to six small subsarcolemmal vesicles are found in each cross-section of a LMBW cell of *Isostichopus badionotus*. These might serve as storage sites for calcium ions that have crossed the plasma membrane, but which might be sequestered until released in the excitation-contraction coupling process. These subsarcolemmal vesicles may also be the sites from which caffeine discharges calcium. If that is the case, the experiments reported above (Hill 1980) indicated that the calcium-storing ability of the vesicles is limited, and that the vesicles are only slowly reloaded across the sarcolemma.

A further study concerned factors that modify the rate of reloading of the depleted vesicles (Hill 1983a).

1. After decline of contractility during a caffeine series, recovery in sea water (9 mM Ca^{++}) increased slowly but steadily with time. A second caffeine series resulted in a second decline of contractility, indicating that Ca^{++}-stores with the same properties had reloaded. However, contractility in the second series was much reduced. In contrast, when recovery occurred in a 100 mM Ca^{++} solution, contractility recovered by 3/5 and the second caffeine series showed a rundown of contractility parallel to the first.

2. Even after caffeine responses in sea water had been made reproducible by treatment with X-537A, caffeine responses in Ca^{++}-free solution diminished sequentially, as before.

3. Brief treatment with 50 mM KCl restored the contractility that had been lost in a caffeine series. Restoration was time-dependent, but was completed in 3 minutes.

4. Brief treatment with 2×10^{-7} M ACh restored the contractility that had been lost in a caffeine series. The extent of a restoration of contractility by ACh was time dependent, and was concentration-dependent between 10^{-8} M and 10^{-6} M.

5. Depolarization and ionophore treatment have strikingly different effects on restoration of contractility. After the calcium ionophore has partitioned into the membrane, the stores (sarcolemmic vesicles?) recharge quickly in 9 mM $[Ca^{++}]_o$ and the isolated muscle responds reproducibly to repeated challenges with caffeine. However, one depolarization only re-

charges the muscle once, and repeated challenges with caffeine then show a renewed rundown of contractility. An unresolved question is the location of the slowly loading and unloading calcium stores. In other words, is the calcium in the secluded extracellular space handled differently from calcium in the abundant extracellular space outside the bundles?

6. Calcium antagonists were used in experiments designed to differentiate between calcium stores used in reproducible contractions and calcium stores used in rapidly extinguishing contractions (Hill 1983b). In caffeine studies, since the untreated LMBW of *Isostichopus badionotus* rapidly loses responsiveness to caffeine, the isolated muscle strips were treated with X-537A to establish a steady state in which Ca^{++} enters quickly to recharge the stores from which caffeine discharges Ca^{++}. Under these conditions, the order of effectiveness of calcium antagonists in blocking caffeine contractures was lanthanum>manganese>dantrolene. The order of reversibility was manganese>dantrolene>lanthanum. Basically, the effects of calcium antagonists indicate simple reduction of Ca-entry in the ionophore-treated preparation. However, in preparations not treated with ionophore, Ca-antagonists do not all act alike. Lanthanum and manganese block tetanic free-loaded contractions or contractions induced by 10^{-8} M or 10^{-7} M ACh, but dantrolene enhances both types of contraction. It could be hypothesized that dantrolene blocks calcium loss from subsarcolemmal vesicles to the extracellular medium, without blocking E-C coupling release of calcium to the cytoplasm (Hill 1983b).

The mechanical responses of the LMBW of *Sclerodactyla briareus* to 50 mM KCl or 10^{-6} M or 10^{-7} M acetylcholine are calcium dependent (Chen 1986). Successive caffeine contractures dwindle to a small amplitude, but contractility is restored after a contraction induced by KCl or ACh. The overall pattern of E-C coupling thus appears identical to that observed in *Isostichopus badionotus.*

Contractions of the LMBW of *Stichopus japonicus* are also calcium-dependent (Sugi et al. 1982). The intracellular calcium storage sites, however, may only be capable of activating about 30% of a maximal contraction. These sites appear to be localized along the inner surface of the plasma membrane and at the flattened subsarcolemmal vesicles (Suzuki & Sugi 1982).

Contractions in response to electrical stimulation, ACh and KCl are calcium-dependent in radial muscles of the echinoid *Asthenosoma ijimai*, but caffeine contractions, which only reach a fraction of a maximal contraction, are not calcium-sensitive (Tsuchiya & Amemiya 1977).

Prosser & Mackie (1980) used treatment with Ca-free artificial sea water (Ca-free ASW) and the Ca-blockers, manganese, cobalt, and verapamil in order to assess calcium-dependence of contractions in pharynx retractor of *Cucumaria minata* and longitudinal retractors (LMBW) of *Leptosynapta*

95

clarki. These agents abolished responses to direct electrical stimulation, neurally-mediated responses to stretch stimulation, and spontaneous rhythmic contractions. Prosser & Mackie suggested these results indicate both the presence of calcium action potentials and a rôle of calcium in E-C coupling.

Tube feet have a similar structure in five extant echinoderm classes (Wood & Cavey 1981). An important functional element is provided by the longitudinal retractor muscle sheet or bands, which appears to be myoepithelial components of the coelomic epithelium. There is no basal lamina between the cells of the adluminal epithelium, which lines the water vascular canal, and the underlying layer of retractor muscle cells, which is penetrated by long basal processes of the epithelial cells. In turn it is retractor muscle cells which provide the main mass of the coelomic lining of the tube feet of *Stylasterias forreri* (Wood & Cavey 1981). The retractor cells interdigitate with each other, in a pattern like that of holothuroid LMBW cells, and it is noteworthy that subsarcolemmal densities of adjacent cells face each other, to form a kind of junction with a cleft of about 50 nm (Fig. 5). Individual cells bifurcate and

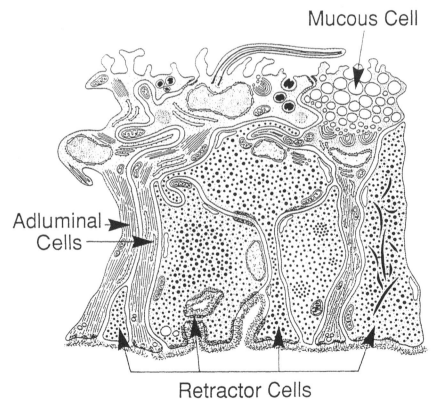

Figure 5. Retractor muscle cells in the coelomic lining of a podium of an asteroid (Wood & Cavey 1981).

branch, in a pattern that allows some branches to be aligned longitudinally. Others which are aligned radially with the basal processes of adluminal cells, run to the basal lamina of the connective tissue which underlies the epidermis. Where these processes reach this basal lamina, they are apposed on 90% of the surface area and are anchored by extensive incursions of the basal lamina faced by subsarcolemmal plaques. Thin (8 nm) myofilaments are associated with subsarcolemmal plaques, both at junctions between muscle cells and junctions with basal lamina. A similar pseudostratified myoepithelium may be the basic pattern for the lining of echinoderm water vascular systems. Cisternae of the sarcoplasmic reticulum, which underlie the sarcolemma of podial retractor cells of *Stylasterias ferreri* (Cavey & Wood 1981), strongly resemble the subsarcolemmal vesicles of LMBW of *I. badionotus* (Hill et al. 1978, 1982). The association of certain cisternae with the sarcolemma suggests a dyadic coupling in podia of *S. ferreri* (Cavey & Wood 1981). Clustered particles in the sarcolemma overlie cisternae. These may be sites for E-C coupling.

5 INNERVATION

Nearly every worker who has studied the nervous system of echinoderms has come to the conclusion that the radial nerve cords have some function other than that of a purely co-ordinating and transmitting system.

(Binyon 1972)

Binyon (1972) pointed out that, for a number of reasons, it should be of particular interest to explore the neuromuscular systems of echinoderms: (1) The extremely small diameter of many echinoderm axons; (2) Simple stimulation of an echinoderm radial nerve cord elicits a complex response which dies out with distance; (3) The anatomical organization in which ectoneural tissue is separated by connective tissue from hyponeural tissue in the radial nerve cord of an asteroid; (4) The apparent presence of pacemakers for spontaneous activity in neuromuscular systems of echinoderms; (5) The problematic nature of synapses in nerve trunks; (6) The problematic nature of intramuscular conduction.

Echinoderms generally have three 'somewhat primitive' neural networks or radiating ganglionated nerve cords (Hyman 1955). These are: (1) the ectoneural (oral) nervous system, which underlies the oral epidermis; (2) the deeper hyponeural nervous system; and (3) the entoneural (aboral) nervous system. The prominence of the three systems varies among classes. For instance, the entoneural system predominates in crinoids but is absent in holothuroids (Hyman 1955). The correlation of anatomical and neurobiological studies of the nervous system of echinoderms has recently been reviewed (Cobb 1987). Cobb felt that it is because of the technical difficulty

97

of working with the echinoderm nervous system that, 'There is less known about the organization of the echinoderm nervous system than that of any other phylum of metazoan animals.' Cobb pointed out that the crinoid nervous systems: ectoneural, hyponeural, and entoneural are not directly comparable to the nervous systems of other classes (1987). In that case, it seems unfortunate that the same names are used, especially in studies of neuromuscular physiology, where Cobb's own work has classified the relationship between the major ectoneural and hyponeural nervous systems in asteroids, echinoids, holothuroids, and ophiuroids. Most simply put, the hyponeural nervous system is the skeletal motor nervous system. A separate visceral plexus innervates muscles of the viscera. This leaves the ectoneural nervous system as the main nervous system, containing sensory neurones, interneurons, and motor neurons.

Cobb (1985) carried out experiments designed to test an hypothesis of ectoneural/hyponeural motor control in the rather enigmatic nervous system of echinoderms. The hypothesis suggested that motor control is exerted through chemical transmission across the basement membrane and layer of collagenous connective tissue which separate the 'main' ectoneural nervous system from the motor hyponeural nervous system, of mesodermal origin. It follows from this hypothesis, and from earlier neurophysiological work, that echinoderm behavior is generated by patterns of neural activity arising in the presynaptic ectoneural system which drives the postsynaptic hyponeural system.

In the ophiuroid *Ophiura ophiura*, and for other eleutherozoan echinoderms, Cobb's scheme suggests that activity of the whole animal is initiated peripherally, when the oral ectoneural region of any of the radial nerve cords becomes active. Patterns of activity originating in one such peripheral locus then spread through the whole ectoneural system, with the circumoral nerve ring acting primarily as a relay between the radial nerve cords of the arms. This hypothesis could be tested in *O. ophiura*, since the large ectoneural interneurons and hyponeural motor neurons were susceptible to dye injection and intracellular recording (Cobb 1985). In each segmental ganglion, of a radial nerve cord, ectoneural cells form an oral layer, while hyponeural motor neurons form aboral swellings on each side of the midline, separated from the main oral ganglion by a partition which ranges in thickness from a connective tissue layer with a basal lamina on each side to a bare 40 nm basal lamina. This partition prevents the ectoneural cells from forming direct synapses with the hyponeural cells, much as a similar partition prevents ectoneural nerve endings from directly innervating muscles of tube feet (Cobb 1970, Florey & Cahill 1977). However, intervertebral skeletal muscles of the articulating calcite ossicles are directly innervated by varicose endings of the hyponeural motor system (Stubbs & Cobb 1981). Experiments with *O. ophiura* (Cobb 1985) tested the hypothesis of Cobb & Pentreath (1976), that a transmitter is

98

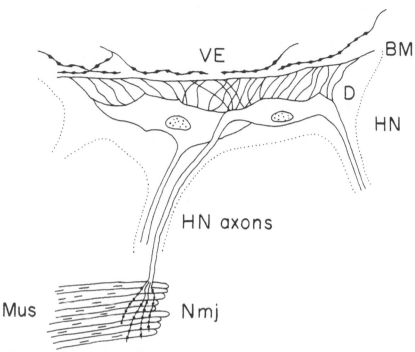

Figure 6. Innervation of intervertebral muscle cells of *Ophiura ophiura* by hyponeural motor neurons, which are excited and inhibited across a basal lamina by ectoneural interneurons (Cobb 1985). Mus-muscle cells, Nmj-neuromuscular junctions, HN = cell bodies of hyponeural neurons, VE-varicose endings, BM-basement membrane, D-dendrites of hyponeural neurons.

released by a continuous layer of vesicular ectoneural axon varicosities, adjacent to the thin basal lamina, and diffuses across the basal lamina to the hyponeural motor nerves. A non-spiking ectoneural cell showed excitatory post-synaptic potentials, as an off-response to interruption of a light beam, directed at peripheral segments of the arm. This photic stimulus induced a burst of spikes in the ectoneural nerve cord, EPSPs in ectoneural ganglion cells, and a correlated burst of EPSPs or IPSPs in hyponeural cells. The hyponeural cell body was found to be electrically inexcitable, but action potentials were initiated in the axon hillock and could be recorded intracellularly from hyponeural motor axons. This indicated chemical transmission across the basal lamina, from ectoneural interneurons to hyponeural motor neurons, which could then excite effectors by trains of action potentials. However, the whole chain of excitation (Fig. 6) remains unusual in that no specialized synaptic regions have been seen on the membranes of ectoneural neurons, hyponeural neurons, or muscle cells.

Cobb (1978) proposed that echinoderms have 'skeletal' smooth muscles and 'visceral' smooth muscles, which are innervated differently. In the dermal papulae of *Asterias rubens*, the 'skeletal' smooth muscle cells are innervated at their proximal ends by simple synapses with ectoneural axons. In contrast, the 'visceral' smooth muscle is closely associated with axons containing large granular vesicles. Thus echinoderm muscles associated with skeletal elements (including the hydro-elastic skeleton) have an innervation reminiscent of vertebrate skeletal muscle, while echinoderm muscles associated with visceral or 'hollow-organ' functions have an innervation reminiscent of vertebrate autonomic fibers (Table 1). 'Skeletal' echinoderm smooth muscles are frequently found to be innervated on long specialized extensions of the muscle cells. For instance, the muscle fibers of the ampullae of the tube feet of *Astropecten irregularis* are innervated by means of a neuromuscular junction, which is formed on a long thin clear extension of the muscle cell, which contains only a core of modified muscle fibers.

Lantern muscles of *Echinus esculentus* are innervated by the hyponeural nervous system (Cobb & Laverack 1966b). This follows the pattern of innervation of skeletal smooth muscle. Each muscle cell is innervated at the proximal end by a single neuron at a point where a dilated region of the axon, filled with synaptic vesicles, is enveloped by an extension of the sarcolemma, at a region where the muscle cell is packed with mitochondria (Cobb & Laverack 1966c). Experiments with electrical and mechanical stimulation of hyponeural tissue suggested the presence of a motor system consisting of a series of pacemakers, with shifting dominance (Cobb & Laverack 1966a).

Neuromuscular transmission in echinoderms may involve a 'stromal pathway', in which neurotransmitters are released into extracellular space, where the nerve endings are not closely approximated to muscle cells (Dolder 1975) or a synaptic pathway, where nerve endings are enveloped by muscle tissue (Cobb & Laverack 1966b) or where a conventional synaptic cleft has

Table 1. Classification of echinoderm smooth muscle (Cobb 1978).

	Skeletal smooth muscle	Visceral smooth muscle
Location:	Echinoid pedicellariae	Echinoid esophagus
	Echinoid lantern	Echinoid gills
	Asteroid ampullae	Echinoid tube feet
		Asteroid ampullae
		Holothurian hemel vessels
Characteristics:	Long cells in bundles	Short cells
	Thick myofilaments	Lack bundles
		Irregular outline
Innervation:	At one end	Associated axons
	On long process	No neuromuscular junctions

been identified (Cobb 1978). Postulated neurotransmitters include 5-hydroxytryptamine (Dolder 1975), acetylcholine (Florey & Cahill 1980; see Pentreath & Cobb 1972 for earlier references) and catecholamines (Cottrell & Pentreath 1970).

6 ELECTRICAL RECORDING FROM ECHINODERM MUSCLE

Electrical responses have been recorded in isolated LMBW preparations, from several species of holothuroids. Prosser et al. (1951) used wick electrodes to detect active responses to direct stimulation in LMBW of *Sclerodactyla briareus*. They observed that the responses were decremental, disappearing over a distance of 2 to 5 mm from the site of stimulation. However, Prosser (1954) concluded that the recorded potentials were not electrotonic, since, although decrementing with distance, they could be detected (in a second study) as far as 12 mm from the site of stimulation. The fact that 'active potentials' died out over a relatively short distance was congruent with earlier evidence that LMBW from other species is neurally activated at many loci, in coordination of contractions of the long muscle (Henri 1903a, b, Tao 1927).

The early work with electrical responses made use of wick electrodes for extracellular recording, since even the decremental, non-propagated action potentials could not be detected from point sources such as silver wire electrodes inserted into an isolated muscle (Prosser 1954). However, the most reliable technique to date seems to be sucrose gap recording, which has now been applied to LMBW preparations from several species of holothuroids. This might have been predicted from the early work with LMBW of *Sclerodactyla briareus*, since with a maximum sucrose gap width of 0.5 mm, electrical continuity across the sucrose gap would be provided by the pathways which allowed responses to be detected 2 to 12 mm from the site of stimulation (Prosser et al. 1951, Prosser 1954). The first published sucrose gap recording from holothuroid LMBW reported work with *Isostichopus badionotus* (Hill et al. 1978). The main technical problems encountered at first, in that work, were those associated with the peculiar visco-elastic nature of holothuroid muscle. (That is, an isolated LMBW breaks during an isometric contraction and electrical continuity across the sucrose gap is lost). It has long been noted that holothuroid LMBW offers rigid resistance to a sudden application of force, but shows plastic compliance and ultimately tears apart under gradually applied force or during an isometric contraction (Levin & Wyman 1927, Van Weel 1955). However, these problems do not arise when isotonic contractions accompany sucrose gap recording, and electrical responses to ACh or to KCl can then be detected with a sucrose gap technique (Hill et al. 1978, Hill 1987). However, electrical responses to stretch-

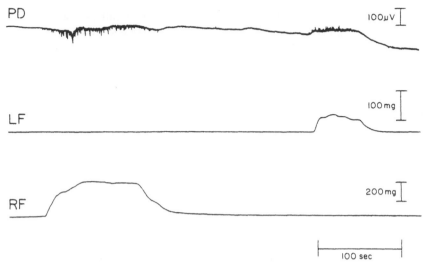

Figure 7. LMBW of *Holothuria cinerascens*. Differential recording across a sucrose gap shows that spontaneous contractions occurring on either side are not propagated across the gap. The direction of pen deflection indicating depolarization (PD) is reversed for a contraction on the left (LF), with reference to a contraction on the right (RF). (original).

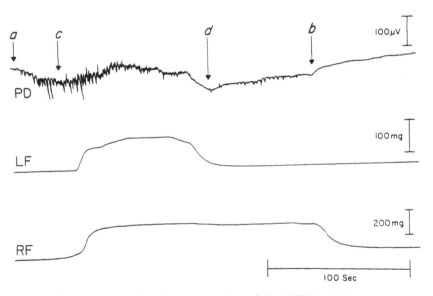

Figure 8. Simultaneous independent contractions of the LMBW of *Holothuria cinerascens* on opposite sides of the sucrose gap. RF corresponds to the electrical signals at ab, while LF corresponds to cd. (original).

102

stimulation are more difficult to detect. Prosser & Mackie (1980) attempted to record electrical correlates to the contraction excited by applying sudden stretches to LMBW of 5 species of holothuroids, or to the pharyngeal retractor muscle of one species. Three types of point-source recording failed to detect action potentials, which could not be picked up with microelectrodes, polyethylene suction-electrodes, or glass-tipped pressure electrodes. Considering these results, Prosser & Mackie (1980) hypothesized that stretch-stimulation may excite a neural reflex which is 'centrally' asynchronous, and thus evokes asynchronous membrane responses, that cannot be detected invididually by means of conventional extracellular electrodes. However, Prosser (1954) suggested that point-source recording might be less effective than the use of wick electrodes in extracellular recording from holothurian muscle. Thus it was perhaps not unexpected to find that sucrose-gap recording (Hill et al. 1978, Hill 1987) was more effective than point-source recording (Prosser & Mackie 1980) for holothuroid LMBW, or at least, for recording electrical responses accomanying spontaneous contraction, as well as contraction driven by treatment with acetylcholine solution or elevated $[K^+]_o$.

Using differential recording across a sucrose gap, an isolated LMBW of *Holothuria cinerascens* was oriented so that a spontaneous contraction on the right side correspond to a downward deflection of the pen recording potential difference (right electrode less positive), while a spontaneous contraction on the left side corresponded to an upward deflection (left electrode less positive). A spontaneous contraction on the right was accompanied by downward spiking which intensified as force developed, while a spontaneous contraction on the left was accompanied by upward spiking (Fig. 7). This spiking is interpreted as a record of asynchronous action potentials in muscle units on either side of the gap. In each case, the spiking was not transmitted across the sucrose gap, although there is evidently electrical continuity across the gap, since the electrode on the quiescent side acts as a reference. However, the segments of the LMBW on opposite sides of the gap are entirely independent in generating spontaneous contractions. Spikes recorded were in the 100 μV range, well within expected values for extracellular recording, but attenuated considering the use of a sucrose gap method. However, even sucrose gap recordings of depolarization with isotonic KCl are attenuated to about 1/5 of expected values in these recordings from LMBW of *H. cinerascens*. This may be attributed to the large amount of extracellular space (Robertson 1980) and the large connective tissue component (Chen 1983), both of which may provide a large longitudinal shunting conductance in parallel with the muscular elements of the LMBW, across the sucrose gap.

At greater amplification (Fig. 8) an inherent instability in PD across the sucrose gap is more evident, but this does not mask a depolarization correlated with force development, which may reflect activation of large numbers

103

of muscle units. These units may well be the bundles, each containing closely apposed cells, but only loosely linked to other bundles, although all richly supplied by the nervous plexus (Chen 1983). A maintained (downward) depolarization from (a) to (b) corresponded to maintained right-hand force (Fig. 8). The depolarization clearly began to develop (a) well before force appeared (RF) and downward spiking increased in amplitude as RF rose to a maximum. The maintained downward potential difference (PD) deflection from (a) to (b) was interrupted by a briefer upward PD deflection from (c) to (d), which corresponded to left-hand force. Upward spiking was correlated with the rise in force (LF). The time relationships in Figure 8 thus indicate that depolarization and spiking induce slowly developing force in the LMBW of *H. cinerascens*.

Direction of depolarization across the sucrose gap was always such that the electrode on the contracting side became less positive with reference to the quiescent side (Fig. 9). LF was thus correlated with upward depolarization and upward spiking, while RF was correlated with downward depolarization and downward spiking.

In contrast to the irregularly spiking electrical corollary to spontaneous contractions, electrical responses of *Holothuria cinerascens* LMBW to ace-

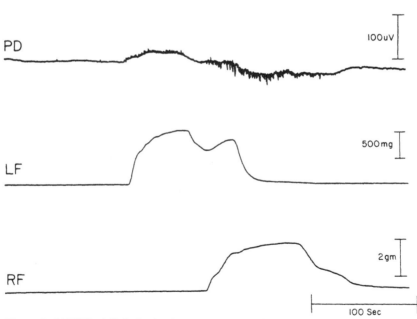

Figure 9. LMBW of *Holothuria cinerascens*. Deflection and spiking upwards clearly correspond to a contraction on the left (LF) while deflection and spiking downwards correspond to a contraction on the right (RF). (original).

104

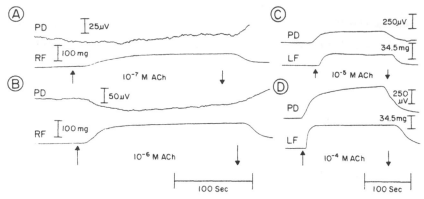

Figure 10. Recordings from an isolated LMBW of *Holothuria cinerascens* in a sucrose gap apparatus. Potential difference and force increase with the concentration of acetylcholine, applied on the right (A, B) or on the left (C, D). (original).

tylcholine were smooth depolarizations (Fig. 10). The threshold for detecting a depolarization in response to ACh (D-response) was around 10^{-7} M. At that concentration, ACh elicited a marked, slowly developing contractile response, but the D-response was difficult to discern from background drift. However, at 10^{-6} M the D-response to ACh was very small, but clearly identifiable, since it corresponded to a contraction, and the PD inflection in response to application of ACh on the right side of the sucrose gap was in the appropriate direction. When 10^{-5} M ACh was applied to the muscle on the left side of the sucrose gap a smooth 'synchronized' response was elicited. Depolarization led to contraction, and repolarization upon washout led to relaxation. The even larger D-response to 10^{-4} M ACh was relatively quick and induced a mechanical response like that due to a high concentration of KCl. The immediate mechanical response may be due to muscle units near the surface of the preparation, while depolarization continues as ACh penetrates the muscle. No spiking appears even in response to high concentrations of ACh, in the same preparation in which spiking accompanies spontaneous contractions.

Lantern retractor muscle (LRM) of holothuroids and echinoids respond to stimulation with a dual electrical or mechanical response, having fast and slow components (Cobb 1968). Stimulation of the hyponeural ganglion of *Echinus esculentus* evokes a rapid twitch, followed by a slow contraction which lasts several seconds, but stimulation of the 'ultimate' motor neurons only evokes a twitch. Cobb (1968) suggested that the delayed slow contraction is due to interaction in the hyponeural ganglion. Lantern retractor muscles of *E. esculentus* must be classified with the 'skeletal' smooth muscles of echinoderms; that is, the fibers run the full length of the muscle,

105

5 sec

Figure 11. The slow component of the mechanical response of an echinoid lantern muscle corresponds to prolonged spiking (Cobb 1968).

innervation is at the proximal end of the LRM, and compound action potentials or motor units can be recorded with external suction electrodes (Cobb 1968). When the motor nerves are stimulated 2 or 3 mm from the 2 cm muscle, only a rapid compound response occurs, which precedes the mechanical response by 75 msec. When hyponeural tissue is stimulated, the quick response in the muscle is followed by repetitive firing of many motor units, as a slow contraction develops (Fig. 11). This slow response of the LRM of echinoids resembles the spontaneous contractions of LMBW, in holothuroids.

REFERENCES

Bianconcini, M.S.C., E.G. Mendes & D. Valente 1985. The respiratory metabolism of the lantern muscles of the sea urchin *Echinometra lucunter* (L.). I. The respiratory intensity. *Comp. Biochem. Physiol.* 80 A: 1-4.

Binyon, J. 1972. *Physiology of Echinoderms*. Pergamon Press, Oxford.

Candia Carnevali, M.D. & A. Saita 1985. Muscle system organization in the echinoderms. III. Fine structure of the contractile apparatus of the arm flexor muscles of the Comatulids (*Antedon mediterranea*). *J. Morph.* 185: 75-78.

Cavey, M.J. & R.L. Wood 1979. Sarcoplasmic reticulum and sarcolemmal couplings in the podial muscle cells of an asteroid echinoderm. *Am. Zool.* 19: 903.

Cavey, M.L. & R.L. Wood 1981. Specializations for excitation-contraction coupling in the podial retractor cells of the starfish *Stylasterias forreri. Cell Tiss. Res.* 218: 475-485.

Chen, Ching-ju 1983. A study of the longitudinal body wall muscle of the sea cucumber *Sclerodactyla briareus*. Ph.D. Thesis, University of Rhode Island.

Chen, Ching-ju 1986. Contractions of holothurian longitudinal muscle. *Chinese J. Physiol.* 29: 43-52.

Cobb, J.L.S. 1968. Observations on the electrical activity within the retractor muscles of the lantern of *Echinus esculentus* using extracellular recording electrodes. *Comp. Biochem. Physiol.* 24: 311-315.

Cobb, J.L.S. 1970. The significance of the radial nerve cord in asteroids and echinoids. *Z. Zellforsch* 108: 457-474.

Cobb, J.L.S. 1978. An ultrastructural study of the dermal papulae of the starfish, *Asterias rubens*, with special reference to innervation of the muscles. *Cell. Tiss. Res.* 187: 515-523.

Cobb, J.L.S. 1985. The neurobiology of the ectoneural/hyponeural synaptic connection in an echinoderm. *Biol. Bull.* 168: 432-446.

Cobb, J.L.S. 1987. Neurobiology of the Echinodermata. In *Nervous Systems in Invertebrates*, M.A. Ali (ed.). Plenum Press, N.Y.

Cobb, J.L.S. & M.S. Laverack 1966a. The lantern of *Echinus esculentus* (L.). I. Gross anatomy and physiology. *Proc. Roy. Soc. Lond.* (B) 164: 624-640.

Cobb, J.L.S. & M.S. Laverack 1966b. The lantern of *Echinus esculentus* (L.). II. Fine structure of hyponeural tissue and its connexions. *Proc. Roy. Soc. Lond.* (B) 164: 641-650.

Cobb, J.L.S. & M.S. Laverack 1966c. The lantern of *Echinus esculentus* (L.). III. The fine structure of the lantern retractor muscle and its innervation. *Proc. Roy. Soc. Lond.* (B) 164: 651-658.

Cobb, J.L.S. & M.S. Laverack 1967. Neuromuscular systems in echinoderms. *Symp. Zool. Soc. Lond.* 20: 25-51.

Cobb, J.L.S. & V.W. Pentreath 1976. The identification of chemical synapses in Echinodermata. *Thalassia Jugol.* 12: 81-85.

Cottrell, G.A. & V.W. Pentreath 1970. Localization of catecholamines in the nervous system of a starfish, *Asterias rubens*, and of a brittlestar, *Ophiothrix fragilis*. *Comp. Gen. Pharmacol.* 1: 73-81.

Dolder, H. 1975. An ultrastructural and cytochemical study of neuromuscular junctions in echinoderms. *Histochem.* 44: 313-322.

Florey, E. & M.A. Cahill 1977. Ultrastructure of sea urchin tube feet. Evidence for connective tissue involvement in motor control. *Cell Tissue Res.* 177: 195-214.

Florey, E. & M.A. Cahill 1980. Cholinergic motor control of sea urchin tube feet: evidence for chemical transmission without synapses. *J. exp. Biol.* 88: 281-292.

Freeman, W.P. & S.E. Simon 1964. The histology of holothuroidean muscle. *J. Cell. Comp. Physiol.* 63: 25-38.

Furukohri, T. 1971a. Some properties of myosin B from sea-cucumber longitudinal muscle. *Sci. Rep. Tôhoku Univ.* (4, Biol.) 36: 31-40.

Furukohri, T. 1971b. Adenosine triphosphatase activity of myosin of sea-cucumber longitudinal muscle. *Sci. Rep. Tôhoku Univ.* (4, Biol.) 36: 41-55.

Galambos, R. 1941. Characteristics of the loss of tension by smooth muscle during relaxation and following stretch. *J. Cell. Comp. Physiol.* 17: 85-95.

Gay, W.S. & S.E. Simon 1964. Metabolic control in holothuroidean muscle. *Comp. Biochem. Physiol.* 11: 183-192.

Hall, A.R. 1927. Histology of the retractor muscle of *Cucumaria miniata*. *Publ. Puget Sound Mar. Biol. Sta.* 5: 205-219.

Hallam, D.C. & R.J. Podolsky 1969. Force measurements in skinned muscle fibers. *J. Physiol. Lond.* 200: 807-819.

Henri, V. 1903a. Etude physiologique des muscles longitudinaux chez le *Stichopus regalis*. *C.R. Séanc. Soc. Biol.* 55: 1194-1195.

Henri, V. 1903b. Etude des réflexes élémentaires chez le *Stichopus regalis*. *C.R. Séanc. Soc. Biol.* 55: 1195-1197.

Herrera, F.C. & M. Plaza 1981. Extracellular cations and respiration of tissues of *Holothuria glaberrima*. *Comp. Biochem. Physiol.* 70A: 27-32.

Hill, A.V. 1926. The viscous elastic properties of smooth muscle. *Proc. Roy. Soc. London* (B) 100: 108-115.

Hill, R.B. 1980. Use of an ionophore to maintain repeated caffeine contractures in holothurian muscle. *Life. Sci.* 27: 1967-1973.

107

Hill, R. B. 1983a. Restoration of contractility by depolarizing agents and by calcium after caffeine treatment of holothurian muscle. *Comp. Biochem. Physiol.* 75C. 5-15.

Hill, R. B. 1983b. Effects of calcium antagonists on contraction of a holothurian muscle. *Comp. Biochem. Physiol.* 76C: 1-8.

Hill, R. B. 1987. Correlation of electrical and mechanical activity of holothurian muscle. *J. Exper. Biol.* 130: 331-339.

Hill, R.B. & J.W. Sanger & C.-J. Chen 1982. Close apposition of muscle cells in the longitudinal bands of the body wall of a holothurian, *Isostichopus badionotus*. *Cell Tissue Res.* 227: 465-473.

Hill, R. B., J.W. Sanger, R. E. Yantorno & C. Deutsch 1978. Contraction in a muscle with negligible sarcoplasmic reticulum: The longitudinal retractor of the sea cucumber *Isostichopus badionotus* (Selenka), Holothurioidea, Aspidochirota. *J. Exper. Zool.* 206: 137-150.

Hyman, L. H. 1955. *The Invertebrates*, Vol. 4, *Echinodermata*. McGraw-Hill, New York, Toronto, London.

Kawaguti, S. 1964. Electron microscopic structures of the podial wall of an echinoid with special references to the nerve plexus and the muscle. *Biol. J. Okayama, Univ.* 10: 1-12.

Kawaguti, S. & Y. Kamishima 1964. Electron microscopy on the spine muscle of the echinoid. *Biol. J. Okayama Univ.* 11: 31-40.

Kerrick, W.G.L. & L.L. Bolles 1982. Evidence that myosin light chain phosphorylation regulates contraction in the body wall muscles of the sea cucumbers. *J. Cell Physiol.* 112: 307-315.

Koizumi, T. 1935. Studies on the exchange and the equilibrium of water and electrolytes in a holothurian, *Caudina chilensis* (J. Müller). V. On the inorganic composition of the longitudinal muscles and the body wall without longitudinal muscles. *Sci. Rep. Tôhoku Imp. Univ.* (4, Biol.) 9: 281-286.

Lehman, W., J. Kendrick-Jones & A.G. Szent-Györgyi 1973. Myosin-linked regulatory systems: Comparative studies. *Cold Spring Harbor Symp. Quant. Biol.* 37: 319-330.

Lehman, W. & A.G. Szent-Györgyi 1975. Regulation of muscular contraction. Distribution of actin control and myosin control in the animal kingdom. *J. Gen. Physiol.* 66: 1-30.

Levin, A. & J. Wyman 1927. The viscous elastic properties of muscle. *Proc. Roy. Soc. London* (B) 101: 218-243.

Madrid, E., I. P. Zanders & F.C. Herrera 1976. Changes in coelomic fluid and intracellular ionic composition in holothurians exposed to diverse sea water concentrations. *Comp. Biochem. Physiol.* 54A: 167-174.

Mattisson, A.G.M. 1959. Cytochrome c, cytochrome oxidase, and respiratory intensity in some types of invertebrate muscles. *Ark. Zool.* 12: 143-163.

Mattisson, A.G.M. 1961. The effect of inhibitory and activating substances on the cellular respiration of some types of invertebrate muscles. *Ark. Zool.* 13: 447-474.

Olson, M. 1938. The histology of the retractor muscles of *Thyone briareus* Lesueur. *Biol. Bull.* 74: 342-347.

Pentreath, V.W. & J.L.S. Cobb 1972. Neurobiology of Echinodermata. *Biol. Rev.* 47: 363-392.

Prosser, C.L. 1954. Activation of a non-propagating muscle in *Thyone*. *J. Cell. Comp. Physiol.* 44: 247-253.

Prosser, C.L., H.J. Curtis & D.M. Travis 1951. Action potentials from some invertebrate non-striated muscles. *J. Cell. Comp. Physiol.* 44: 299-319.

Prosser, C.L. & G.O. Mackie 1980. Contractions of holothurian muscles. *J. Comp. Physiol.* 136: 103-112.

Robertson, J.D. 1980. Osmotic constituents of some echinoderm muscles. *Comp. Biochem. Physiol.* 67A: 535-543.

Shida, H. 1971. Electron microscopic studies on the longitudinal muscle of the sea-cucumber (*Stichopus japonicus*). *Sci. Rep. Tôhoku Univ.* (4, Biol.) 35: 175-187.

Simon, S.E., S. Edwards & P.J. Dewhurst 1963. Potassium exchange in holothuroidean muscle. *J. Cell Comp. Physiol.* 63: 89-100.

Simon, S.E., J. Muller & D.J. Dewhurst 1963. Ionic partition in holothuroidean muscle. *J. Cell Comp. Physiol.* 63: 77-84.

Steinbach, H. 1937. Potassium and chloride in thyone muscle. *J. Cell Comp. Physiol.* 9: 429-435.

Steinbach, H.B. 1940. Electrolytes in thyone muscle. *J. Cell Comp. Physiol.* 15: 1-9.

Stubbs, T. & J.L.S. Cobb 1981. The giant neurone system in ophiuroids. II. The hyponeural motor tracts. *Cell Tissue Res.* 220: 373-385.

Sugi, H., S. Suzuki, T. Tsuchiya, S. Gomi & N. Fujieda 1982. Physiological and ultrastructural studies on the longitudinal retractor muscle of a sea cucumber *Stichopus japonicus*. I. Factors influencing the mechanical response. *J. Exper. Biol.* 97: 101-111.

Suzuki, S. 1982. Physiological and cytochemical studies on activator calcium in contraction by smooth muscle of a sea cucumber, *Isostichopus badionotus*. *Cell Tissue Res.* 222: 11-24.

Suzuki, S. & H. Sugi 1982. Physiological and ultra-structural studies on the longitudinal retractor muscle of a sea cucumber (*Stichopus japonicus*). II. Intracellular localization and translocalization of activator calcium during mechanical activity. *J. Exper. Biol.* 97: 113-119.

Tao, L. 1927. Physiological characteristics of *Caudina* muscle, with some accounts on the innervation. *Sci. Rep. Tohôku Imp. Univ.* (4, Biol.) 2: 265-291.

Tsuchiya, T. 1985. The maximum shortening velocity of holothurian muscle and effects of tonicity changes on it. *J. Exp. Biol.* 119: 31-40.

Tsuchiya, T. & S. Amemiya 1977. Studies on the radial muscle of an echinothuriid sea-urchin, *Asthenosoma* – I. Mechanical responses to electrical stimulation and drugs, *Comp. Biochem. Physiol.* 57C: 69-73.

VanWeel, P.B. 1955. The problem of the smooth muscle. *Pubbl. Staz. Zool. Napoli* 26: 10-16.

Wood, R.L & M.J. Cavey 1981. Ultrastructure of the coelomic lining in the podium of the starfish *Stylasterias forreri*. *Cell Tissue Res.* 218: 449-473.

Zanders, I.P. & F.C. Herrera 1974. Ionic distribution and fluxes in holothurian tissues. *Comp. Biochem. Physiol.* 74A: 1153-1170.

Pharmacological effects of compounds from echinoderms

J.F. VERBIST

Substances Marines à Activité Biologique (SMAB), Faculté de Pharmacie, Nantes, France

Final manuscript acceptance: October 1990.

CONTENTS

1 INTRODUCTION

During the last 20 years, marine invertebrates have become the victims of a new race of predators, the chemists, who in alliance with pharmacologists consider them, like land plants, as a potential source of new molecules for the creation of novel drugs. It was readily noted that marine invertebrates synthesize substances not previously identified in other organisms. It was thus logical to suppose that these substances must play a biological role either in the functioning of the organism itself or in its interspecific relations (e.g., roles of defense, attack, attraction, repulsion, recognition), and that they could be of pharmacological interest.

However, the approach in this research often remained empirical. With the exception of marine organisms known for their toxicity, little evidence suggested that it would be more advantageous to study one species than another. Biological 'leads' remained rare, and invertebrates, contrary to plants, rarely had an ethnopharmacological reputation. Thus the only approach at first consisted in blind screening. On the basis of successive results, the screening became more refined for certain preferential phyla, classes, families or genera. This empirical method was also justified because of the novelty of the field of exploration and the high statistical probability of finding not only an interesting active substance but also a new substance either in the chemical or pharmacological sector which could serve as a leader for a whole series of derived drugs.

In this way, many interesting substances were identified (Kaul & Daftari 1986, Gerwick 1987), including, for example, many novel neurotoxins whose effects on ion channels make them quite valuable as pharmacological reagents (Wu & Narahashi 1988), and many potentially anticancer com-

112

pounds (Munro et al. 1987). Several of these substances are at the stage of clinical trials, particularly two potential antineoplastic compounds: didemnin B, a cyclic depsipeptide from a tunicate of the *Trididemnum* genus which in 1984 became the first natural marine product to enter clinical trials and is now in phase II (Suffness & Thompson 1988); and Girolline, a new guanidine derivative from the sponge *Pseudaxinyssa cantharella* (Ahond et al. 1988). However, no marine substance from this research has yet been used in therapy.

This review defines our knowledge about the active substances of echinoderms and considers the potential interest of this phylum from that point of view.

2 HEMOLYTIC AND HEMAGGLUTINATING SUBSTANCES

Few pharmacological substances normally use the bloodstream to produce their effect. Thus it is especially important to know their behavior relative to blood and particularly its formed elements. This aspect is fundamental in the case of echinoderms since one of their characteristics is the frequent presence of hemolytic substances. Hemagglutinating properties have sometimes been detected.

2.1 *Hemolytic substances*

The first observations of hemolytic activity in echinoderms were made by Yamanouchi in 1942 and 1943 (Yamanouchi 1955) during studies concerning the toxicity of holothuroids, particularly against fish. He showed that this property is a nearly constant feature of this class since it was found in aqueous extracts of 22 out of the 27 species studied. These hemolytic properties have been observed also in aqueous extracts from other species (Rogers et al. 1980). They have been localized in aqueous extracts from Cuvierian tubules (Poscidio 1983b) and in the coelomic fluid (Canicatti, 1987). In echinoids, this same activity has been noted for aqueous extract of the entire body and of gonads (Rogers et al. 1980) the homogenate of globiferous pedicellariae (Alender et al. 1965) and coelomic fluid (Brown et al. 1968, Ryoyama 1973, Bertheussen 1983). Observations concerning asteroids are less frequent [total isopropanolic extract (Owellen et al. 1973); aqueous extract from gonads (Rogers et al. 1980)] and non-existent for other classes. The results obtained with extracts or non-purified preparations are shown in Table 1. The activity observed depends on the origin of the erythrocytes (Table 2). The principles responsible for this activity are currently attributed to two groups of compounds: saponosides and proteins.

Table 1. Hemolytic activity of crude or purified echinoderm extracts.

Species	Extract[1]	Erythrocyte	Results[2, 3]	Reference
Crinoidea				
Antedon bifida	Purified BuOH and EtOH extracts	Sheep	±	Mackie et al. 1975
Asteroidea				
Asterias forbesi	Body fluid	Human	–	Brown et al. 1968
Asterias rubens	H₂O (gonads)	Human	+	Rogers et al. 1980
	Purified BuOH and EtOH extracts	Sheep	+	Mackie et al. 1975
	Body fluid	Human	+	Mackie et al. 1968
Asterias vulgaris	i-PrOH	Mouse	+	Owellen et al. 1973
Astropecten irregularis	Purified BuOH and EtOH extracts	Sheep	+	Mackie et al. 1975
Coscinasterias tenuispina	Purified BuOH and EtOH extracts	Sheep	+	Mackie et al. 1975
Crossaster papposus	Purified BuOH and EtOH extracts	Sheep	+	Mackie et al. 1975
Diplasterias brucei	Purified BuOH and EtOH extracts	Sheep	+	Mackie et al. 1975
Hippasterias phrygiana	Purified BuOH and EtOH extracts	Sheep	+	Mackie et al. 1975
Luidia ciliaris	Purified BuOH and EtOH extracts	Sheep	+	Mackie et al. 1975
Marthasterias glacialis	Purified BuOH and EtOH extracts	Sheep	+	Mackie et al. 1968
	Body fluid	Human	+	Brown et al. 1968
Oreaster reticulatus	Body fluid	Human	–	Brown et al. 1968
Porania pulvillus	Purified BuOH and EtOH extracts	Sheep	+	Mackie et al. 1975
Ophiuridae				
Ophiocomina echinata	Body fluid	Human	+	Brown et al. 1968
Ophiocomina nigra	Purified BuOH and EtOH extracts	Sheep	±	Mackie et al. 1975
Ophiura texturata	Purified BuOH and EtOH extracts	Sheep	±	Mackie et al. 1975
Echinoidea				
Anthocidaris crassispina	Body fluid	See Table 2	+	Ryoyama 1973
Arbacia punctulata	Body fluid	Human	–	Brown et al. 1968
Diadema antillarum	Body fluid	Human	–	Brown et al. 1968

114

Species	Material	Test organism	Result	Reference
Echinarachnus parma	Body fluid	Human	–	Brown et al. 1968
Echinocardium cordatum	Purified BuOH and EtOH extracts	Sheep	±	Mackie et al. 1975
Echinometra lucunter	Body fluid	Human	–	Brown et al. 1968
Echinus esculentus	Globiferous pedicellariae homogenate	?	+	Levy 1935 (in Alender et al. 1965)
	H_2O (gonads)	Human	+	Rogers et al. 1980
	Purified BuOH and EtOH extracts	Sheep	±	Mackie et al. 1975
Hemicentrotus pulcherrimus	Body fluid	See Table 2	+	Ryoyama 1973
Lytechinus variegatus	Body fluid	Human	+	Brown et al. 1968
Paracentrotus lividus	Globiferous pedicellariae homogenate	?	+	Levy 1935 (in Alender et al. 1965)
Psammechinus miliaris	Globiferous pedicellariae homogenate	?	+	Levy 1935 (in Alender et al. 1965)
	H_2O (gonads)	Human	+	Rogers et al. 1980
Pseudocentrotus lividus	Body fluid	See Table 2	+	Ryoyama 1973
Strongylocentrotus droebachiensis	Body fluid	Rabbit	+	Bertheussen 1983
		Human	–	Bertheussen 1983
		Human	+	Brown et al. 1968
		Mouse	–	Bertheussen 1983
		Sheep	–	Bertheussen 1983
Tripneustes esculentus	Body fluid	Human	–	Brown et al. 1968
Tripneustes gratilla	Globiferous pedicellariae homogenate	?	+	Levy 1935 (in Alender et al. 1965)
Holothuroidea				
Aslia levefrei	H_2O	Human	+	Rogers et al. 1980
Actinopyga lecanora	H_2O (Cuvierian tubules)	Human	+	Poscidio 1983b
Caudina chilensis	H_2O	Rabbit	–	Yamanouchi 1955
Cucumaria echinata	H_2O	Rabbit	+	Yamanouchi 1955
Cucumaria japonica	H_2O	Rabbit	+	Yamanouchi 1955
Holothuria arenicola	Body fluid	Human	–	Brown et al. 1968
Holothuria argus	H_2O	Rabbit	+	Yamanouchi 1955
Holothuria atra	H_2O	Rabbit	+	Yamanouchi 1955
Holothuria axiologa	H_2O	Rabbit	+	Yamanouchi 1955

Table 1 (continued).

Species	Extract[1]	Erythrocyte	Results[2,3]	Reference
Holothuria bivittata	H_2O	Rabbit	+	Yamanouchi 1955
Holothuria cubana	Body fluid	Human	+	Brown et al. 1968
Holothuria fuscocinerea	H_2O	Human	+	Poscidio, 1983b
Holothuria lecanora	H_2O	Rabbit	+	Yamanouchi 1955
Holothuria lubrica	H_2O	Rabbit	+	Yamanouchi 1955
Holothuria mexicana	Body fluid	Human	+	Brown et al. 1968
Holothuria forskali	H_2O	Human	+	Rogers et al. 1980
Holothuria monacaria	H_2O	Rabbit	+	Yamanouchi 1955
Holothuria nobilis	H_2O	Rabbit	−	Yamanouchi 1955
Holothuria parvula	Body fluid	Human	+	Brown et al. 1968
Holothuria pervicax	H_2O	Rabbit	+	Yamanouchi 1955
Holothuria polii	Body fluid[4]	pig	+	Parrinello et al. 1979
		Rabbit	+	Parrinello et al. 1979
				Caniccati 1987
Holothuria pulla	H_2O (Cuvierian tubules)	Human	+	Poscidio 1983b
Holothuria scabra	H_2O	Human	+	Yamanouchi 1955
Holothuria surinamensis	Body fluid	Human	+	Brown et al. 1968
Holothuria vagabunda	H_2O	Rabbit	+	Yamanouchi 1955
Leptosynapta ooplax	H_2O	Rabbit	+	Yamanouchi 1955
Opheodesoma grisea	H_2O (Cuvierian tubules)	Human	±	Poscidio 1983b
Molpadia sp.	H_2O	Rabbit	−	Yamanouchi 1955
Parastichopus nigripunctatus	H_2O	Rabbit	+	Yamanouchi 1955
Pentacta australis	H_2O	Rabbit	+	Yamanouchi 1955
Polycheira rufescens	H_2O	Rabbit	+	Yamanouchi 1955
Pseudocucumis africana	H_2O	Rabbit	+	Yamanouchi 1955
Stichopus badionotus	Body fluid	Human	+	Brown et al. 1968
Stichopus chloronotus	H_2O	Rabbit	+	Yamanouchi 1955

Stichopus japonicus	H_2O	Rabbit	+	Yamanouchi 1955
Stichopus oshsimae	H_2O	Rabbit	–	Yamanouchi 1955
Stichopus variegatus	H_2O	Rabbit	+	Yamanouchi 1955
Thelenota ananas	H_2O	Rabbit	+	Yamanouchi 1955

1. When unspecified, the extracts was prepared from the body wall. H_2O: aquous extracts, BuOh: butanol extract, EtOH: ethanol extract, i-PrOH: isopropanol extract.

2. The results have been specified as positive or negative. When positive, the activity can be more or less important according to the species (Yamanouchi 1955).

3. The sampling date, which is seldom given by the authors, has not been considered in this table. This can have a great importance because the hemolytic substances content, e.g. echinoderm saponins content, changes significantly during the year.

4. Hemolytic activity on chicken, horse, sheep or human erythrocytes appears when $CaCl_2$ (20-100 mM) is added. In this same conditions, the activity on pig or rabbit erythrocytes is increased.

117

Table 2. Relative hemolytic activity of coelomic fluid preparation obtained from three different species of echinoids against erythrocytes from various species of animals (Ryoyama 1973).

Erythrocyte origin*	Hemolytic titer**		
	Anthocidaris crassispina	*Pseudocentrotus depressus*	*Hemicentrotus pulcherrimus*
Human:			
Group A	2^4	2^4	2^8
Group B	2^4	2^4	2^8
Group O	2^4	2^4	2^7
Rabbit	2^9	2^9	2^{10}
Mouse	2^2	2^2	2^6
Rat	<2	<2	2^4
Guinea pig	<2	<2	2^5
Dog	<2	<2	–
Sheep	<2	<2	<2
Goat	<2	–	–
Pig	<2	–	–
Hen	<2	<2	<2
Snake	<2	<2	<2
Frog	<2	–	–

* Final concentration of erythrocytes is 6.10^7-7.10^7 cells per ml of reaction mixture.
** Represented as the reciprocal of the highest dilution effecting hemolysis.

2.1.1 *Hemolytic saponins*

Yamanouchi (1955) noted that the hemolytic (and ichthyotoxic) product isolated from *Holothuria vagabunda* Selenka was similar to plant saponins, which are tensio-active substances characterized by hemolytic properties. These animal saponins were identified with triterpenic glycosides in holothuroids and named holothurins. Similar substances of steroid type, *viz.* asterosaponins, were found in asteroids. These holothurins and asterosaponins are mixtures of saponosides which vary qualitatively and quantitatively according to the species. The other classes of echinoderms contain few or no saponins (Mackie et al. 1977). Two steroid saponosides from the ophiuroid *Ophioderma longicauda* have recently been described (Riccio et al. 1986). The chemistry of echinoderm saponins has been reviewed by Burnell & Apsimov (1983), Stonik (1986), Minale et al. (1986) and Stonik & Elyakov (1988). Saponins whose hemolytic properties have been studied *in vitro* are shown in Table 3. The hemolytic activity of holothurin from *Actinopyga agassizi* has been observed *in vivo* by injection into the dorsal lymph spaces of the frog *Rania pipiens*: an injection of 0.2 ml of a 50% solution led to hemolysis of 50% of red blood cells. Hemolysis was followed by an intense hematopoietic activity (Jakowska et al. 1958). Saponins of the asteroid

118

Table 3. Hemolytic activity of echinoderm saponins.

Saponin	Organism	Erythrocyte species	Activity[1]	Reference
Asterosaponins				
Asterosaponin[2]	Asterias vulgaris	Human, mouse, rat	ECmin: 530.10^{-3} mg/ml	Owellen et al. 1973
Asterosaponin[2]	Asterias amurensis	Rabbit	EC100: 32.10^{-3} mg/ml	Yasumoto et al. 1964
Asterosaponin[2]	Asterina miniata	Human	EC100: 4.10^{-3} mg/ml	Rio et al. 1965
Asterosaponin[2]	Asterina pectinifera	Rabbit	EC100: 142.10^{-3} mg/ml	Hashimoto & Yasumoto 1960
Asterosaponin[2]	Pycnopodia helianthoides	Human	EC100: 25.10^{-3} mg/ml	Rio et al. 1965
Asterosaponin[2]	Pisaster ochraceus	Human	EC100: $>20.10^{-3}$ mg/ml	Rio et al. 1965
Asterosaponin[2]	Pisaster brevispinus	Human	EC100: 3.10^{-3} mg/ml	Rio et al. 1965
Asterosaponin XI		Unspecified	EC100: 46.10^{-3}%	Fusetani et al. 1984
Asterosaponin XX		Unspecified	EC100: 45.10^{-3}%	Fusetani et al. 1984
Asterosaponin XXI	from Acanthaster planci	Unspecified	EC100: $>392.10^{-3}$%	Fusetani et al. 1984
Asterosaponin XXII	or Luidia maculata	Unspecified	EC100: 1196.10^{-3}%	Fusetani et al. 1984
Asterosaponin XXIII	or Asterias amurensis	Unspecified	EC100: 87.10^{-3}%	Fusetani et al. 1984
Asterosaponin XXIV		Unspecified	EC100: 56.10^{-3}%	Fusetani et al. 1984
Asterosaponin XXXV		Unspecified	EC100: $>413.1.^{-3}$%	Fusetani et al. 1984
Asterosaponin XLI		Unspecified	EC100: 31.10^{-3}%	Fusetani et al. 1984
Asterosaponin XLII		Unspecified	EC100: 93.10^{-3}%	Fusetani et al. 1984
Asterosaponin LXVII		Unspecified	EC100: $>440.10^{-3}$%	Fusetani et al. 1984
Holothurins				
Holothurin[2]	Actinopyga agassizi	Rabbit	EC50: 40.10^{-3}%	Nigrelli & Jakowska 1960
Holothurin[2]	Actinopyga agassizi	Mouse	ECmin: 1.10^{-3} mg/l	Cairns & Olmsted 1973
Holothurin[2]	Holothuria vagabunda	Rabbit	EC100: 25.10^{-3} mg/ml	Yamanouchi 1955
Holothurin H	Holothuria fuscocinerea	Human	ECmin: 15.10^{-3} mg/ml	Poscidio 1983b
Holothurin H	Holothuria pulla	Human	ECmin: 60.10^{-3} mg/ml	Poscidio 1983b
Holothurin H	Actinopyga lecanora	Human	ECmin: 60.10^{-3} mg/ml	Poscidio 1983b
Holothurin H	Opheodesoma grisea	Human	ECmin: 400.10^{-3} mg/ml	Poscidio 1983b
Holothurin A (I)	Actinopyga agassizi	Human	EC50: 0192.10^{-3} mg/ml	Thron 1964
Holothurin A (I)	Unidentified holothuroid	Human	EC100: 0039.10^{-3} mg/ml	Rio et al. 1965
Griseogenin glucoside (III)	Holothuria floridana	Unspecified	Powerful activity	Kaul & Daftari 1986

1. For an easier comparison we have calculated concentration in mg/ml or in % according to the units chosen by the authors. EC50: concentration required to give 50% hemolysis; EC100: threshold concentration to give complete hemolysis; ECmin: minimal hemolytic concentration. However these results depend on the methods of testing.

2. Unidentified.

119

Pycnopodia helianthoides produced hemolysis *in vivo* in erythrocytes from *Fundulus heterocyclus* when injected IP (Rio et al. 1963), as did that of *Asteria vulgaris* by the same route in the mouse (Owellen et al. 1973).

Echinoderm saponins generally are more hemolytic than plant saponins (Thron 1964). Yamanouchi (1955) reported that 'holothurin' from *Holothuria vagabunda* was 6 to 7 times more active than the 'saponin' from the bark of *Quillaya smegmadermos*.'Holothurin' from *Actinopyga agassazi* has been reported to be 2 (Nigrelli & Jakowska 1960) to 250 to 500 (Cairns & Olmsted 1973) times as active. Compared to quillajasaponin. Holothurin A (I)*, a pure saponin representing around a third of the 'holothurin' of *Actinopyga agassazi* (Kitagawa et al. 1979, 1981b, 1982), showed the same activity as quillajasaponin at a concentration 500 times lower (Rio et al. 1965). However, Thron (1964), using partially purified fractions of this same 'holothurin,' found a ratio of only 6 to 15 with respect to quillajasaponin and 3.8 with respect to digitonin. These differences in results may be due to differences in test conditions. The saponins contained in asteroids are 7 times more active than the 'saponin' of Quillaja wood (Rio et al. 1965). However, great differences exist between the substances concerned. Fusetani et al. (1984) emphasized the importance of a long carbon 17 side chain on the steroid nucleus (compare XX and LXVII) and the presence of a methyl on carbon 24 in this chain (ergostane structure) (compare XX and XVIII, and XXIII and XLII) and of a sulfate group on the hydroxyl on carbon 3 (compare XXI and XXIV). Plant saponins are not generally sulfated.

The mode of hemolytic action of echinoderm saponins is similar to that of plant saponins. For both, the addition of cholesterol reduces or eliminates hemolysis.This has been demonstrated for holothuroids (Yamanouchi 1955, Nigrelli & Jakowska 1960) and asteroids (Mackie et al. 1975, 1977). This suggests that saponins react with cholesterol of the erythrocyte membrane, thereby creating holes or cavities of 50 to 250 Å on the outer surface of this membrane (Seeman et al. 1973, Canicatti 1987) to allow or facilitate the hemolysis**. The affinity for cholesterol is such that holotoxin A1 (XV) from *Stichopus japonicus* has been proposed instead of digitonin for assaying free blood cholesterol (Ivanov et al. 1986). Interaction of plant saponins with

---

*The Roman numerals refer to formulae of substances cited in the Annex.

**This action on membrane cholesterol is of course not limited to red blood cells. The resistance of holothuroids and asteroids to their own saponosides is sometimes explained by the fact that the predominant membrane sterol is not cholesterol as in orphiuroids, echinoids, crinoids and many other invertebrates and vertebrates but 5-cholest-7-ene-3β-ol, which interacts very little with saponins, particularly of asteroids (Mackie et al. 1977). The high sulfate cholesterol content (Goodfellow & Goad 1973) in asteroid tissues may also play a protective role (Burnell & Apsimov 1983).

120

membrane proteins and phospholipids has been suggested (Assa et al. 1973).

2.1.2 *Hemolytic proteins*
The hemolytic activity observed with holothuroid and echinoid coelomic fluid has not been related to saponins but to proteins (Ryoyama 1973, Bertheussen 1983, Canicatti 1987). This activity is abolished by heating aqueous extracts, and an optimal pH has generally been determined. For the coelomic fluid of several echinoids, hemolytic activity was reduced by the action of trypsin and 2-mercaptoethanol, a reagent that splits disulfide bonds (Ryoyama 1973). Various authors have noted that the presence of Ca^{2+} ions is essential and excludes in particular a mechanism of action of saponoside type. However, Canicatti & Ciulla (1987) noted one of the two hemolysins of coelomocytes from *Holothuria polii* was heat-stable and calcium-independent. These two hemolytic proteins (MW: 76-80 KDa) are serologically cross-reactive (Canicatti & Ciulla 1988) and are produced by two distinct populations of coelomocytes (Canicatti et al. 1988). Hemolysis caused by coelomic fluid from *Strongylocentrotus droebachiensis,* unlike that caused by coelomic fluid from *Holothuria polii* (Canicatti 1987), would seem to be rather similar to that induced by human complement. In particular, it is inhibited in a similar way by substances such as hydrazine or NH_4^+ ions (Bertheussen 1983).

2.2 *Hemagglutinating substances*

Compared to other invertebrate phyla, echinoderms have not often been studied with respect to their content of hemagglutinating substances (Table 4), probably because potential hemolytic properties are likely to mask the presence of such substances. Ryoyama (1974) and Bertheussen (1983) avoided this difficulty by denaturing the hemolysins of the coelomic fluid of several echinoid species by heat before performing the hemagglutination test. Parrinello et al. (1979) used the same procedure with the coelomic fluid of *Holothuria polii.* However, this method can only be used with protein hemolysins that are more heat sensitive than agglutinins. Bertheussen (1983) based his approach on the fact that the lytic activity of a complement-like substance from echinoid coelomic fluid was significantly inhibited when Ca^{2+} was decreased to 6 mM, whereas the agglutination reaction was still very strong even at 1 mM Ca^{2+}. Though some preparations possessed hemagglutinating properties when these procedures were used, these properties generally lacked specificity. The experiments generally used unpurified or slightly purified extracts which could have corresponded to hemagglutinin mixtures with different specificities.

The activity of hemagglutinin from *Holothuria polii*, a protein-like

Table 4. Hemagglutinating substances from echinoderms.

Species	Extract	Erythrocyte											Hemagglutinim structure	Reference
		a	b	c	d	e	f	g	h	i	j	k		
Asteroidea														
Asterias forbesi	Body fluid		+	+	+	(+)								Brown et al. 1968
Asterias rubens	Aqueous extract		+	+	+				+	+	+	−[1]	Lectin	Andersson et al. 1986
Asterina miniata	Body fluid	+												Tyler 1946
Oreaster reticulatus	Body fluid		+	+	+	(+)			+		+			Brown et al. 1968
Echinoidea														
Anthocidaris crassispina	Body fluid	+	+	+	+	+	+	+					Glycoprotein	Ryoyama 1974
Arbacia punctulata	Body fluid	+	+	−	+	(+)	+	+	−	−	−			Brown et al. 1968
Diadema antillarum	Body fluid	−	−	−	−	−								Brown et al. 1968
Echinarachnius parna	Body fluid	+	+	+	+	(+)								Brown et al. 1968
Echinometra lucunter	Body fluid	−	−	−	−	−								Brown et al. 1968
Echinus esculentus	Haemolymph			+		+	+	+		−	−	+[2]		Uhlenbruck et al. 1970
	Aqueous extract			−	−				−	−	−	−[1]		Andersson et al. 1986
Hemicentrotus pulcherrimus	Body fluid	+	+	+	+	+	+	+	+	−	−		Polysaccharide?	Ryoyama 1974
	Seminal fluid	+	+			+	+						Glycoprotein and protein	Yumada & Aketa 1982
Lytechinus pictus	Seminal fluid					−								Tyler 1946
Psammechinus miliaris	Haemolymph							+				+[2]		Uhlenbruck et al. 1970
Pseudocentrotus depressus	Body fluid	−	+	+	−								Protein	Ryoyama 1974
Strongylocentrotus droebachiensis	Body fluid*	+	+	+	+	+	−	+	+	−	−			Brown et al. 1968
Strongylocentrotus purpuratus	Body fluid					+	+		−				Lectin	Bertheussen 1983
	Body fluid							−						Tyler 1946
Tripneustes esculentus	Seminal fluid	−				−					−			Brown et al. 1968

< unusual reaction** >

122

Holothuroidea								
Holothuria arenicola	Body fluid							Brown et al. 1968
Holothuria atra	Haemolymph	+	+	+			$-^3$	McKay et al. 1969
Holothuria leucospilota	Haemolymph	+	+	−			$+^3$	McKay et al. 1969
Holothuria polii	Coelomic fluid from Polian vesicles	+ + +	+ + +	+	+	$+^{1,3,4,5}$ Protein like substances	Parrinello et al. 1976
Holothuria tubulosa	Coelomic fluid from Polian vesicles	+ + +	+	+		$-^{1,3,4,5}$	Parinello et al. 1976

Erythrocytes: a = human unspecified, b = human A, c = human B, d = human O, e = human Oh, f = rabbit, g = dog, h = guinea-pig, i = mouse, j = sheep, k = other: 1: Horse erythrocytes; 2: Pig erythrocytes; 3: Chicken erythrocytes; 4: Star-gather fish (*Uranoscopus scaber*) erythrocyte, cat erythrocyte, calf erythrocyte, rat erythrocyte, 5: Toad (*Discoglossus pictus*) erythrocyte.
*Likely anti-H activity.
**Possible anti-M activity.

substance, and from *Holothuria tubulosa* was not inhibited by any of 11 sugars (Parrinello et al. 1976). Two agglutinins were purified from seminal plasma of *Hemicentrotus pulcherrimus* (Yamada & Aketa 1982). One, a glycoprotein (MW 14 000) whose action on Group A human erythrocytes or rabbit erythrocytes decreased more or less in the presence of several sugars, had lectin characteristics. The other, a protein (MW 22 000), was not inhibited by any sugar and was resistant to trypsin. The lectin responsible for the agglutination caused by coelomic fluid of *Anthocidaris crassispina* echinodin, is a chain of 147 amino acids (the complete sequence is known) associated with a carbohydrate chain (Giga et al. 1987). The hemagglutinating effect of another lectin found in *Strongylocentrotus droebachiensis* is prevented by D-fucose (Bertheussen 1983). The aqueous extract of *Asterias rubens* is particularly active (0.1 µg/ml) on mouse erythrocytes but much less active (100 µg/ml) or inactive on those of other species (Andersson et al. 1986). Its lectin-like activity is inhibited by glucosamine and to a lesser degree by glucuronic acid. It also agglutinates murine spleen lymphocytes (10 µg/ml) and murine bone-marrow cells (1 µg/ml). The proliferation of the latter was stimulated at that concentration. However, murine thymus cells or human peripheral blood lymphocytes were not affected.

3 CYTOTOXIC AND ANTITUMOR SUBSTANCES

Cytotoxic (*in vitro*) and thus potentially antitumoral (*in vivo*) substances are common in echinoderms. Weinheimer et al. (1978) detected this activity on human KB cancer cells and on P 388 murine ascitic leukemia (T/C* ≥ 125%) in 9.6% of isopropanolic extracts prepared from 187 species. However, this percentage is slightly lower than that (11.4%) noted overall by these authors in 1700 species of invertebrates tested. Nemanich et al. (1978) obtained higher results on these same KB cells: 26% of 1:3 toluene-methanol extracts prepared from 71 echinoderm species were active vs 13.2% overall for 507 other marine species. This difference can be attributed to the particularly high detection threshold (IC50 ≤ 200 µg/ml) used by these authors. On the same KB cells, 35% of 95% alcoholic extracts from 17 species of echinoderms from French or Tunisian coasts were active at 100 µg/ml (Verbist et al. unpubl. res.). However, murine leukemic cells are less sensitive than KB cells as none of 10 extracts prepared with 1:3 toluene-methanol and tested at 10 µg/ml were active on L 1210 cells (Rinehart et al. 1981). Likewise, the percentage of activity on P 388 cells in the conditions cited above dropped to 17% (Verbist et al. unpubl. res.). Asteroids, followed by holothuroids, are the

---

*T/C = mean survival time of the test group divided by mean survival time of the control group expressed as a percentage.

124

classes with the highest percent of species with active components with 40 and 35% of extracts respectively prepared from these species giving a positive response (Nemanich et al. 1978). These results do not concern aqueous extracts likely to contain active macromolecules for which no systematic studies were found in the literature. Table 5 lists the major results published about crude or partially purified extracts.

Some echinoderms have cytotoxins (Table 6). These are related to holo-thurins for holothuroids and to asterosaponins, polyhydroxylated sterols, benzyltetraisoquinoline alkaloids and perhaps polysaccharides for asteroids (Heilbrunn et al. 1954). However, few of these substances have an antitumor activity, except for the crude holothurin of *Actinopyga agassizi*, the imbricatine (an alkaloid) from *Dermasteria imbricata*, and a polysaccharide from *Asterias amurensis*. To these must be added a mucopolysaccharide from *Stichopus japonicus* and echinoid glycoproteins, that have no or very low cytotoxic activity *in vitro*.

3.1 Holothurin from Actinopyga agassizi

Nigrelli (1952) showed that the crude holothurin, referred to as sun-or oven-dried powdered Cuvierian tubules of the holothuroid *Actinopyga agassizi* and considered as a very toxic agent against fish and various invertebrates, inhibits the growth of mouse sarcoma 180 when injected at the inoculation site of the sarcoma. Sullivan et al. (1955, 1956) obtained survival times (T/C) of over 250% on Krebs-2 ascitic tumor at a dose of 0.1 mg/mouse IP every other day for 21 days. Nigrelli & Jakowska (1960) reported results in the same conditions ranging from 168 to 273%. Cairns & Olmsted (1973) found a significant activity on mouse sarcoma 180 used in the form of ascitic tumor (T/C: 145% at 0.15 mg IP/mouse every other day), although no effect was obtained on melanoma B 16. All of these tests were carried out with crude holothurin. Chanley et al. (1959) isolated the glycosidic part representing 40% of the crude material which was designated as 'holothurin A.' This substance is a mixture of two triterpene-glycosides [holothurin A (I) and 24-dehydroechinoside A (XIV)] at an approximate ratio of 1:2 (Kitagawa et al. 1982). The 'holothurin A' of Chanley, which was consistently cytotoxic against KB cells (Nigrelli et al. 1967; Colon et al. 1976), was inactive on Krebs-2 ascitic tumor at 1 to 1.5 mg/mouse (Nigrelli & Jakowska 1960) in the same conditions described above, and inactive or relatively inactive on three other transplanted mouse tumors: sarcoma 180 (tested in the form of solid tumor), mammary adenocarcinoma 755 and leukemia L 1210 (Leiter et al. 1962). 'Holothurin A' was also much less toxic than crude holothurin since the lethal dose (LD 100) of the latter was 0.2 mg/mouse IP (Sullivan et al. 1955, Nigrelli & Jakowska 1960). The antitumor substances of the crude holothurin from *Actinopyga agassizi* unidentified.

Table 5. Principal published results about cytotoxic and antitumoral assays of crude and semipurified extracts of Echinodermata.

Species	Extract[1]	KB[9] IC50[2]	P 388[9] IC50[2]	M.S.37[9] IC100[3]	AP[9]	P 388[9] T/C%[5]	Reference
Crinoidea							
Antedon bifida	EtOH-CH$_2$Cl$_2$	Inact. at 100	Inact. at 100				Verbist et al. (unpubl.)
Nemaster rubignosa	H$_2$O	+ (100)[6]					Ruggieri & Nigrelli 1974
Asteroidea							
Acanthaster planci	H$_2$O	+ (50-100)[6]					Ruggieri & Nigrelli 1974
	H$_2$O				+		Ruggieri 1965
Asterias forbesi	EtOH-CH$_2$Cl$_2$	> 100	> 100				Verbist et al. (unpubl.)
Asterias rubens	H$_2$O[7]	Inact. at 100					Verbist et al. (unpubl.)
Asterina miniata	H$_2$O				+		Ruggieri 1965
Chaetaster longipes	EtOH-CH$_2$Cl$_2$	Inact. at 100	Inact. at 100				Verbist et al. (unpubl.)
Echinaster sepositus	EtOH-CH$_2$Cl$_2$	10	90				Verbist et al. (unpubl.)
Heliaster kubinji	Toluene-MeOH 1:3	100-120					Nemanich et al. 1978
Henricia leviuscula	Toluene-MeOH 1:3	50-100					Nemanich et al. 1978
Leptychaster sp.	Toluene-MeOH 1:3	50-100					Nemanich et al. 1978
Linckia columbiae	Toluene-MeOH 1:3	50-100					Nemanich et al. 1978
Luidia ciliaris	EtOH-CH$_2$Cl$_2$	28	> 100				Verbist et al. (unpubl.)
	H$_2$O	17/inact. at 100					Verbist et al. (unpubl.)
Luidia superba	Toluene-MeOH 1:3	50-100					Nemanich et al. 1978
Masthasterias glacialis	EtOH-CH$_2$Cl$_2$	Inact. at 100	> 100				Verbist et al. (unpubl.)
	H$_2$O	Inact. at 100					Verbist et al. (unpubl.)
Mediaster aequalis	Toluene-MeOH 1:3	< 10					Nemanich et al. 1978
Mithrodia bradleyi	Toluene-MeOH 1:3	100-200					Nemanich et al. 1978
Ophidiaster ophidianus	EtOH-CH$_2$Cl$_2$	10-13	Inact. at 100				Verbist et al. (unpubl.)
	EtOH-H$_2$O	Inact. at 100					Verbist et al. (unpubl.)
Oreaster reticulatus	EtOH 50%					+	Sigel et al. 1970
	Toluene-MeOH 1:3	100-200					Nemanich et al. 1978

Species	Extract / Solvent	Conc.	Conc.	Activity	Reference
Pharia pyramidata	Toluene-MeOH 1:3	100-200			Nemanich et al. 1978
Pisaster brevispinus	H_2O			+	Ruggieri 1965
Pisaster ochraceus	H_2O			+	Ruggieri 1965
Pycnopodia helianthoides	H_2O			+	Ruggieri 1965
Ophiocentrotus brachiata	Toluene-MeOH 1:3	50-100			Nemanich et al. 1978
Ophiocomina echinata	EtOH-CH_2Cl_2	> 100	> 30		Verbist et al. (unpubl.)
	EtOH 50%			+	Sigel et al. 1970
Ophiocomina nigra	EtOH-CH_2Cl_2	100	100		Verbist et al. (unpubl.)
Ophioderma variegatum	Toluene-MeOH 1:3	100-200			Nemanich et al. 1978
Ophiothrix fragilis	EtOH-CH_2Cl_2	> 100	Inact. at 100		Verbist et al. (unpubl.)
Echinoidea					
Arbacia lixula	EtOH-CH_2Cl_2	Inact. at 100	Inact. at 100		Verbist et al. (unpubl.)
Echinocardium cordatum	EtOH-CH_2Cl_2	> 100	> 100		Verbist et al. (unpubl.)
Lytechinus variegatus	EtOH			150%	Pettit et al. 1981b
	H_2O			127%	Pettit et al. 1981b
Paracentrotus lividus	EtOH-CH_2Cl_2	85	> 100		Verbist et al. (unpubl.)
Psammechinus miliaris	EtOH-CH_2Cl_2	Inact. at 100	Inact. at 100		Verbist et al. (unpubl.)
Sphaerechinus granularis	EtOH-CH_2Cl_2	Inact. at 100	Inact. at 100		Verbist et al. (unpubl.)
Strongylocentrotus droebachiensis	H_2O			145%	Pettit et al. 1981b
Toxopneustes roseus	Toluene-MeOH 1:3	10-50			Nemanich et al. 1978
Holothuroidea					
Astichopus multifidus	H_2O	+ (100-500)[6]		+	Ruggieri & Nigrelli 1974
Actinopyga agassizi	EtOH 50%			+	Sigel et al. 1970
Actinopyga mauritiana	H_2O			112%	Allen et al. 1986
	EtOH			104%	Allen et al. 1986
Actinopyga obesa	G.E. (holothurin A)	25			Kuznetsova et al. 1982
Bohadschia argus	G.E. (holothurin A)	12.5			Kuznetsova et al. 1982
	G.E. (bivittosides type)	12.5			Kuznetsova et al. 1982

Table 5. (continued).

Species	Extract[1]	KB[9] IC50[2]	P 388[9] IC50[2]	M.S.37[9] IC100[3]	AP[9]	P 388[9] T/C%[5]	Reference
Bohadschia marmorata	G. E. (bivittosides type)			12.5			Kuznetsova et al. 1982
Bohadschia vitiensis	G. E. (bivittosides type)			25			Kuznetsova et al. 1982
Brandtothuria arenicola	Toluene-MeOH 1:3	50-100					Nemanich et al. 1978
Cucumaria montagui	EtOH-CH$_2$Cl$_2$	10-18	2.6				Verbist et al. (unpubl.)
Cucumaria sp.	Toluene-MeOH 1:3	50-100					Nemanich et al. 1978
Holothuria atra	G. E. (holothurin A)			25			Kuznetsova et al. 1982
	G. E. (holothurin B)			50			Kuznetsova et al. 1982
	G. E. (hol. A and B)						Kuznetsova et al. 1982
Holothuria coluber	G. E. (holothurin A)			12.5			Kuznetsova et al. 1982
Holothuria discrepans	G. E. (hol. A and B)			6.2			Kuznetsova et al. 1982
Holothuria edulis	G. E. (hol. A and B)			6.2			Kuznetsova et al. 1982
Holothuria forskali	EtOH-CH$_2$Cl$_2$	>100	>100				Verbist et al. (unpubl.)
	H$_2$O	Inact. at 100					Verbist et al. (unpubl.)
Holothuria hilla	H$_2$O					116%	Allen et al. 1986
	EtOH					110%	Allen et al. 1986
	G. E. (hol. A and B)			50			Kuznetsova et al. 1982
Holothuria impatiens	G. E. (holothurin A)			100			Kuznetsova et al. 1982
Holothuria leucospilota	G. E. (hol. A and B)			25			Kuznetsova et al. 1982
Holothuria mexicana	H$_2$O		Inact. at 100				Ruggieri & Nigrelli 1974
Holothuria tubulosa	EtOH-CH$_2$Cl$_2$	+ (750)					Verbist et al. (unpubl.)
Neothyone gibbosa	H$_2$O	>100					Nemanich et al. 1978
Pearsonothuria graeffei	G. E. (holothurin A)	50-100		6.2			Kuznetsova et al. 1982
Pentamera chierchia	Toluene-MeOH 1:3	50-100					Nemanich et al. 1978
Psolidium dorsipes	Toluene-MeOH 1:3	50-100					Nemanich et al. 1978
Selenkothuria lubrica	Toluene-MeOH 1:3	50-100					Nemanich et al. 1978
Stichopus chloronotus	G. E. (stichoposides C and D)			12.5			Kuznetsova et al. 1982

Stichopus variegatus	G. E. (stichoposides C and D)	6.2	Kuznetsova et al. 1982
Stichopus horrens	G. E. (stichoposides D and ?)	50	Kuznetsova et al. 1982
Stichopus sp.	G. E. (unknown stichoposides)	6.2	Kuznetsova et al. 1982
Thelenota ananas	G. E. (thelothurins A and B)	6.2	Kuznetsova et al. 1982
Thelenota anax	G. E. (thelothurins)	25.0	Kuznetsova et al. 1982

1. EtOH-CH_2Cl_2 and EtOH-H_2O: ethanol extract separated between CH_2Cl_2 and H_2O; G.E.: semipurified glycosydic extract with the main compounds in brackets.

2. IC50: concentration which inhibits 50% of the cell proliferation in vitro expressed in μg/ml.

3. IC100: concentration which inhibit 100% of the cell proliferation in vitro. The authors expressed this value in mg/ml (?). We supposed this means μg/ml.

4. Various aspects are considered: sperm immobilization; cytolytic effects on unfertilized eggs; larval animalization.

5. T/C: mean survival time (in days) of the test group divided by the mean survival time of the control group expressed as a percent.

6. Cytolytic effect at the concentration specified in brackets.

7. Anderson et al. (1986) published the activity of aqueous extract of *Asterias rubens* on human T-cell Lymphoma (MOLT 4), human burkitt lymphoma (Raji & Daudi), Mouse T-cell lymphoma (**RBL** 5) and mouse leukemia L 1210 at 01 μg/ml. The observed activity is not related to a cytotoxic effect because at 1 μg/ml a stimulating effect can be observed on the human peripheral blood lymphocytes.

8. Pettit et al. (1970) also mentioned the antitumoral activity of unspecified extracts of asteroids (*Astropecten scoparius, Luidia clathrata, Pisaster ochraceus*) and echinoids (*Lytechinus variegatus, Mellita quinquiesperforata*) on Walker 256 carcinoma or on L 1210 or P 388 leukemia.

9. KB = KB human epithelial carcinoma cells, P 388 = P 388 mouse lymphoid leukemia cells, M.S. 37 = mouse sarcom – 37 cells, AP = *Arbacia punctulata* eggs assay.

129

Table 6. Principal published results about isolated cytotoxic compounds.

Compound	Species	AP* 1	HP* 2	KB* 3	P388* 3	HLC* 11	MTC* 11	BF* 12	Other	Reference
Holothurins from Holothuroidea										
Crude holothurin	Actinopyga agassizi				5-50					Nigrelli et al. 1967
Holothurin mixture with 40% holothurin A	Actinopyga agassizi								40[4]	Lasley & Nigrelli 1970
Holothurin A (I)	Actinopyga agassizi				5-50					Nigrelli et al. 1967
Holothurin A (I)	Holothuria mexicana	0.78								Anisimov et al. 1980
Holothurin B (II)	Holothuria mexicana	6.25								Anisimov et al. 1980
Stichoposide A (IV)	Stichopus japonicus	1.55								Anisimov et al. 1980
Stichoposide A1 (desulfated stichoposide A)	Stichopus japonicus	2.5-5.0								Anisimov et al. 1972a
Stichoposide C (V)	Stichopus japonicus	1.55								Anisimov et al. 1980
Stichostatin 1[8]	Stichopus chloronotus				2.9					Pettit et al. 1976
Thelenostatin 1[8]	Thelonota ananas				1.5					Pettit et al. 1976
Actinostatin[8]	Actinopyga mauritana			2.6					2.1[5]	Pettit et al. 1976
Cucumarioside G (XI)	Cucumaria fraudatrix	0.72								Anisimov et al. 1980
Saponin-like substance	Holothuria edulis	≈1								Ruggieri & Nigrelli 1966b
Asterosaponins and sterols from Asteroidea										
Saponin-like substance	Pycnopodia helianthoides	+								Rio et al. 1963
Saponin-like substance	Pycnopodia sp.		50							Nigrelli et al. 1967
Saponin-like substance	Pisaster sp.		50-500							Nigrelli et al. 1967
Saponin-like substance	Acanthaster sp.		5							Nigrelli et al. 1969
Saponin-like substance	Acanthaster planci	10-50								Ruggieri & Nigrelli, 1966a
Thornasteroiside A (XX)	A.p. or L.m. or A.a.[10]		10							Fusetani et al. 1984
	Acanthaster planci				0	1+	2+			Andersson et al. 1989
XXI	A.p. or L.m. or A.a.[10]		50							Fusetani et al. 1984
XXII	A.p. or L.m. or A.a.[10]		30							Fusetani et al. 1984

130

Marthasteroside A1 (XXIII)	A.p. or L.m. or A.a.[10]	15					Fusetani et al. 1984
	Marthasterias glacialis		0	0	1+		Andersson et al. 1989
XXIV	A.p. or L.m. or A.a.[10]	19					Fusetani et al. 1984
XXVI	A.p. or L.m. or A.a.[10]	15					Fusetani et al. 1984
XXVII	A.p. or L.m. or A.a.[10]	15					Fusetani et al. 1984
Marthasteroside A2 (XXVIII)	A.p. or L.m. or A.a.[10]	15					Fusetani et al. 1984
	Marthasterias glacialis				1+		Andersson et al. 1989
Pectinioside A (XXIX)	Asterina pectinifera	10.0				10.0[5]	Dubois et al. 1988
Pectinioside E (XXX)	Asterina pectinifera	11.5				8.8[5]	Dubois et al. 1988
Pectinioside F (XXXI)	Asterina pectinifera	†				†[5]	Dubois et al. 1988
XXXIII	A.p. or L.m. or A.a.[10]	50	0	0	1+		Fusetani et al. 1984
Sepitoside A (XXXIV)	Echinaster sepositus		0	0	1+		Andersson et al. 1989
XXXV	A.p. or L.m. or A.a.[10]	5					Fusetani et al. 1984
XXXVI	A.p. or L.m. or A.a.[10]	10					Fusetani et al. 1984
XXXVII	A.p. or L.m. or A.a.[10]	15					Fusetani et al. 1984
XXXVIII	A.p. or L.m. or A.a.[10]	10					Fusetani et al. 1984
Marthasteroside B (XXXIX)	Marthasterias glacialis	15	1-	0	1+		Andersson et al. 1989
XL	A.p. or L.m. or A.a.[10]	15					Fusetani et al. 1984
XLI	A.p. or L.m. or A.a.[10]	10					Fusetani et al. 1984
XLII	A.p. or L.m. or A.a.[10]	> 51					Fusetani et al. 1984
Nodoside (XLIII)	Protoreaster nodosus		2+	0	1+		Andersson et al. 1989
Halityloside D (XLIV)	Halityle regularis		0		0		Andersson et al. 1989
XLV	Asterina pectinifera	14				23[5]	Higuchi et al. 1988
Crossasteroside A (XLVI)	Crossaster papposus		0	1+	1+		Andersson et al. 1989
Crossasteroside B (XLVII)	Crossaster papposus		0	0	0		Andersson et al. 1989
Crossasteroside C (XLVIII)	Crossaster papposus				1+		Andersson et al. 1989
XLIX	Archaster typicus		1-		1+		Andersson et al. 1989
L	Archaster typicus		1-		0		Andersson et al. 1989
LI	Protoreaster nodosus					+(?)	Minale et al. 1984
LII	Poraster superba				0		Andersson et al. 1989

Table 6 (continued).

Compound	Species	AP* (1)	HP* (2)	KB* (3)	P388* (3, 11)	HLC* (11)	MTC* (11)	BF* (12)	Other	Reference
LIII	Protoreaster nodosus								+(?)	Minale et al. 1984
	Asterina pectinifera		38						46[5]	Higuchi et al. 1988
LIV	Protoreaster nodosus								+(?)	Minale et al. 1984
	Asterina pectinifera		36						−[5]	Higuchi et al. 1988
LV	Asterina pectinifera		34						26[5]	Higuchi et al. 1988
Halityloside A (LVI)	Halityle regularis							1+		Andersson et al. 1989
LVII	Hacelia attenuata								+(?)	Minale et al. 1982
LVIII	Protoreaster nodosus				1+	1+		1+		Andersson et al. 1989
LIX	Protoreaster nodosus				0	0		1+		Andersson et al. 1989
LX	Protoreaster nodosus				1−	0		1+		Andersson et al. 1989
Ophidianoside C (LXI)	Ophidiaster ophidianus							1+		Andersson et al. 1989
LXVII	A.p. or L.m. or A.a.[10]									Fusetani et al. 1984
Asterosaponin L (LXVIII)	Linckia guildingi	12.5								Anisimov et al. 1980
Regularoside A (LXX)	Halityle regularis			>50	1−	0		1+		Andersson et al. 1989
Pectinioside D (LXIX)	Asterina pectinifera			†					†[5]	Dubois et al. 1988
Pectinioside B (LXXI)	Asterina pectinifera			†					†[5]	Dubois et al. 1988
Pectinioside C (LXXII)	Asterina pectinifera			10.8					11.0[5]	Dubois et al. 1988
Unidentified substance from Asteroidea										
$C_{27}H_{40}O_8$	Coscinasterias sp.								+[6]	Institute of Physical and Chemical Research 1984
Saponins and sterols from Ophiuridea										
LXII	Ophioderma longicaudum				1+	1+				Andersson et al. 1989
LXIII	Ophioderma longicaudum					1+	0			Andersson et al. 1989
LXIV	Ophiocoma insularis				0	1+				Andersson et al. 1989
LXV	Ophioderma longicaudum				0	0	0			Andersson et al. 1989
LXVI	Ophioderma longicaudum				1+	0		1+		Andersson et al. 1989

Polysaccharides from Asteroidea		
Heparine-like substance	*Asterias forbesi*	+[7] Heilbrunn et al. 1954
Glycoproteins from Echinoidea		
Several glycoproteins	*Anthocidaris crassispina*	100^9 Suntory 1983
Alkaloids from Asteroides		
Imbricatine (LV)	*Dermasterias imbricata*	< 1 Pathirana & Andersen 1986

† No activity at 20.
(?) (Unprecised).
*AP = *Arbacia punctulata* eggs assay, HP = *Hemicentrotus pulcherrimus* eggs assay, KB = KB human epithelial carcinoma cells, P388 = P 388 mouse lymphoid leukemia cells, HLC = human lymphoma cells (JURCAT), MTC = mouse T-cell lymphoma cells (YAC-1), BF = turbinate cells from bovine foetus.

1. The activity is expressed as the minimum concentration in $\mu g/ml$ which ceases eggs division at the first blastomere stage.
2. The activity is expressed as the concentration in $\mu g/ml$ which inhibit 50% of the first egg cleavage.
3. IC50 expressed in $\mu g/ml$.
4. IC50 in $\mu g/ml$ on human leucocytes.
5. IC50 in $\mu g/ml$ on L 1210 mouse lymphoid leukemia cells.
6. IC50 in $\mu g/ml$ on mouse myeloide leukemia cells M1.
7. Effect on eggs cleavage of *Chaetopterus pergamentaceus* (Annelidea).
8. Unidentified.
9. IC50 in $\mu g/ml$ on mouse sarcome 180 cells.
10. *Acanthaster planci* or *Luidia maculata* or *Asterias amurensis* (Cf.) *versicolor*.
11. Activity at 5 $\mu g/ml$: 0 = 1-25%, 1 = 26-50%, 2 = 51-75%, 3 = 76-100%, – = inhibition, + = stimulation.
12. Cytotoxic activity at 10 $\mu g/ml$: 3+ = total lysis of the cells, 2+ = monolayer broken, 1+ = some cells with different forms, 0 = normal cells.

133

Table 7. Antitumoral activity of glycoproteins from Echinoids.

Glycoproteins	M.W.[1]	Species	Antitumoral activity		References
			Mouse leukemia P 388 T/C (i.p. dose)[2]	Mouse sarcome 180 TWI (i.p. dose)[3]	
Strongylostatine 1 (Unnamed)	$> 40.10^6$	*Strongylocentrotus droebachiensis*	153% (10 mg/kg)		Pettit et al. 1979
	$> 40.10^6$	*Strongylocentrotus droebachiensis*		72.9% (10 mg/mouse)	Shimizu et al. 1985
Strongylostatine 2	$2 > 65.10^3$	*Strongylocentrotus droebachiensis*	142% (4.5 mg/kg)		Pettit et al. 1981b
SUP II, IV, 4, 6 and 8	$> 5.10^3$	*Strongylocentrotus nudus*		from 58 to 100% (1/10 mg/mouse)	Suntory 1985
Lytechinastatine SU 100-110-120-130-140	$> 2.10^6$	*Lytechinus variegatus*	+		Pettit et al. 1981a
SU 210-220-230-240	$> 10^4$	*Anthocidaris crassispina*		from 42.8 to 99.9% (10 mg/mouse)	Suntory 1983
(Unnamed)	$> 40.10^6$	*Anthocidaris* sp.		77.0% (10 mg/mouse)	Shimizu et al. 1985

1. Expressed in Daltons.
2. T/C: Mean survival time of the test group divided by the mean survival time of the control group.
3. TWI: Tumor weight inhibition.

3.2 *Imbricatine*

Imbricatine (LXXV) is an unusual sulfur benzyltetrahydroquinoline alkaloid isolated from an asteroid, *Dermasterias imbricata*. It is one of only a few nonsaponin metabolites from echinoderms and has a moderate activity on P388 mouse leukemia (T/C = 139% at a dose of 0.5 mg/kg IP).

3.3 *Glycoproteins from echinoids*

The major results regarding echinoid glycoproteins are given in Table 7. The level (Ca, 10 mg/kg) of toxicity exhibited by strongylostatin I suggests that it may be a component of the venom of pedicellariae (Pettit et al.1979). Antitumoral glycoproteins have been isolated from echinoid tests (Suntory 1983, and Shimizu et al. 1985).

3.4 *Mucopolysaccharide from Stichopus japonicus*

Fan and Chen (1981, 1982) have isolated an antitumor acid mucopolysaccharide from *Stichopus japonicus*.

3.5 *Polysaccharide from Asterias amurensis*

A polysaccharide (designated NRP-1) from the asteroid *Asterias amurensis* has antitumor properties against sarcoma 180 (ILS* > 367% at 100 mg/kg IP), carcinoma IMC (ILS > 229% at 100 mg/kg IP) and fibrinosarcoma Meth A (37.3% reduction in tumor size at 50 mg/kg SC) grafted into the mouse (Ito & Masuda 1986; Masuda et al. 1987). This compound does not exhibit direct cytocidal activity on HeLa cells at 1 mg/ml but induces cytolytic activity of peritoneal exsudate cells and provokes interferon induction. Therefore, its antitumor action might be mediated by an immunological response. The molecular weight of this polysaccharide is about 10 000. It is composed of several kinds of sugar, including uronic acid (1.5%) and sialic acid (9.2%).

3.6 *Mechanism of cytotoxic action of steroid derivatives*

The cytotoxic action of various holothurins on echinoid eggs involves an interruption of cell multiplication (1st cleavage), cytolysis and animalization of larvae (differentiation effect) (Ruggieri & Nigrelli 1960, 1964, 1966b,

---

*ILS = increased life span:

$$\frac{\text{median survival days (MSD) of treated group - MSD of control group}}{\text{MSD of control group}} \%$$

135

Ruggieri 1965, Anisimov et al. 1972a, 1973, 1974a). These effects result from a disruption of biosynthesis of proteins, DNA and RNA (Anisimov et al. 1971, 1974a, 1977, Elyakov et al. 1972, Baranova et al. 1973). Mitosis is arrested and DNA synthesis inhibited in onion root bulbs (Santhakumari & Stephen 1988). These authors compared the action of holothurin from *Holothuria vagabunda* with that of adriamycin. It seems likely that membrane cholesterol is affected (mentioned above for the study of hemolytic properties) associated with a complexing of glycolipids residing in the plasma membrane (Anisimov et al. 1979). This would lead to changes in the selective permeability of the membrane and interference with the biosynthesis of macromolecules (Anisimov et al. 1978).

Asteroid saponins produced the same cytotoxic effects as holothurins on echinoid eggs (Rio et al. 1963, Ruggieri & Nigrelli 1966a, Anisimov et al. 1980). With respect to inhibitory activity on cell division in fertilized eggs, Fusetani et al. (1984) emphasized that the cytologic modifications observed are similar to those produced by cytochalasin, thus concluding that asterosaponins inhibit active polymerization during embryogenesis. Both for XXII (asterosaponin group) (Fusetani et al. 1984) and IV (holothurin group) (Anisimov et al. 1973), the presence of a sulfate group seems to slightly weaken the cytotoxic effect.

4 ANTIMICROBIAL SUBSTANCES

Toluene-methanol extracts of 1020 marine animal species were examined against five microorganisms: a gram-negative bacterium (*Escherichia coli*), a gram-positive bacterium (*Bacilus subtilus*), a yeast (*Saccharomyces cerevisiae*) and a virus (*Herpes simplex* type I) (Shaw et al. 1976, Rinehart et al. 1981) (Table 8). These authors emphasize the habitual and preferential presence of antibacterial and/or antifungal substances in the echinoderms. On the average, 47% of the species tested were active. In this respect, only the sponges had a greater activity, with a mean percentage of 69%. The percentage of antiherpetic species (18% on average) was also among the highest. The distribution of antibacterial and antifungal activity within the echinoderms is specified in Table 9. Extracts from no species was active on *Escherichia coli*; 50% of the ophiuroids gave positive results on *Bacillus subtilis;* holothuroids commonly contain antifungal substances, mainly against yeasts. However, these results do not imply that other types of extracts different from the ones tested lack such activity. Table 10 gives the results of antibacterial or antifungal tests performed from more or less purified extracts of various species, and Table 11 those of published antiviral tests.

136

Table 8. *In vitro* antimicrobial activity of toluene-methanol extracts of marine animal species.

Phylum	% species active (number of species examined)							
	Overall antibacterial and antifungal[1]				(HSV-1)[2] antiviral			
	Shaw[3]		Rinehart[4]		Shaw[3]		Rinehart[4]	
Porifera	37	(71)	82	(187)	61	(13)	14	(180)
Cnidaria	21	(72)	35	(70)	0	(1)	17	(69)
Ctenophora	0	(3)	–		–		–	
Platyhelminthes	25	(4)	–		–		–	
Nemertina	0	(4)	–		–		–	
Annelida	16	(37)	33	(3)	–		0	(3)
Mollusca	16	(199)	16	(20)	–		0	(21)
Arthropoda	1	(98)	0	(6)	–		0	(6)
Sipunculida	0	(4)	–		–		–	
Entoprocta	0	(1)	–		–		–	
Ectoprocta	23	(13)	100	(1)	–		0	(1)
Chaetognatha	0	(1)	–		–		–	
Echinodermata	43	(83)	58	(36)	67	(3)	16	(36)
Chordata	6	(81)	68	(27)	–		23	(26)
Overall	19	(670)	63	(350)	58	(17)	14	(342)

1. Species: *Escherichia coli, Bacillus subtilis, Sacchararomyces carlsbergiensis, Penicillium atrovenetum.*
2. HSV-1: Herpes Simplex Virus type l.
3. Shaw et al. 1976.
4. Rinehart el al. 1981.

Table 9. Distribution of antibacterial and antifungal activity in the Echinodermata (Shaw et al. 1976).

Echinodermata (number of species)	% active species			
	Escherichia coli	*Bacillus sub-tilis*	*Saccharomyces cerevisiae*	*Penicillium atrovenetum*
Ophiuroidea (14)	0	10	13	6
Asteroidea (31)	0	50	0	0
Echinoidea (15)	0	0	0	0
Holothuroidea (22)	0	9	83	48

4.1 *Antibacterial substances*

Although echinoderm extracts sometimes show antibacterial activity, as in the case of ophiuroids on *Bacillus subtilus*, few substances have been isolated which cause this effect. *Asterias rubens* contains a lysozyme (an N-acetylmuramylhydrolase mucopeptide with 15 000 ± 1000 MW) (Jolles &

137

Table 10. *In vitro* antibacterial and antifungal activity of crude or semipurified extracts of echinoderms.

Organism	Extract*	Gram+ bacteria			Gram- bacteria		Yeasts						Other fongi		Reference
		1	2	3	4	5	6	7	8	9	10	11	12	13	
Crinoidea															
Nemaster rubiginosa	Purified EtOH-H$_2$O 8:2 E						−					−			Ruggieri & Nigrelli 1974
Asteroidea															
Asterias rubens	H$_2$O	−			+										Andersson et al. 1983
	Petroleum ether	−			−										Andersson et al. 1983
	CH$_2$Cl$_2$	−			−										Andersson et al. 1983
	Purified EtOH-H$_2$O 1:1 E											+			Ruggieri & Nigrelli 1974
Acanthaster planci	Toluene-MeOH 1:3	+					−					−	−		Nemanich et al. 1978
Astrometis sertulifera	EtOH or EtOH-H$_2$O 1:1		+		−		−						+		Constantine et al. 1975
Dermasterias imbricata	EtOH or EtOH-H$_2$O 1:1		+		+		+						+		Constantine et al. 1975
Evasterias troschelli	EtOH or EtOH-H$_2$O 1:1		+		+		+						+		Constantine et al. 1975
Heliaster kubinji	Toluene-MeOH 1:3		−		−							+		−	Nemanich et al. 1978
Henricia leviuscula	Toluene-MeOH 1:3		−		−							−		−	Nemanich et al. 1978
Hippasterias spinosa	EtOH or EtOH-H$_2$O 1:1		+		+		−						+		Constantine et al. 1975
Leptychaster sp.	Toluene-MeOH 1:3		−		−							−		−	Nemanich et al. 1978
Linckia columbiae	Toluene-MeOH 1:3	+			−							+		−	Nemanich et al. 1978
Luidia superba	Toluene-MeOH 1:3	−			−							+	+		Nemanich et al. 1978
Mediaster aequalis	EtOH or EtOH-H$_2$O		+		+		+						+		Constantine et al. 1975
Mithrodia bradleyi	Toluene-MeOH 1:3		−		−							−		−	Nemanich et al. 1978
Pharia pyramida	Toluene-MeOH 1:3		−		−							−		−	Nemanich et al. 1978
Pisaster brevespinus	EtOH or EtOH-H$_2$O 1:1		+		−		+						+		Constantine et al. 1975
Pycnopodia helianthoides	EtOH or EtOH-H$_2$O 1:1		−		−		−						−		Constantine et al. 1975
	Toluene-MeOH 1:3		−		−							−		−	Nemanich et al. 1978

Species	Sample / Solvent						Reference
Solaster dawsoni	EtOH or EtOH-H_2O 1:1	+	−	+			Constantine et al. 1975
Solaster stimpsoni	EtOH or EtOH-H_2O 1:1	+	−	+			Constantine et al. 1975
Stylasterias forreri	EtOH or EtOH-H_2O	+	+	−			Constantine et al. 1975
Ophiuroidea							
Gorgonocephalus caryi	EtOH or EtOH-H_2O 1:1	+	+	−			Constantine et al. 1975
Ophiocoma echinata	Purified EtOH-H_2O 8:2 E	−		−			Ruggieri & Nigrelli 1974
Ophioderma variegatum	Toluene-MeOH 1:3	+		−			Nemanich et al. 1978
Echinoidea							
Brissopsis lyrifera	H_2O	−	+				Andersson et al. 1983
	Petroleum ether	−	−				Andersson et al. 1983
	CH_2Cl_2	−	+				Andersson et al. 1983
Echinus esculentus	Body fluid	Div: 12+/12					Wardlaw & Unkles 1978
	Body fluid	*Ps.*+ Div: 11+/11					Service & Wardlaw 1985
	H_2O	−	−				Andersson et al. 1983
	Petroleum ether	−	−				Andersson et al. 1983
	MeOH	−	−				Andersson et al. 1983
	CH_2Cl_2	−	−				Andersson et al. 1983
Mellita quinquiesperforata	Purified EtOH-H_2O 8:2 E	−					Ruggieri & Nigrelli 1974
Toxpneustes reseus	Toluene-MeOH 1:3	−	−				Nemanich et al. 1978
Holothuroidea							
Actinopyga lecanora	G.F.	+	+	+	+	+	Anisimov et al. 1982b
Actinopyga mauritiana	G.F.	+	+	+	+	+	Anisimov et al. 1982b
	G.F. (holothurin A)	+	+	+			Kuznetsova et al. 1982
Actinopyga obesa	G.F. (holothurin A)	+	+	+			Kuznetsova et al. 1982
Astichopus multifidus	Purified EtOH-H_2O 8:2 E	+	+	+	+	+	Ruggieri & Nigrelli 1974
Bohadschia argus	G.F.	+	+	+	+	+	Anisimov et al. 1982b
	G.F. (bivittosides)	+	+	+			Kuznetsova et al. 1982

139

Table 10. (continued).

Organism	Extract*	Gram+ bacteria			Gram− bacteria		Yeasts						Other fongi		Reference
		1	2	3	4	5	6	7	8	9	10	11	12	13	
Bohadschia marmorata	G.F.						+	+	+	+	+				Anisimov et al. 1982b
	G.F. (bivittosides)										+				Kuznetsova et al. 1982
Bohadschia vitiensis	G.F. (bivittosides)		−		−		+	+	+	+	+				Kuznetsova et al. 1982
Bohadschia sp.	G.F.						+	+	+	+	+				Anisimov et al. 1982b
Brandtothuria arenicola	Toluene-MeOH 1:3	−			−						+		+		Nemanich et al. 1978
Cucumaria fraudatrix	G.F.						+	+	+	+	+		+		Anisimov et al. 1982b
Cucumaria miniata	EtOH or EtOH-H_2O 1:1		+		−		+							+	Constantine et al. 1975
Cucumaria sp.	Toluene-MeOH 1:3	−			−							+	+		Nemanich et al. 1978
Holothuria atra	G.F.						+	+	+	+	+				Anisimov et al. 1982b
	G.F. (holothurin A)							+	+	+	+				Kuznetsova et al. 1982
	G.F. (holothurin B)							+	+	+	+				Kuznetsova et al. 1982
Holothuria cinerescens	G.F.						+	+	+	+	+				Anisimov et al. 1982b
Holothuria coluber	G.F. (holothurin A)							+	+	+	+				Kuznetsova et al. 1982
Holothuria discrepans	G.F. (holothurins A and B)		−		−		+	+	+	+	+				Kuznetsova et al. 1982
Holothuria edulis	G.F.						+	+	+	+	+				Anisimov et al. 1982b
	G.F. (holothurins A and B)							+	+	+	+				Kuznetsova et al. 1982
Holothuria hilla	G.F.						+	+	+	+	+				Anisimov et al. 1982b
	G.F. (holothurins A and B)		−		−		+	+	+	+	+				Kuznetsova et al. 1982
Holothuria impatiens	G.F. (holothurin A)		−		−			+	+	+	+				Kuznetsova et al. 1982
Holothuria leucospilota	G.F.						+	+	+	+	+				Anisimov et al. 1982b
	G.F. (holothurins A and B)											+			Kuznetsova et al. 1982
Holothuria mexicana	Purified EtOH-H_2O 8:2 E						−					+			Ruggieri & Nigrelli 1974
Holothuria parvula	Purified EtOH-H_2O 8:2 E						−					+			Ruggieri & Nigrelli 1974
Labidodemas americanum	Toluene-MeOH 1:3	−			−									−	Nemanich et al. 1978

140

Species	Extract/Fraction	Test results	Reference
Neothyone gibbosa	Toluene-MeOH 1:3	– ... + –	Nemanich et al. 1978
Pearsonothuria graeffei	G.F.	+ + + + + + +	Anisimov et al. 1982b
	G.F. (holothurin A)	+	Kuznetsova et al. 1982
Pentamera chierchia	Toluene-MeOH 1:3	– + + + + +	Nemanich et al. 1978
Psolidium dorsipes	Toluene-MeOH 1:3	– + + + +	Nemanich et al. 1978
Selenkothuria lubrica	Toluene-MeOH 1:3	– + + + +	Nemanich et al. 1978
Stichopus chloronatus	G.F.	+ + + + + +	Anisimov et al. 1982b
	G.F. (stichoposides C and D)	+	Kuznetsova et al. 1982
Stichopus horrens	G.F. (stichoposide D)	– + + +	Kuznetsova et al. 1982
Stichopus japonicus	G.F.	+ + + + + +	Anisimov et al. 1982b
Stichopus tremulus	Body fluid	G.h.+ Div.: 5-/5 S.m.–	Johnson & Chapman 1971
Stichopus variegatus	G.F. (stichoposides C and D)	– + +	Kuznetsova et al. 1982
Stichopus sp.	G.F. (stichoposides)	– + +	Kuznetsova et al. 1982
Thelonota ananas	G.F.	+ + + + +	Anisimov et al. 1982b
	G.F. (thelothurins A and B)	+	Kuznetsova et al. 1982
Thelonota anax	G.F. (thelothurins)	+	Kuznetsova et al. 1982

* E = extract; G.F = glycoside fraction (when specified, the name of the main glycoside is put in brackets); EtOH = ethanol; MeOH = methanol; H_2O = water, CH_2Cl_2 = methylene chloride.

** + = Active extract (or, when specified, minimum inhibitory concentration < 100 µg/ml); – = inactive extract.

1 = *Bacillus subtilis*; 2 = *Staphylococcus aureus*; 3 = other Gram+ bacteria; 4 = *Escherichia coli*; 5 = other Gram- bacteria; 6 = *Candida albicans*; 7 = *Candida krusei*; 8 = *Candida tropicalis*; 9 = *Candida utilis*; 10 = *Saccharomyces carlsbergiensis*; 11 = *Saccharomyces cerevisiae*; 12 = *Penicillium atrovenetum*; 13 = *Trichophyton mentagrophytes*; P.s. = *Pseudomonas* sp.; G.h. = *Gaffkya homari*; S.m. = *Serratia marcescens*; Div. = Various marine bacteria, the number of which being specified.

Table 11. *In vitro* antiviral activity of crude or semipurified extracts of echinoderms.

Organism	Extract	Antiviral activity			Reference
		Influenza virus	Herpes virus type 1	Other viruses*	
Asteroidea					
Acanthaster planci	Glycoside fraction	+			Shimizu 1971
Asterias forbesi	Glycoside fraction	+			Shimizu 1971
Asterias rubens	Petroleum ether	–	–		Andersson et al. 1983
Asterina pectinifera	Glycoside fraction	+			Shimizu 1971
Luidia senegalensis	CH_3OH		–	–	Brito et al. 1981
	CH_2Cl_2		–	–	Brito et al. 1981
Echinoidea					
Brissopsis lyrifera	H_2O	–	–		Andersson et al. 1983
	Petroleum ether	–	–		Andersson et al. 1983
	CH_2Cl_2	–	–		Andersson et al. 1983
	CH_3OH	–	–		Andersson et al. 1983
Echinus esculentus	H_2O	–	–		Andersson et al. 1983
	Petroleum ether	–	–		Andersson et al. 1983
	CH_2Cl_2	–	+		Andersson et al. 1983
	CH_3OH	–	+		Andersson et al. 1983

*Other viruses: Adeno virus type 1, Polio virus type 1, *Vesicular stomatitis* Alagoas virus.

Table 12. Effect of echinochrome A against marine bacteria (Johnson & Chapman 1971).

Bacteria	Results
Gram-	
Vibrio fischeri	++
Alternomonas citrea	−
Phytobacterium phosphoreum	+
Pseudomonas sp.	++
Gram+	
Micrococcus sp.	+
Planococcus citreus	++

++ = Bactericidal activity; + = Bacteriostatic activity; − = Inactive at 50 µg/ml.

Jolles 1975) whose bacteriological activity could account for aqueous extract properties relative to *Escherichia coli*. According to Johnson & Chapman (1971), the slightly bacteriostatic activity of coelomic fluid from *Stichopus tremulus* could be related to the phagocytic properties of coelomocytes. Holotoxin from *Stichopus japonicus* showed no antibacterial activity on the 11 different bacteria on which it was tested (Shimada 1965). Asterosaponins (XX, XXII, XXVIII, XXXIV, XXXIX, LXI) and some hydroxylated sulfated steroidal glycosides (LXII) or polyhydroxylated sulfated sterols (LXIII, LXIV) from ophiuroids are inactive against *Staphylococcus aureus* and *E. coli*. However, polyhydroxylated steroidal glycosides (XLIII, XLIV, XLVI, XLVII, XLVIII) and polyhydroxylated sterols (L, LVIII, LIX, LX) from asteroids, and desulfated sterols from ophiuroids (LXV, LXVI), although inactive against *E. coli*, showed moderate activity against *S. aureus*. Only one compound (LXV) was active at the lower concentration tested (20 µg/disc). The only other antibacterial substance clearly identified is echinochrome A (LXXIV), a naphthoquinone pigment from red spherule coelomocytes of *Echinus esculentus* (Johnson & Chapman 1971). Although the activity of echinochrome A (Table 12) might suggest that it can play a role in the organism's defense against bacteria, it seems inadequate for therapeutic application.

4.2 Antifungal substances

It is noteworthy (Table 9) that holothuroids, and to a lesser extent asteroids – the only two classes containing saponosides in appreciable quantity (Mackie et al. 1977) – are by far the echinoderms with the highest activity. Holothurin activity has been widely studied (Table 13). D'Auria et al. (in press) tested 40 steroidal glycosides from asteroids against a pathogenic fungus, *Cladospo-*

Table 13. Antifungal activity of holothurins.

Substance[1]	Organism	Antifungal activity[2] (minimum growth inhibitory concentration in μg/ml)[2]									Reference
		A.n.	A.o.	C.a.	C.u.	C.h.	H.p.	M.g.	M.s.	P.ch.	
Cucumarioside I[3]	Cucumaria japonica			30							Batrakov et al. 1980
Cucumarioside II[3]	Cucumaria japonica			60							Batrakov et al. 1980
Echinoside A (XII)	Actinopyga echinites	3.12	3.12	12.5	6.25	25.0			50.0	3.12	Kitagawa et al. 1985
Desulfated echinoside A	Actinopyga echinites	1.56	1.56	50	3.12	$>10.0^2$			$>10.0^2$	1.56	Kitagawa et al. 1985
Echinoside B (XIII) I)	Actinopyga echinites	3.12	3.12	$>10.0^2$	12.5	$>10.0^2$			$>10.0^2$	3.12	Kitagawa et al. 1985
Desulfated echinoside B	Actinopyga echinites	3.12	3.12	12.5	6.25	25.0			50.0	3.12	Kitagawa et al. 1985
Holothurin A (I)	Actinopyga agassizi			10.0^4							Ruggieri & Nigrelli 1974
Desulfated holothurin A	Actinopyga agassizi			10.0^3							Ruggieri & Nigrelli 1974
Holotoxin A (mixture)	Stichopus japonicus	16.7		16.7						16.7	Shimada 1965
Holotoxin A (XV)	Stichopus japonicus		6.25	6.25	3.12			3.12		3.12	Kitagawa et al. 1976
Holotoxin A1 (XVI)	Stichopus japonicus	6.25		12.5			6.25				Mal'tsev et al. 1985
Holotoxin B (XVII)	Stichopus japonicus		12.5	6.25	3.12			1.50		6.25	Kitagawa et al. 1976
Holotoxin C (XVIII)	Stichopus japonicus		25.0	25.0	12.5			12.5		12.5	Kitagawa et al. 1976
Stichoposide A (IV)	Stichopus chloronotus	$>10.0^2$		$>10.0^2$			$>10.0^2$				Mal'tsev et al. 1985
Stichoposide C (V)	Stichopus chloronotus	6.25		6.25			12.5				Mal'tsev et al. 1985
Stichoposide D (VI)	Stichopus chloronotus	12.5		25.0			50.0				Mal'tsev et al. 1985
Stichoposide E (VII)	Stichopus chloronotus	25.0		12.5			25.0				Mal'tsev et al. 1985
Thelenotoside A (IX)	Stichopus chloronotus	0.75		6.25			12.5				Mal'tsev et al. 1985
Thelenotoside B (X)	Stichopus chloronotus	6.25		25.0			25.0				Mal'tsev et al. 1985
Various vegetale saponins				$>10.0^2$				$>10.0^2$			Kitagawa et al. 1976

		P.ci.	P.n.	R.r.	S.ca.	T.a.	T.r.	T.u.	Other	
Cucumarioside I[3]	Cucumaria japonica								C.t.: 40	Batrakov et al. 1980
Cucumarioside II[3]	Cucumaria japonica								C.t.: 60	Batrakov et al. 1980
Echinoside A (XII)	Actinopyga echinites	3.12		6.25		6.25	6.25			Kitagawa et al. 1985
Desulfated echinoside A	Actinopyga echinites	3.12		3.12		6.25	3.12			Kitagawa et al. 1985

144

Compound	Source							Reference
Echinoside B (XIII) [1])	*Actinopyga echinites*	3.12		12.5		12.5	6.25	Kitagawa et al. 1985
Desulfated echinoside B	*Actinopyga echinites*	3.12		6.25		6.25		Kitagawa et al. 1985
Holothurin A (I)	*Actinopyga agassizi*		10.0[3]					Ruggieri & Nigrelli 1974
Desulfated holothurin A	*Actinopyga agassizi*		10.0[2]					Ruggieri & Nigrelli 1974
Holotoxin A (mixture)	*Stichopus japonicus*				1.56	2.78	6.25	*F.l., G.sa., G.si., H.a., O.m.* = 16.7; *P.o.* = 2.78; *S.ce* = 2.78; *T.a.* = 6.25 Shimada 1965; *T.i.* = 1.56/6.25 Shimada 1969
Holotoxin A (XV)	*Stichopus japonicus*		25.0		1.56	0.78	3.12	Kitagawa et al. 1976
Holotoxin A1 (XVI)	*Stichopus japonicus*			3.12	3.12	3.12	3.12	Mal'tsev et al. 1985
Holotoxin B (XVII)	*Stichopus japonicus*				1.56	0.78	3.12	Kitagawa et al. 1976
Holotoxin C (XVIII)	*Stichopus japonicus*				12.5	6.25	12.5	
Stichoposide A (IV)	*Stichopus chloronotus*	>10.0[2]		12.5	>10.0[2]		12.5	Mal'tsev et al. 1985
Stichoposide C (V)	*Stichopus chloronotus*	12.5		1.55	1.55		1.55	Mal'tsev et al. 1985
Stichoposide D (VI)	*Stichopus chloronotus*	6.25		12.5	6.25		12.5	Mal'tsev et al. 1985
Stichoposide E (VII)	*Stichopus chloronotus*	12.5		6.25	25.0	6.25	6.25	Mal'tsev et al. 1985
Thelenotoside A (IX)	*Stichopus chloronotus*	6.25		1.55	0.75		1.55	Mal'tsev et al. 1985
Thelenotoside B (X)	*Stichopus chloronotus*	25.0		6.25	6.25		6.25	Mal'tsev et al. 1985
Various vegetale saponins				>10.0[2]	>10.0[2]			Kitagawa et al. 1976

1. Six other antifungal holothurins have been isolated from *Stichopus*: Stichlorosides A1, A2, B1, B2, C1 and C2, the antifungal activity of which have not yet been determined (Kitagawa et al. 1981a).

2. *A.n.: Aspergillus niger; A.o.: Aspergillus orizae; C.t.: Candida tropicalis; C.u.: Candida utilis; C.h.: Cladosporium herbaceum; F.l.: Fusarium lini; G.sa.: Gibberella saubinetii; G.si.: Glomerella singulata; H.a.: Helminthosporium avenae; H.p.: Homodendro pedrosoi; M.g.: Microsporum gypseum; M.s.: Mucor spinescens; O.m.: Ophiobolus miyabeanus; P.ch.: Penicillium chrysogenum; P.ci.: Penicillium citrinum; P.n.: Penicillium niger; P.o.: Pyricularia orizae; R.r.: Rhodotorula rubra; S.ca.: Saccharomyces carlsbergiensis; S.ce.: Saccharomyces cerevisiae; T.a.: Trichophyton asteroides; T.i.: Trichophyton interdigitale; T.r.: Trichophyton rubrum; T.u.: Torula utilis.*

3. Unidentified.

rium cucumerinum. The most active compound (at 1 μg in the bioauthography using thin layer chromatography) was laeviusculoside A (LXXIII).

The discovery of this property, which is active on different sorts of fungi (e.g., yeasts, dermatophytes, *Aspergillaceae*), was expected since antifungal properties have generally been considered to be characteristic of saponosides (Tschesche 1971, Pinkas et al. 1972, Kapoor & Chawla 1986). As for hemolysis, the activity was greater than that of plant homologues (Kitagawa et al. 1976).

It seems difficult to establish structure-activity relationships between the different saponosides. Although the nature of the aglycone is important, the number and type of monosaccharide units, as well as the presence of a sulfate group in the carbohydrate chain, play a role (Mal'tsev et al. 1985). Holothurin A (4 monosaccharide units) is less antifungal than holothurin B which differs only in lacking the first two carbohydrates (Kitagawa et al. 1979). However, stichoposide C (6 monose units) is more active than its triose derivative obtained by Smith degradation (Malt'sev et al. 1985). Stichoposide C (V) and D (VI) each contain 6 monose units. In stichoposide D a quinovose has replaced a glucose residue found in stichoposide C, which is 4 to 8 times as active. Thus even small changes in the carbohydrate chain affect antifungal properties (Malt'sev et al. 1985). Similarly, desulfated echinoside B is much less active than echinoside B, although echinoside A-desulfated echinoside A and holothurin A-desulfated holothurin A produce an inverse change (Ruggieri & Nigrelli 1974, Kitagawa et al. 1985).

The mechanism of saponoside antifungal action is uncertain. Anisimov et al. (1981) showed that two triterpene glycosides from holthuroids increase the membrane permeability of *Saccharomyces carlsbergiensis*. RNA biosynthesis is inhibited in this same species (Baranova et al. 1973), and addition of cholesterol to the medium prevents stichoposide A antifungal action (Anisimov et al. 1974b). It is thus tempting to suppose that the mechanism of action is similar to that of hemolytic or cytotoxic effects (see above), especially since the action of these saponosides on yeasts and tumor cells corresponds closely (Kuznetsova et al. 1982). However, Anisimov et al. (1981) called attention to certain differences between membrane effects of these saponosides on fertilized eggs of the echinoid *Strongylocentrus nudus* and on the yeast *Saccharomyces carlsbergiensis* . Moreover saponoside antifungal activity is not related to either hemolytic capability or the complexing power of cholesterol (Tschesche & Wulff 1965). Finally, the plasma membrane of fungi contains ergosterol instead of the cholesterol of animal cells. When added to medium, ergosterol does not prevent the antifungal action of any of the triterpene saponosides, particularly not stichoposide A (Anisimov et al. 1974b). Thus, though there is general agreement about the membrane effect of these substances, the site of their action at this level is uncertain.

146

4.3 *Antiviral substances*

Echinoderm extracts have antiviral properties. Shimizu (1971, 1973) noted the presence of pregnane glycoside derivatrives active against influenza virus in *Asterias forebesi, Acanthaster planci* and *Asterina pectinifera* . Andersson et al. (1989) studied the moderate antiviral activity of saponins and other steroidal compounds from asteroids and ophiuroids (Table 14) on pseudorabies virus (Suid herpesvirus 1, SHV-1). Saponosides produce this antiviral property (Tschesche & Wulff 1965, Rao et al. 1974).

Table 14. Antiviral activity of saponins and stereoidal compounds from echinoderms (Andersson et al. 1989).

Substance	Organism	Antiviral activity[1] Dose in µg/ml	
		10	1
Asterosaponins			
Thornasteroside A (XX)	*Acanthaster planci*	9	5
Marthasteroside A1 (XXIII)	*Marthasterias glacialis*	22	17
Marthasteroside A2 (XXVIII)	*Marthasterias glacialis*	17	14
Marthasteroside B (XXXIX)	*Marthasterias glacialis*	11	14
Ophidianoside C (LXI)	*Ophidiaster ophidianus*	0	6
Cyclic steroidal glycosides			
Sepitoside A (XXXIV)	*Echinaster sepositus*	4	10
Polyhydroxylated steroidal glycosides			
Halityloside A (LVI)	*Halityle regularis*	17	20
Halityloside D (XLIV)	*Halityle regularis*	8	2
Crossasteroside A (XLVI)	*Crossaster papposus*	6	0
Crossasteroside B (XLVII)	*Crossaster papposus*	24	43
Crossasteroside XLVIII)	*Crossaster papposus*	8	27
(LII)	*Poraster superbus*	18	24
Polyhydroxylated sterols			
(LVIII)	*Protoreaster nodosus*	2	6
(LIX)	*Protoreaster nodosus*	0	0
(XLIX)	*Archaster typicus*	13	18
Polyhydroxulated sulfated sterols			
(L)	*Archaster typicus*	20	15
Disulfated sterols			
(LXV)	*Ophioderma longicaudum*	19	12
(LXVI)	*Ophioderma longicaudum*	0	11

1. The activity is expressed as per cent reduction of pseudorabies virus plaque formation.

5 ANTIPARASITIC SUBSTANCES

As holothurin is toxic *in vitro* on various species including protozoans (Nigrelli & Zahl 1952, Nigrelli & Jakowska 1960), its affects *in vivo* on parasites, is of interest. Rats intraperitoneally inoculated with crude holothurin and previously or simultaneously infected by *Tripanosoma lewisi* had lower parasitemia than the controls (Styles 1970). This result is of very limited interest since, on the contrary, a higher level of parasitemia was observed in rats treated after infection. These results were confirmed in mice inoculated with *Trypanosoma duttoni* (Sen & Lin 1975, 1977). This kind of effect had already been noted when bacterial endotoxins were administered to mice with trypanosomal infections. However, it is difficult to find a reasonable explanation for the observed effects unless two antagonistic influences, one inhibitory and the other stimulatory, are ascribed to the holothurin (Styles, 1970). Supposedly, the antiparasitic action of holthurins is counteracted by their interference in the formation of antibodies directed against the parasite (Sen & Lin 1977). Holothurin is active *in vitro* against *Amoeba proteus* and nemathelminths (Nigrelli & Jakowska 1960), and holotoxins A, B and C are active against *Trichomonas vaginalis* (Kitagawa et al. 1976). However, the value of these substances has not been tested in animals with parasites.

6 IMMUNE FACTORS AND SUBSTANCES REACTING WITH THE IMMUNE SYSTEM

An immune defense system in invertebrates resembling that of mammals. is particularly apparent in echinoderms. Asteroid immune defense factors have been especially studied, and phagocytic B-like and T-like cells identified. Some immune factors have also been described in coelomic fluid of echinoids. These immune factors, as well as the effect of holothurin on immunological reactions, are considered below.

6.1 *Holothurins*

Pretreating polynuclear leukocytes with small quantities of holothurin (1 to 25 µg/ml, 5 h) from *Actinopyga agassizi* increases phagocytosis of *Staphylococcus aureus*. Holothurin apparently decreases surface energy at the leukocyte cell membrame, thereby enhancing bacteria uptake during the ingestion period. This causes an increase in lactic acid production which subsequently has an effect on intracellular granules, causing lysis and release of lytic enzymes active against bacteria. Higher concentrations (50 to 100 µg/ml, 5 h) enhance phagocytic activity but also cause leukocyte death. It appears that the cell membrane is altered, allowing holothurin to penetrate the cell and act on intracellular granules and provoking the release of phagocytic and hydrolytic enzymes (Lasley & Nigrelli 1970).

Small concentrations of crude holothurin (0.1 to 0.4 μg/ml blood), holothurin A (I) (0.02 μg/ml) or desulfated holothurin (1.0 μg/ml) cause stimulation of leukocyte migration. Holothurin may act by stimulating aerobic and anaerobic metabolism as a result of its membrane effect. At higher concentrations (16 to 22 μg crude holothurin/ml, or 0.1 μg holothurin A/ml, or 300 μg desulfated holothurin/ml), but lower than those inducing leukocyte death, these saponins inhibit cell movement, probably by altering leukocyte metabolism. Ouabain, another nonhemolytic saponin-like substance produces the same effects (Lasley & Nigrelli 1971).

Of particular interest is the effect of cucumarioside (unidentified) from *Cucumaria japonica* on the development of immune response to corpuscular pertussis vaccine (*Bordetella pertussis*) in mice. This substance has pronounced immunomodulating properties, depending on concentration, administration route, and the dose of pertussis vaccine: e.g., oral administration of 4 μg/mouse or IP injection of O.04 and 0.0004 μg/mouse has a positive adjuvant effect. However, no IP route effect is observed at 0.004 μg/mouse. In the same way, IP injection of 0.001 μg/mouse abolishes the suppressive action of large doses of pertussis vaccine in the background rosette-formation test 7 and 14 days after oral immunization, but stimulates the effect on days 21 and 28 (Sedov et al. 1984). This variability in response is not favorable for the therapeutic use of these compounds.

6.2 *Lymphokine and interleukin-like factors from asteroids*

A protein (MW 32,000) isolated from coelomocytes (macrophages) of the asteroid *Asterias forbesi*, referred to as 'sea-star factor' (SSF), produces an erythematous indured lesion indistinguishable from that of cutaneous delayed tuberculin hypersensitivity when injected intracutaneously into various mammalian species (Prendergast & Susuki 1970). This protein, like nonspecific migration inhibiting factor (MIF), inhibits guinea pig peritoneal macrophage migration directly *in vitro* and produces macrophage activation (Prendergast 1971). It also possesses chemotactic properties for mononuclear cells (Prendergast 1971) and abolishes primary immune response to T-dependent antigens (Prendergast et al. 1974) and enhances susceptibility of SSF-pretreated mice to infection with a normally sublethal dose of *Listeria monocytogens* (Willenborg & Prendergast 1974). SSF-activated macrophages exert a powerful nonimmunologically-specific suppressive effect on proliferation of tumor cells in vitro (Liu et al. 1983). None of these effects are related to cytotoxic properties (Prendergast et al. 1974 Liu et al. 1983). These results emphasize the similarity of SSF to lymphokines produced by stimulated immunocompetent T lymphocytes in vertebrates (Prendergast et al. 1974, Liu et al. 1983).

An interleukin-1-like protein (MW 29,500) from coelomocytes and coelomic fluid of the same asteroid species has been identified. This substance stimulates murine thymocyte proliferation directly (to a greater degree in the presence of submitogenic concentrations of concanavilin A),and fibroblast proliferation. Its activity can be inhibited by an antibody to human interleukin-1 (Beck & Habicht 1986).

The asteroid *Asterias rubens* possesses an immune system able to induce specific cellular and humoral responses against a foreign antigen (Leclerc et al. 1986). Three subpopulations have been identified in the axial organ: B-like cells, T-like cells and phagocytic cells. Leclerc et al. (1981) showed that the T-like cells release lymphokine-like mediator with mitogenic properties such as vertebrate interleukin-2. They demonstrated that cultured axial organ cells produce a stimulatory factor, apparently a protein (MW > 10 000), when stimulated with *Nocardia* delipidated cell mitogen. This last factor is not T-like-dependent and has a mitogenic effect only on B-like cells (Leclerc et al. 1987).

These findings suggest an important analogy between the immune system of highly developed invertebrates and that of mammals.

6.3 *Complement-like factor from echinoids*

As noted in the section on hemolytic substances, complement-like activity has been detected in coelomic cell-free fluid from *Strongylocentrotus droebrachiensis* through lytic action on rabbit erythrocytes and an opsonic effect on echinoid coelomic phagocytes and mouse peritoneal macrophages. This activity greatly resembles that of mammalian complement activated via the alternative pathway (Bertheussen 1983).

7 SUBSTANCES ACTIVE ON NERVE AND MUSCLE

Many experiments have been performed on biological excitable systems: nerves, striated muscle and synapses, smooth muscle and cardiac muscle. Holothurins, asterosaponins and protein echinoid toxins are implicated as well as nonprotein echinoid substances, tetrodotoxin and prostaglandins.

7.1 *Holothurins*

Intravenous injection of holothurin A (I) from *Actinopyga agassizi* into mice (9 mg/kg) produces acute intoxication characterized by thrashing convulsions, taut muscle, total body contortions, splayed and stiff legs, humping of the back and considerable respiration difficulty (Friess et al. 1960). All these symptoms could involve interaction with muscle and/or nerve elements.

Different experimental models have been used to investigate this effect of holothurin A. On a single fiber-single node preparation from frog (or toad) sciatic nerve, holothurin A produces a conduction blockade along the fiber which is irreversible at 10^{-5} M and reversible at very low concentrations (10^{-7} M) (Friess et al. 1959, 1960, *Thron et al.* 1963, Dettbarn et al. 1965). On a rat phrenic nerve diaphragm preparation, it blocks twitch response via a direct (muscle stimulation) or indirect (nerve stimulation) pathway. These effects are irreversible at higher concentrations (Friess et al. 1959, 1960, 1967, Thron et al. 1964). Holothurin A produces a direct contractile effect on muscle, which is partially released with time (Friess et al. 1959, 1960). On a frog sartorious nerve/muscle preparation contracture is not propagated when holothurin A is applied topically to muscle (Thron et al. 1963). The substance irreversibly blocked (2.10^{-4} M) a neurally and directly evoked response on a stimulated monocellular electroplax preparation of *Electrophorus electricus* (Dettbarn et al. 1965) and produced irreversible inactivation of a cat superior cervical ganglion preparation (Friess et al. 1970). Holothurin A thus causes a destructive interaction with excitable nerves (medullary, peripheral or cervical) and muscle tissues. It should be noted that related phenomena have been observed for other saponin-like products, such as *Quillaya* saponin and digitonin (Thron et al. 1964).

Friess et al. (1959) noted that effects on propagation of action potential resemble those of cholinesterase inhibitors (e.g., veratridine, eserine) which are, however, reversible. In other respects, preconditioning of a rat phrenic nerve diaphragm preparation with tiny concentrations of physostigmine well below those producing a direct effect on tissue response (for instance, 5.10^{-10} M) prevents irreversible destruction of twitch response when stimulation is indirect. This protective action is readily overridden by concentration increments of anticholinesterasic agents, leading to complete disappearance of the effect (Thron et al. 1964). Other anticholinesterasic inhibitors, such as neostigmine and galanthamine produce the same phenomena (Friess et al. 1965). The value of the optimal concentration of all these protective agents is linearly related to the anticholinesterasic potencies of the agents as measured on the anticholinesterase-acetylcholine system *in vitro*. These findings, which have been interpreted in terms of possible drug chemoreceptor interaction, have led to the formulation of a receptor model at the neuromuscular synapse (Friess et al. 1965, Friess & Durand 1965, Friess 1972). This model takes into account the fact that holothurin A does not produce any inhibitory activity against the acetylcholinesterase-acetylcholine system *in vitro* (Thron et al. 1963).

Holothurin A produces irreversible depolarization in a monocellular electroplax preparation of *Electrophorus electricus* (Dettbarn et al. 1965) and, by external application or internal infusion, in squid (*Loligo peali*) axon membrane (De Groof & Narahashi 1976). These effects on the resting potential have been interpreted in terms of passive ionic transport and attributed to an initial increase of K^+ efflux which then decreases steadily (Dettbarn et al. 1965) as well as to an increase in resting sodium permeability which is not antagonized by tetrodotoxin (De Groof & Narahashi 1976). These authors suggested that membrane permeability modification is related to the hemolytic effect of saponins on erythrocytes. Moreover, holothurins Increase Ca^{2+} membrane permeability of an artificial membrane formed from liposomes from the bulk fraction of egg lecithin (Rubtzov et al. 1980).

The neuromuscular effects of holothurin can also be attributed to a modification of active transport of ions. This substance decreases the activity *in vitro* of rat brain Na^+-K^+ ATPase (Gorshkov et al. 1982). Nevertheless, this effect is nonspecific since holothurin also inhibits rat brain Mg^{2+}-ATPase activity (Gorshkov et al. 1982) as well as that of Ca^{2+} ATPase from sarcoplasmic reticulum vesicles (Rubtzov et al. 1980). The effect of ATPase activity disappears when holothurins are preincubated with cholesterol (Gorshkov et al. 1982). As the functioning of membrane Na^+-K^+ ATPase depends on its lipidic environment, particularly on the presence of cholesterol (Järnefelt 1972), it is likely that the effects obtained on active cation transport are due to interaction with this substance. It thus seems, as in the case of hemolytic properties, that the different neuromuscular effects of holothurin A are mainly attributable to interaction with membrane lipids, leading to modification of ionic conductance at this level and disturbance of ATPase functioning. This mode of action could account for the observed loss of basophilic macromolecular material from axoplasm in and near the nodes of Ranvier caused by holothurin A (Thron et al. 1964).

Substances from other holothuroids such as holothurin B (II) from *Holothuria vagabunda* or *H. lubrica,* are also active, on a phrenic nerve diaphragm preparation (Friess et al. 1968). Friess et al. (1965) considered that the observed effect was lower, but Gorshkov et al. (1982) found no real activity difference on rat brain Na^+-K^+ ATPase between holothurin A (I) from *Actinopyga agassizi* , holothurin B (II) from *H. atra* , holothurin C (XIX) from *Bohadschia argus* , cucumarioside G (XI) from *Cucumaria fraudatix* , astichoposide C (= stichloroside C2) (VIII) from *Astichopus multifidis* and thelenotoside B (X) from *Thelenota ananas.* Likewise, cucumarioside (unidentified) from *Cucumaria japonica* is as active as the holothurin of *H. grissi* on the ionic permeability of sarcoplasmic reticulum vesicles (Rubtzov et al. 1980).

Loss of the half-esterified sulfate function of holothurin A results in an abrupt decrease in activity (Friess et al. 1965). On a rat phrenic nerve-diaphragm preparation, desulfated holothurin was 6 to 10 times less active than holothurin on directly induced muscle contraction and blockade of directly and indirectly elicited twitch response. The blocking activity of desulfated holothurin is largely reversible upon washing. This compound offers significant protection against irreversible destruction of the twitch response normally evoked by holothurin (Friess et al. 1967).

Holothurin is 10 times as potent as its desulfated homologue on cat superior cervical ganglion preparation. However, at levels below its own effective blockade concentration, desulfated holothurin markedly extends the survival of ganglionic excitability against the rapid destruction normally wrought by holothurin (Friess et al. 1970). Nevertheless, on IV administration in the intact mouse, desulfated holothurin does not counteract the irreversibility of holothurin A on central nervous system receptor (Friess et al. 1968). Syndromes leading to death* are quite similar, except for their relative courses (Friess et al. 1967).

Although desulfated holothurin is less potent than the parent toxin and produces effects that are less irreversible, the presence of a sulfate group (or the number or polarity of monosaccharides) does not influence the extent of Na^+-K^+ ATPase inhibition. Holothurins A, B and C and cucumarioside G are as active as two desulfated saponins, astichoposide C and thelenotoside B. The latter nevertheless inhibit Mg^{2+}-ATPase activity much less (Gorshkov et al. 1982). Therefore, the total structure of glycosides determines the specificity and intensity of their activity.

7.2 *Asterosaponins*

Fänge (1963) noted that the aqueous extracts of some asteroids (*Asterias rubens, Asterias glacialis* and *Henricia sanguinolenta*) cause a strong, nearly irreversible contraction of isolated radula muscle of *Buccinum* sp. *in vitro*. The fact that this activity disappears upon boiling makes it uncertain that potentially present asterosaponins are actually responsible, even though they are capable of exerting such activity. Friess et al. (1968), using a rat phrenic nerve-diaphragm preparation, showed that asterosaponin A and B (unidentified) of *Asterias amurensis*, like holothurin A, cause powerful direct muscle contraction. These substances block the twitch response elicited by indirect or direct stimulation. On Na^+-K^+ ATPase activities of the microsomal fraction of rat brain *in vitro*, asterosaponin L (LXVIII) from *Linckia guildingi* and asterosaponin II (unidentified) from *Linckia laevigata* are much less active than holothurin. They seem to be inactive on Mg^{2+} ATPase activity (Gorshkov et al. 1982). These results corroborate the lack of significant activity of other asterosaponins such as marthasteroside A

---

*IV LD50 value of holothurin on mice: 9 mg/kg wt; LD50 of desulfated holothurin: 11.0 mg/kg wt (Friess et al. 1967).

(XXIII) and B (XXXIX) or sepositoside A (XXXIV) against *Neurospora* ATPase (Minale & Riccio 1987). Except for sepositoside A, all these asteroid saponins are sulfated.

Smooth muscles are also affected. Polyhydroxylated steroidal glycoside crossasteroside (XLVI) from the asteroid *Crossaster papposus* inhibits activity (80%) on electrically-induced contractions of a guinea pig ileum preparation at 100 µg/ml (Anderson et al. 1985). No mechanisms of action have been proposed.

7.3 *Protein toxins from echinoids*

Toxins are of interest to pharmacologists not only for their toxic effect but also for their mechanism of action. If this mechanism concerns a clearly determined target of the organism, the toxin can sometimes either be used as a medicine at low doses or serve as a pharmacological reagent for studying the mechanism of action of other substances. Different protein toxins from echinoid pedicellariae have an effect on junction transmitter release on smooth or cardiac muscle. One is a high molecular weight protein from pedicellariae of *Lytechinus pictus* which produces a reduction in the amplitude of the excitatory junction potential recorded from a single muscle fiber of the leg of the crab *Pachygrapsus crassipes*. The resting membrane potential of these fibers experiences little change. This suggests that the toxin acts primarily upon the transmitter release mechanism (Biedebach et al. 1978).

Studies on toxins have involved smooth and cardiac muscle. Feigen et al. (1966, 1974) noted that crude toxin and the nondialyzable purified fraction (Alender et al. 1965) from pedicellariae of *Tripneustes gratilla* cause contractions in isolated guinea pig ileum. The results suggest that histamine and other agents (e.g., serotonin, bradykinin) might be released. Chemical evidence indicates the release of histamine from ileal, cardiac and lung tissues of the guinea pig, and from colon and lung tissues of the rat. The toxin acts as a proteolytic enzyme producing, together with the globulin fraction of mammalian plasma, active dialyzable kinin-like substances (Feigen et al. 1968). One of these substances seems to be bradykinin which can also be destroyed by echinoid toxin once it is formed (Feigen et al. 1974). Kimura et al. (1975) also emphasized the kinin-like activity of the crude toxin from *Toxopneustes piloleus*. This toxin is a mixture of several substances that are basic peptides or proteins and induce or stimulate contractions of various isolated smooth muscles. Like bradykinin, they increase capillary permeability, an effect which has been attributed to dilatation of capillaries secondary to constriction of peripheral veins. They also stimulate the cardiac movement of isolated frog heart or rabbit atrium. Nevertheless, this effect is overridden in the crude toxin by a fraction having strong inhibitory action on cardiac movement. The contractile response in longitudinal muscle of isolated guinea pig ileum was not affected by atropine, tetrodotoxin, hexamethonium or methysergide. However, it was partially decreased by H1-receptor antagonist tripelennamine, which suggests that the response may be at least partly due to histamine release. A slight but significant release of this compound was detected, but only at higher concentrations than those causing muscle contraction. Tripelennamine-resistant contraction might be due to the direct action of the extract on muscle (Kimura & Nakagawa 1980).

7.4 Nonprotein substances from echinoids

A dialyzable hemolytic substance different from both proteins and saponins has been extracted from the spine and test of an echinoid (unspecified species) belonging to *Diadematidae*. This substance increases reversibly and dose-dependently miniature endplate potential (MEPP) frequency in frog sartorius muscle. It is of particular interest because of its difference from other toxins. Its action has been interpreted as a result of increased permeability of presynaptic membrane to both divalent cations (Ca^{2+} and Mg^{2+}) and Na^+. The membrane potential of the muscle is not changed appreciably (Kihara et al. 1983, Anraku et al. 1984).

Mendes et al. (1963) found a dialyzable acetylcholine-like substance from homogenates of pedicellariae from *Lytechnius variegatus*. This was tested by responses of guinea pig ileum, rat uterus, dog blood pressure, amphibian heart, holothuroid longitudinal muscle, and protractor muscle of the echinoid lantern.

7.5 Tetrodotoxin

Tetrodotoxin, one of the most potent neurotoxins, is found mainly in several vertebrates including various species of pufferfish (Tetraodontidae) and amphibians, and mollusks (Maruyama et al. 1984). It has also been detected in two asteroid species: *Astropecten latespinosus* (Maruyama et al. 1984) and *Astropecten polyacanthus* (Noguchi et al. 1982, Miyazawa et al. 1987). Tetrodotoxin affects excitable membranes by selectively blocking the sodium channel, thus leading to muscular paralysis, arterial hypotension and respiratory depression. Tetrodotoxin has been reviewed (World Health Organization 1984).

7.6 Prostaglandins

Prostaglandin group B has been identified in the viscera of ten species of echinoderms: five asteroids (*Asturias amurensis*: 1200 ng/g tissue; *Disto-*

lasterias nippon: 700 ng/g; *Evasterias r. tabulata*: 700 ng/g; *Lysastrosoma anthosticta*: 800 ng/g; *Patiria pectinifera*: 2000 ng/g), two echinoids (*Strongylocentrotus nudus* and *S. intermedius*), and three holothuroids (*Stichopus japonicus*: 70 ng/g; *Cucumaria fraudatrix*; and *C. japonica*). PGE$_2$ and PGF$_2$ were identified on the basis of their chromatographic behavior and a bioassay of extracts from the corresponding zone. These prostaglandins produced contractions of the uterine horns of the rat and of smooth muscle of the holothuroid *Cucumaria japonica* (Korotchenko et al. 1983).

8 SUBSTANCES ACTIVE ON THE CARDIOVASCULAR SYSTEM

A few preliminary experiments have been carried out on the cardiovascular properties of echinoderm extracts (Table 15).

Goldsmith & Carlson (1976) identified the substances from *Astirias forbesi* that decrease blood pressure as asterosaponin (XXV). The effect is dose-related in the anesthetized cat. The product caused a sudden 30% drop at 0.5 mg/kg IV, and about 50% at 2 mg/kg. In all cases, the blood pressure drop was only transient: 33 s for the lowest dose and 126 s for the highest. The rat was less sensitive to asterosaponin than the cat. The drug had a negative chronotropic effect but altered the normal rat ECG. Hypotensive activity was due to a direct effect on the vasculature. The Wakunaga Pharmaceutical Co. (1984) has patented asterosaponins (XX, desulfated XX, XXI and XXII) from *Asterias amuransis* for their hypotensive properties.

For echinoids, the kinin-like action of a toxin from *Toxopneustes piloleus*, responsible for stimulation of cardiac movement of isolated frog heart or rabbit atrium has been considered previously. However, the effect of crude toxin is overridden by a strong inhibitory fraction. Crude echinoid toxin constricts peripheral blood vessels (Kimura et al. 1975). The acetylcholine-like substance from homogenates of globiferous pedicellariae of *Lytechinus variegatus* has been mentioned. This product is responsible for slowing the heart rhythm of the toad *Bufo ictericus* and lowering the blood pressure of the dog.

To our knowledge, the only observation about the cardiovascular activity of products from holothuroids was reported by Ruggieri & Nigrelli (1974). These authors, on the basis of an erroneous reference, indicate that holothurin A increases the conduction time through the atrioventricular node and decreases the automatic rate of spontaneous beating Purkinje cells. They suggested a possible use of the substance in cases of sinus node arrhythmia and tachycardia.

Table 15. Effect (increase: ↑ or decrease: ↓) of extracts of echinoderms on inotropic (I), chronotropic (C), coronary flow (CF) and blood pressure.

Species	Extract	Cardiac activity (% baseline)			Effect on blood pressure	Reference
		I	C	CF		
Asteroidea						
Astropecten duplicatus	a	–	–	11 ↓	–	Kaul et al. 1977
Culcita grex	a	–	0	11 ↑	40 ↓	Kaul et al. 1977
Culcita novaeguinae	a	0	0	13 ↑	8 ↓	Kaul et al. 1977
Oreaster reticulatus	a	67 ↑	–	–	–	Kaul et al. 1977
Echinoidea						
Echinometra diadema	a	–	–	–	70 ↓	Kaul et al. 1977
Echinometra mathei	a	–	–	14 ↓	–	Kaul et al. 1977
Lytechinus variegatus	a	–	–	–	80 ↓	Kaul et al. 1977
Mellita quinquiesperforata	a	19 ↓	–	–	–	Kaul et al. 1977
Strongylocentrotus sp.	a	–	–	–	9 ↑	Kaul et al. 1977
Holothuroidea						
Actinopyga mauritiana	a	50 ↑	71 ↓	81 ↓	15 ↓	Kaul et al. 1977
Holothuria floridana	a	–	–	–	8 ↓	Kaul et al. 1977
Holothuria mexicana	a	–	–	22 ↓	30 ↑	Kaul et al. 1977
	a	5 ↑	5 ↑	15 ↓	20 ↓	Kaul et al. 1977
Stichopus japonicus	b	40 ↓	↓	–	–	Lee et al. 1984
Stichopus tropicalis	a	15 ↓	8 ↓	28 ↓	–	Kaul et al. 1977

a = Isopropanol or methanol extracts; 10 mg/kg I.V. for dog blood pressure, 100 µg for isolated guinea pig heart.

b = Dialysable fraction of water extract; 100 g/ml for isolated spontaneous beating guinea pig atria.

9 OTHER SUBSTANCES OF PHARMACOLOGICAL INTEREST

Other pharmacological activities or other substances of interest in this field have been reported but have not been studied experimentally in detail.

9.1 Anti-inflammatory substances

Goldsmith & Carlson (1976) reported the anti-inflammatory properties of asterosaponin (XXV) from *Asterias forbesi* and compared them with those of phenylbutazone. At 120 µg/ml *in vitro*, this substance inhibits the heat denaturation of bovine serum albumin by over 80%, and at 500 µg/ml by 85%. In carraghennin-induced rat-paw edema, phenylbutazone reduced edema formation by 52%; at 75 mg/kg, asterosaponin produced an identical level of inhibition. This activity is common to several saponins (Shibata 1977, Jaecker et al. 1982).

9.2 Analgesic substances

The same asterosaponin (XXV) has analgesic properties which have been assessed using the mouse acetic acid writhing test. At a dose of 25 mg/kg IP, saponin showed 99% reduction in writhings, whereas morphine at 10 mg/kg eliminated all writhings (Goldsmith & Carlson 1976).

9.3 Insulin from asteroids

An acid-isopropanol extraction of digestive tissue from the asteroid *Pisaster ochraceous* yielded material with properties resembling those of insulins from vertebrates (Wilson & Falkmer 1965). This was the first identification of an insulin-like substance in invertebrates.

9.4 Substances active on the central nervous system

Ninety-five percent isopropanol or methanol echinoderm extracts have been tested (Dose: 100 mg/kg IP) on mouse spontaneous motor activity to assess central nervous system response. Chlorpromazine (3 mg/kg) was used as standard (Table 16) (Kaul et al. 1977). The active extracts have not been studied.

9.5 Hypothermic substances

A griseogenin-derived saponin (III) from *Holothuria floridana* has a powerful, dose-related hypothermic activity (Kaul & Schmitz, unpublished data; cited in Kaul & Daftari 1986).

9.6 Anticoagulant substances

Anticoagulant acid mucopolysaccharides have been isolated from holothuroids. One, from *Holothuria leucospilota* , is active in rabbit blood both *in*

158

Table 16. Effect of echinoderm extracts on mouse spontaneous motor activity (Kaul et al. 1977)

Species	Depressant activity (% CPZ)[1]
Asteroidea	
Astropecten duplicatus	72
Culcita grex	71
Culcita novaeguineae	94
Oreaster reticulatus	63
Echinoidea	
Diadema antillarum	43/34
Echinometra lucunter	57
Echinometra mathei	51
Echinothrix diadema	94
Lytechinus variegatus	63
Mellita quinquiesperforata	85
Strongylocentrotus sp.	53
Tripneustes esculentus	45
Tripneustes ventricosus	69
Holothuroidea	
Actinopyga agassizi	67
Actinopyga mauritiana	177
Holothuria difficilis	66
Holothuria floridana	71
Holothuria glaberrima	66
Holothuria mexicana	90/94
Unidentified holothuroid	59
Stichopus badionotus	60
Stichopus tropicalis	94

1. Dose: 100 mg/kg IP; results are expressed as percent of chlorpromazine (CPZ) activity (3 mg/kg) IP.

vitro and *in vivo* and is as potent as heparin (Zhang et al. 1988). A second, from *Stichopus japonicus*, identified as a chondroitin sulfate, contains D-glucuronic acid, D-galactosamine, sulfur and fucose in concentrations of 18.4, 16, 11.8 and 18%, respectively. Its molecular weight is about 43 000 (Hashimoto et al. 1988).

10 DISCUSSION

Three groups of pharmacologically-active substances are predominant among the echinoderms: saponosides (holothurins and asterosaponins), proteins, and polysaccharides.

Holothurins are probably the most studied echinoderm chemical group. The particular structure of their aglycones gives them a certain originality among the saponosides and provides for an often greater pharmacological activity. Their most notable features include antifungal properties which have justified the implementation of clinical trials in Japan. Shimada (1969) reported the effect of holotoxin for external treatment of dermatophytosis. (Improvement was noted in 77 out of 87 cases treated (88%), with no harmful side effects.) Perhaps these substances may not compete with amphotericin B, nystatin or griseofulvin which also act on membranes. Certain other properties require complementary studies to be better understood. For example, the substance responsible for the antitumor properties of crude holothurin has not been identified. It cannot be holothurin A which is only active *in vitro* but might be another holothurin, or possibly an impurity. The immunostimulating properties of this same preparation are also of interest. But we may ask whether these effects are useful for therapy despite the fact that they seem to depend on the membrane toxic properties of holothurin. Ruggieri & Nigrelli (1974) suggested that the holothurin A might be used for the treatment of sinus node arhythmia and tachycardia. Powerful antithermic properties have been attributed to a griseogenin-derived saponin (Kaul & Daftari 1986). The mechanism of action of these two properties has not been studied. It may be related to the toxic membrane properties of this group of substances that have already been implicated in hemolytic and cytotoxic effects and the destructive interaction observed in excitable nerve and muscle tissues. Moreover, crude holothurins from the body wall of *Actinopyga lecanora, Holothuria pulla* and *H. fuscocinerea* have indirect mutagenic and clastogenic activity. Although no action was noted *in vitro* (*Rec-* assay and Ames' test) after metabolic activation, these holothurins produced aberration scores in Swiss albino mice (micronucleum test) when administered IP. Nevertheless, *per os* they exhibited a 35-fold reduction in activity, which indicates inactivation in the alimentary canal (Poscidio 1983a). Thus, except for external usage, the future therapeutic potential of holothurins seems necessarily limited by their toxicity.

Asterosaponins, though chemically interesting because of their frequently polyhydroxyl nature and the customary presence of a sulfated group, have been studied considerably less. Although they have the same type of toxicity (cytotoxicity, hemolysis, contractile effects on smooth and striated muscle), it would be of interest to examine some of their properties more thoroughly, such as the hypotensive activity of asterosaponins from *Asterias amurensis* and the analgesic and inflammatory effect of asterosaponins from *A. forbesi*. The latter property is common to several plant saponins (Shibata 1977, Jaecker et al. 1982) which have only been used for minor therapeutic applications.

Echinoid pedicellariae contain toxic protein derivatives whose mechanism of action on nerve and muscle has been studied (reduced transmitter release function for *Lytechinus pictus*; kinin-like activity and release of histamine for *Tripneustes gratilla* and *Toxopneustes piloleus*). Not only do these substances have toxic-related effects but they are also likely to be poorly tolerated by foreign organisms because of their protein nature. In fact, they have not attracted the attention of therapists. Two glycoproteins of these pedicellariae have been patented because of their antitumor properties (Suntory 1983, 1985), but their potential therapeutic value has not been considered.

Neither holothuroid nor asteroid proteins have been the subject of detailed pharmacological study. The most notable result has been the detection of lymphokine and interleukin-like factors in asteroids. However, despite the fundamental interest of this discovery, it is doubtful that these substances have therapeutic value.

Antitumor properties (sometimes very intense) have been noted for various polysaccharides from the holothuroid *Stichopus japonicus* and the asteroid *Asterias amuriensis*. Many polyholosides of different origin possess such properties in IP applications in animals. The major one is lentinane, a potent immunostimulant extracted from the mushroom *Lentinus edodes* (Chihara 1981), which in association with Tegafur permits an increased survival rate, as compared to Tegafur alone, for treatment of inoperable gastric cancer (Tsukagoshi 1988). Echinoderm polysaccharides may play a role.

The study of holothuroid acid mucopolysaccharides is also of potential interest for heparin-like substances.

These results are meager. The potential value of marine invertebrates depends on the detection of substances with an original structure that could be of pharmacological interest. This would not seem to be the case for echinoderms. Except for saponins, the only chemical originality that these invertebrates offer is the presence of carotenoid-protein complexes and naphthoquinone (or anthraquinone) pigments found particularly in the testes and spines of echinoids. Among these substances, only echinochrome A has proved of pharmacological value for its slight antibacterial properties. Nonproteic nitrogenous compounds, such as alkaloids, which often have an activity of pharmacological interest, are rather rare (Palumbo et al. 1982, Pettit et al. 1973, Nakamura et al. 1981, Pathirana & Andersen 1986). Between 1977 and 1985, only 3 of 85 compounds isolated from echinoderms were nitrogenous metabolites (Ireland et al. 1988). Among these substances, only imbricatine, an alkaloid from an asteroid, has been studied from a pharmacological point of view (it possesses moderate antitumoral activity).

These conclusions are not definitive. Certain properties have been observed in various echinoderm extracts for which active substances have not yet been identified. This is the case for gram-positive antibacterial substances of ophiuroids, for a nonproteic, nonsaponin hemolytic substance from an echinoid (because of its effect on miniature endplate potentials) and for hypotensive or sedative substances of the central nervous system. Moreover, few studies have been carried out on ophiuroids and crinoids, and only a limited number of echinoderms have attracted the attention of researchers. In any event, echinoderms are less promising for pharmacological purposes than other marine invertebrate phyla such as sponges, cnidarians or urochordates.

ACKNOWLEDGEMENTS

The author wishes to thank Mrs C. Cuenca for her information about taxonomy and Mrs J. Philippot and Mr Y. F. Pouchus for their help in preparing the manuscript.

REFERENCES

Ahond, A., M.B. Zurita, M. Colin, C. Fizames, P. Laboute, F. Lavelle, D. Laurent, C. Poupat & J. Pusset 1988. La Girolline, nouvelle substance antitumorale extraite de l'éponge *Pseudaxinyssa cantharella* n. sp. (Axinellidae). *C.R. Acad. Sci.* Ser. 2, 307: 145-148.

Alender, C.B., G.A. Feigen & J.T. Tomita 1965. Isolation and characterization of sea urchin toxin. *Toxicon* 3: 9-17.

Allen, T.M., A. Sharma & R.E. Dubin 1986. Potential new anticancer drugs from marine organisms collected at Enewtewak atoll. *Bull. mar. Sci.* 38: 4-8.

Andersson, L., S. Bano, L. Bohlin, R. Riccio & L. Minale 1985. Studies of Swedish marine organisms. Part 6 – A novel bioactive steroidal glycoside from the starfish *Crossaster papposus. J. Chem. Research* 1985: 3873-3885.

Andersson, L., L. Bohlin, M. Iorizzi, R. Riccio, L. Minale & W. Moreno-Lopez 1989. Biological activity of saponins and saponin-like compounds from Starfish and Brittle-stars. *Toxicon* 27: 179-188.

Andersson, L., G. Lidgren, L. Bohlin, L. Magni, S. Ögren & L. Afzelius 1983. Studies of Swedish marine organisms. I – Screening of biological activity. *Acta Pharm. Sued.* 20: 401-414.

Andersson, L., G. Lidgren, L. Bohlin, P. Pisa, H. Wigzell & R. Kiessling 1986. Studies of Swedish marine organisms. V – A screening for lectin- like activity. *Acta Pharm. Sued.* 23: 91-100.

Anisimov, M.M., E.B. Fronert, T.A. Kuznetsova & G.B. Elyakov 1973. The toxic effect of triterpene glycosides from *Stichopus japonicus* Selenka on early embryogenesis of the sea urchin. *Toxicon* 11: 109-111.

Anisimov, M.M., N.N. Gafurov, S.I. Baranova, V.V. Shcheglov, A.I. Parsnitskaya & V.A. Rasskarov 1977. The influence of triterpenoid glycosides on activity of some enzymes of subcellar fractions of the rat liver and sea urchin embryo. *Izn. Akad. Nauk SSSR, (Biol.)* 2: 301-303.

162

Anisimov, M.M., A.S. Ivanov, A.M. Popov, M.I. Kiseleva, I.G. Sebko, L.Y. Korotkikh, A.S. Antonov, V.A. Stonik & V.F. Antonov 1981. Effect of triterpene glycosides and polyene antibiotics on cell membrane permeability for potassium and UV-absorbing agents. *Prikl. Biokhim. Mikrobiol.* 17: 890-895.

Anisimov, M.M., T.A. Kuznetsova, V.P. Shirokov, N.G. Prokofyeva & G.B. Elyakov 1972a. The toxic effect of a stichoposide A from *Stichopus japonicus* Selenka on early embryogenesis of the sea urchin. *Toxicon* 10: 187-188.

Anisimov, M.M., A.M. Popov & C.N. Dzizenlo 1979. Effect of lipids from sea urchin embryos on cytotoxic activity of certain triterpene glycosides. *Toxicon* 17: 319-321.

Anisimov, M.M., N.G. Prokofieva, L.Y. Korotkikh, I.I. Kapustina & V.A. Stonik 1980. Comparative study of cytotoxic activity of triterpene glycosides from marine organisms. *Toxicon* 18: 221-223.

Anisimov, M.M., N.G. Prokofieva, T.A. Kuznetsova & N.V. Peretolchin 1971. Effect of some triterpene glycosides on protein synthesis in tissue cultures of rat bone marrow. *Izv. Akad. Nauk SSSR (Biol.)* 1: 137-140.

Anisimov, M.M., V.V. Shcheglov, S.N. Dzizenko, L.I. Strigina, N.I. Uvarova, G.I. Oshitok, T.A. Kuznetsova, N.S. Chetyrina & I.N. Sokolvsky 1974b. Effect of some sterines on antimicrobial activity of triterpen glycosides of plant and animal origin. *Antibiotiky* 7: 625-629.

Anisimov, M.M., V.V. Shcheglov, V.A. Stonik, E.B. Fronert & G.B. Elyakov 1974a. The toxic effect of cucumarioside from *Cucumaria fraudatrix* on early embryogenesis of the sea urchin. *Toxicon* 12: 327-329.

Anisimov, M.M., V.V. Shcheglov, V.A. Stonik, A.L. Kul'ga, E.V. Levina, U.S. Levin & G.B. Elyakov 1972b. A comparative study of the antifungal activity of triterpene glycosides from pacific holothurians. *Dokl. biol. Sci.* 207: 666-668.

Anisimov, M.M., E.B. Shentsova, V.V. Shcheglov, Y.N. Shumilov, V.A. Rasskazov, L.I. Strigina, N.S. Chetyrina & G.B. Elyakov 1978. Mechanism of action of some triterpene glycosides. *Toxicon* 16: 207-218.

Anraku, M., H. Kihara & S. Hashimura 1984. A new sea urchin toxin and its effect on spontaneous transmitter release at frog neuromuscular functions. *Jap. J. Physiol.* 34: 839-847.

Assa, Y., S. Shany, B. Gestetner, Y. Tencer, Y. Birk & A. Bondy 1973. Interaction of alfalfa saponins with components of the erythrocyte membrane in hemolysis. *Biochim. Biophys. Acta* 307: 83-91.

Baranova, S.I., A.L. Kul'ga, M.M. Anisimov, V.A. Stonik, E.V. Levina, V.S. Levin & G.B. Elyakov 1973. Comparative study of the effect of the glycoside fractions of Pacific Ocean holothuria on RNA biosynthesis in a yeast cell culture. *Izv. Akad. Nauk. SSSR (Biol).* 2: 284-286.

Batrakov, S.G., E.S. Girshovich & N.S. Drozhzhina 1980. Triterpenic glycosides with antifungal activity isolated from the sea cucumber *Cucumaria japonica*. *Antibiotiki* 25: 408-411.

Beck, G. & G.S. Habicht 1986. Isolation and characterization of a primitive interleukine-1-like protein from an invertebrate, *Asterias forbesi*. *Proc. natl. Acad. Sci USA* 83: 7429-7433.

Bertheussen, K. 1983. Complement-like activity in sea urchin coelomic fluid. *Develop. compar. Immunol.* 7: 21-31.

Biedebach, M.C., G. Jacobs & S.W. Langjahr 1978. Muscle membrane potential effects of a toxin extract from the sea urchin *Lytechinus pictus* (Verril). *Comp. Biochem. Physiol.* 95C: 11-12.

163

Brito, M.A., M.H. Lagrota & R.D. Machaou 1981. Use of microtechnique for screening of antiviral substances. *Rev. Microbiol.* 12(3): 65-69.

Brown, R., L.R. Almodovar, H.M. Bihatia & W.C. Boyd 1968. Blood group specific agglutinins in invertebrates. *J. Immunol.* 100(1): 214-216.

Burnell, D.J. & J.W. Apsimov 1983. Echinoderm saponins. In P.J. Scheuer (ed.). *Marine Natural Products: Chemical and Biological Perspectives.* 5: 287-389. New York: Acad. Press.

Cairns, S.D. & C.A. Olmsted 1973. Effect of Holothurin on sarcoma 180 and B16 melanoma tumors in mice. *Gulf. Res. Rep.* 4: 205-213.

Canicatti, C. 1987. Membrane damage by coelomic fluid from *Holothuria polii* (Echinodermata). *Experientia* 43: 611-614.

Canicatti, C. & D. Ciulla 1987. Studies on *Holothuria polii* (Echinodermata) coelomocyte lysate I. Hemolytic activity of coelomocyte hemolysins. *Dev. comp. Immunol.* 11: 705-712.

Canicatti, C & D. Ciulla 1988. Studies on *Holothuria polii* (Echninodermata) coelomocyte lysate II. Isolation and characterization of coelomycte hemolysins. *Dev. comp. Immunol.* 12: 55-64.

Canicatti, C., D. Ciulla & E. Farina-Lipari 1988. The hemolysin-producer coelomocytes in *Holothuria polii*. *Dev. comp. Immunol.* 12: 729-736.

Chanley, J.D., S.K. Kohn, R.F. Nigrelli & H. Sobotka 1959. Holothurin. I. The isolation, properties and sugar components of Holothurin A. *J. Am. chem. Soc.* 81: 5180-5183.

Chihara, G. 1981. The antitumor polysaccharide Lentinan. An overview. *Int. Congr. Ser. – Excerpta Med.* 576: 1-16.

Colon, R.L., G.D. Ruggieri, S.J. Nigrelli & R.F. Nigrelli 1976. Comparison of the biological activities of crude and purified holothurin from the tubule of the Bahamian Sea Cucumber *Actinopyga agassizi* Selenka. In H.H. Webber & G.D. Ruggieri (ed.). *Food drugs from the sea. Proceed. 1974.* 366-374. Washington: Marine Technol. Soc.

Constantine Jr, G.H., P. Catalfomo & C. Chou 1975. Antimicrobial activity of marine invertebrates extracts. *Aquaculture* 5: 299-304.

D'Auria, M.V., A. Fontana, L. Minale & R. Riccio (in press). *Gazz. Chim. It.*

De Groof, R.C. & T. Narahashi 1976. The effects of Holothurin A on the resting membrane potential and conductance of squid axon. *Eur, J. Pharmacol.* 36: 337-346.

Dettbarn, W.D., H.B. Higman, E. Bartels & T. Podleski 1965. Effect of marine toxins on electrical activity and K+ efflux of excitable membranes. *Biochem. Biophys. Acta* 94: 472-478.

Dubois, M.A., R. Higuchi, T. Komori & T. Sasaki 1988. Biologically active glycosides from asteroidea. XVI – Steroid oligoglycosides from the starfish *Asterina pectinifera* Müller et Troschel, 3. Structure of two new oligoglycoside sulfates, Pectinioside E and F, and biological activity of the six new pectiniosides. *Liebigs Ann. Chem.* 9: 845-850.

Elyakov, G.B., M.M. Anisimov, N.G. Prokofyeva, T.A. Kuznetsova & E.B. Fronert 1972. Sensitivity of rat marrow cells in culture to the toxic effect of Stichoposide A, from *Stichopus japonicus* Selenka. *Toxicon* 10: 299-300.

Fan, H.F. & C. Chen 1981. A new method for the isolation and purification of an acid mucopolysaccharide from *Stichopus japonicus*. *Yao Hsweh Tung Pao* 16(4): 57-58.

Fan, H.F. & J. Chen 1982. A new method for the isolation and purification of an acidic mucopolysacchardie from *Stichopus japonicus*. *Zhongyao Tongbao* 7(4): 27-29.

Fänge, R. 1963. Toxic factors in starfishes. *Sarsia* 10: 19-21.

Feigen, G.A., L. Hadji & J.E. Cushing 1974. Modes of action and identities of protein constituents in sea urchin toxin. In H.J. Hum & C.E. Lane (eds). *Bioactive Compounds from the Sea*. 37-97. New-York: Dekker Inc.

Feigen, G.A., E. Sanz & C.B. Alender 1966. Study of the mode of action of sea urchin toxin. I. Conditions affecting release of histamine and other agents from isolated tissues. *Toxicon* 4: 161-175.

Feigen, G.A., E. Sanz, J.T. Tomita & C.B. Alender 1968. Studies on the mode of action of sea urchin toxin. II. Enzymatic and immunological behavior. *Toxicon* 6: 17-43.

Friess, S.L. 1972. Mode of action of marine saponins on neuromuscular tissue. *Fed. Proceed.* 31: 1146-1149.

Friess, S.L., J.D. Chanley, W.V. Hudak & H.E. Weems 1970. Interaction of the echinoderm toxin Holothurin A and its desulfated derivatives with the cat superior cervical ganglion preparation. *Toxicon* 8: 211-219.

Friess, S.L. & R.C. Durant 1965. Blockade phenomena at the mammalian neuromuscular sinapse. Competition between reversible anticholinesterases and an irreversible toxin. *Toxicol. Appl. Pharmacol.* 7: 373-381.

Friess, S.L., R.C. Durant & J.D. Chanley 1968. Further studies on biological actions of steroidal saponins produced by poisonous echinoderms. *Toxicon* 6: 81-92.

Friess, S.L., R.C. Durant, J.D. Chanley & F.J. Fash 1967. Role of the sulfate charge center in irreversible interactions of Holothurin A with chemoreceptors. *Bioch. Pharmacol.* 16: 1617-1625.

Friess, S.L., R.C. Durant, J.D. Chanley & T. Mezzetti 1965. Some structural requirements underlying holothurin A interactions with sinaptic chemoreceptors. *Biochem. Pharmacol.* 14: 1237-1245.

Friess, S.L., F.G. Standaert, E.R. Whitcomb, R.F. Nigrelli, J.D. Chanley & H. Sobotka 1959. Some pharmacologic properties of Holothurin, an active neurotoxin from the sea cucumber. *J. Pharmacol. Exp. Theor.* 126: 323-329.

Friess, S.L., F.C. Standaert, E.R. Whitcomb, R.F. Nigrelli, J.D. Chanley & H. Sobotka 1960. Some pharmacologic properties of Holothurin A, a glycoside mixture from the sea cucumber. *Ann. N.Y. Acad. Sci.* 90: 893-901.

Fusetani, N., Y. Kato, K. Hashimoto, T. Kumori, Y. Itakura & T. Kowasaki 1984. Biological activities of asterosaponins with special reference to structure-activity relationships. *J. nat. Prod.* 47: 997-1002.

Gerwick, W.H. 1987. Drugs from the sea, the search continues. *J. Pharm. Technol.* 3: 136-141.

Giga, Y., A. Ikai & K. Takahashi 1987. The complete amino acid sequence of echinoidin, a lectin from the coelomic fluid of the sea urchin *Anthocidaris crassispina*. Homologies with mammalian and insect lectins. *J. biol. Chem.* 262: 6197-6203.

Goldsmith, L.A. & G.P. Carlson 1976. Pharmacological evaluation of an asterosaponin from *Asterias forbesi*. In H.H. Webber & G.D. Ruggieri (eds). *Food-drugs from the sea, Proceed. 1974.* 354-365. Washington: Marine Technol. Soc.

Goodfellow, R. & L.J. Goad 1973. Observations on the sterol sulfates and sterol esters of the echinoderm *Asterias rubens. Biochem. Soc. Trans.* 1: 759-761.

Gorshkov, B.A., I.A. Gorshkova, V.A. Stonik & G.B. Elyakov 1982. Effect of marine glycosides on adeninetriphosphate activity. *Toxicon* 20: 655-658.

Hashimoto, T., S. Watabe, T. Harada & T. Muto 1988. Extraction of chondroitin sulfates from sea cucumbers. *Jpn Kokai Tokkyo Koho* JP 63 10, 601 [88 10, 601] (Cl. C08 B 37/08), 18 janv. 1988, Appl. 86 58, 678, 12 mars 1986; 5 pp.

Hashimoto, Y. & T. Yasumoto 1960. Conformation of saponin as a toxic principle of starfish. *Bull. Jap. Soc. scient. Fish.* 26: 1132-1138.

165

Heilbrunn, L.V., A.B. Chaet, A. Dunn & W.L. Wilson 1954. Antimitotic substances from ovaries. *Biol. Bull.* 106: 158-168.

Higuchi, R., Y. Noguchi, T. Komori & T, Sasaki 1988. Biologically active glycosides from Asteroidea. XVIII. Proton-NMR spectroscopy and biological activities of polyhydroxylate of steroids from the starfish *Asterina pectinifera* Müller et Troschel. *Liebigs Ann. Chem.* 1185-1189.

Institute of physical and chemical research 1984. Antitumor agent SFE-I from starfish. *Jpn Kokai Tokkyo Koho* JP 58, 148, 825 [84 137, 418] (Cl. A 61 k 35/56), 07 Aug. 1984, Appl. 83/10, 344, 24 Janv. 1983; 6 pp.

Ireland, C.M., D.M. Roll, T.F. Molinski, T.C. McKee, T.M. Zabriskie & J.C. Swersey 1988. Uniqueness of the marine chemical environment: categories of marine natural products from invertebrates. In D.G. Fautin (ed.). *Biomedical importance of marine organisms.* 41-56. San Francisco: California Acad. Sc.

Ito, K. & M. Masuda 1986. Preparation of antitumor polysaccharide from starfish. *Jpn Kokai Tokkyo Koho* JP 61, 22, 021 [86 22, 021] (Cl. A 61 K 35/36), 30 Jan. 1986, Appl. 84/142, 015, 09 Jul. 1989; 7 pp.

Ivanov, A.S., E.M. Khalikov, V.A. Solov'va, I.I. Mal'tsev & G.B. Elyakov 1986. Assay of free cholesterol in human blood serum using the triterpene glycoside Holotoxin A1. *Vopr. Med. Khim.* 32(5): 132-134.

Jaecker, H.J., G. Voigt & K. Hiller 1982. Zum Antiexudation verhalten einiger Triterpen-saponine. *Pharmazie* 37: 380-382.

Jakowska, S., R.F. Nigrelli, P.M. Murray & A. Veltri 1958. Hemopoietic effects of Holothurin, a steroid saponin from the sea cucumber, *Actinopyga agassizi. Anat. Rec.* 132: 459.

Järnefelt, J. 1972. Lipid requirements of functional membrane structure as indicated by reversible inactivation of $(Na^+ -K^+)$-ATPase. *Biochim. Biophys. Acta* 266: 91-102.

Johnson, P.T. & F.A. Chapman 1971. Comparative studies on the in vitro response of bacteria to invertebrate body fluids. III. *Stichopus tremulus* (sea cucumber) and *Dendraster excentricus* (sand dollar). *J. Invert. Pathol.* 17: 94-106.

Jolles, J. & P. Jolles 1975. The lysosyme of *Asterias rubens. Eur. J. Biochem.* 54: 19-23.

Kapoor, V.K. & A.S. Chawla 1986. Biological significance of triterpenoids. *J. Sc. Ind. Res.* 45: 503-511.

Kaul, P.N. & P. Daftari 1986. Marine pharmacology: bioactive molecules from the sea. *Ann. Rev. Pharmacol. Toxicol.* 26: 117-142.

Kaul, P.N., S.K. Kulkarni, A.J. Weinheimer, F.J. Schmitz & T.K.B. Karns 1977. Pharma-cologically active substances from the sea. II. Various cardiovascular activities found in the extracts of marine organisms. *J. nat. Prod.* 40: 253-268

Kihara, H., M. Anraku & S. Hashimura 1963. Extraction and characterization of the biologically active substances from sea urchins in Fiji. *J. Physiol. Soc. Jpn.* 45: 468.

Kimura, A., H. Kayashi & M. Kuramoto 1975. Studies of urchi-toxins: separation, purification and pharmacological actions of toxinic substances. *Japan. J. Pharmacol.* 25: 109-120.

Kimura, A. & H. Nakagawa 1980. Action of an extract from the sea urchin *Toxopneustes pileolus* on isolated smooth muscle. *Toxicon* 18: 689-693.

Kitagawa, I., M. Kobayashi, T. Inamoto, M. Fuchida & Y. Kyogoku 1985. Marine natural products. XIV. Structure of echinosides A and B, antifungal lanostane-oligosides from the sea cucumber *Actinopyga echinites* (Jaeger). *Chem. Pharm. Bull.* 33: 5214-5224.

Kitagawa, I., M. Kobayashi, T. Inamoto, T. Yasuzawa & Y. Kyogoku 1981a. The structure of six antifungal oligoglycosides, stichlorosides A1, A2, B1, B2, C1 et C2 from the sea cucumber *Stichopus chloronotus* (Brandt). *Chem. Pharm. Bull.* 29: 2387-2391.

166

Kitagawa, I., M. Kobayashi & Y. Kyogoku 1982. Marine natural products. IX. Structural elucidation of triterpenoidal oligoglycosides from the Bahamean sea cucumber *Actinopyga agassizi* Selenka. *Chem. Pharm. Bull.* 30: 2045-2050.

Kitagawa, I., T. Nishino & Y. Kyogoku 1979. Structure of Holothurin A, a biologically active triterpene- oligoglycoside from the sea cucumber *Holothuria leucospilota* Brandt. *Tetrahedron Lett.* 16: 1419-1422.

Kitagawa, I., T. Nishino & Y. Kyogoku 1981b. Marine natural products. VIII. Bioactive triterpene-oligoglycosides from the sea cucumber *Holothuria leucospilota* Brandt. Structure of Holothurin A. *Chem. Pharm. Bull.* 29: 1951-1956.

Kitagawa, I., T. Sugawara & I. Yosioka 1976. Saponin and sapogenol. XV. Antifungal glycosides from the sea cucumber *Stichopus japonicus* Selenka (2). Structure of Holotoxin A and Holothoxin B. *Chem. Pharm. Bull.* 24: 275-284.

Korotchenko, O.D., T.Ya. Mishchenko & S.V. Isay 1983. Prostaglandins of Japan sea invertebrates. I. The quantitation of group B prostaglandins in echinoderms. *Comp. Biochem. Physiol.* 14C: 85-88.

Kuznetsova, T.A., M.M. Anisimov, A.M. Popov, S.I. Baranova, S.S. Afiyatullov, I.I. Kapustina, A.S. Antonov & G.B. Elyakov 1982. A comparative study *in vitro* of physiological activity of triterpene glycosides of marine invertebrates of echinoderm type. *Comp. Biochem. Physiol.* 37C: 41-43.

Lasley, B.J. & R.F. Nigrelli 1970. The effect of crude holothurin on leucocyte phagocytose. *Toxicon* 8: 301-306.

Lasley, B.J. & R.F. Nigrelli 1971. The effect of holothurin on leucocyte migration. *Zoologica N.Y.* 56: 1-12.

Leclerc, M., C. Brillouet & R. Barot-Ciobaru 1987. Evidence for a B-like cells stimulatory factor in *Asterias rubens. Med. Sci. Res.* 15: 211-212.

Leclerc, M., C. Brillouet & F. Luquet 1981. Properties of cell subpopulations of starfish axial organ: in vitro effect of pokeweed mitogen and evidence of lymphokine-like substances. *Scand. J. Immunol.* 14: 281-284.

Leclerc, M., C. Brillouet, G. Luquet & R. Binaghi 1986. The immune system of invertebrates: the sea star *Asterias rubens* (Echinoderma) as a model of study. *Bull. Inst. Past. (Paris)* 84: 311-330.

Lee, S.J., J.M. Shin, B.U. Im & Y.H. Kim 1984. Toxicity and biological activity of extracts from *Stichopus japonicus. Arch. Pharmacol. Res.* 7: 61-62.

Leiter, J., A.R. Bourke, D.B. Fitzgerald, M.M. Mac Donald, S.A. Schepartz & I. Wodinsky 1962. Screening data from the Cancer Chemotherapy National Service Center screening laboratories. XI. *Cancer Res.* 22(11): 1002.

Liu, S.H., M.B. McChesney & R.A. Prendergast 1983. Kinetics of tumor cell cytostasis by sea star *(Asterias forbesi)* factor-activated macrophages. *Dev. Comp. Immunol.* 7: 545-554.

Mackie, A.M., R. Lasker & P.T. Grant 1968. Avoidance reactions of a mollusc *Buccinum undatum* to saponin-like surface-active substances in extracts of the starfish *Asterias rubens* and *Marthasterias glacialis. Comp. Biochem. Physiol.* 26: 415-428.

Mackie, A.M., H.T. Singh & T.C. Fletcher 1975. Studies on the cytolytic effects of seastars *(Marthasterias glacialis)* saponins and synthetic surfactants in the plaice *Pleuronectes platessa. Mar. Biol.* 29: 307-314.

Mackie, A.M., H.T. Singh & J.M. Owen 1977. Studies on the distribution, biosynthesis and function of steroidal saponins in echinoderms. *Comp. Biochem. Physiol. B* 56: 9-14.

Mal'tsev, I.I., S.I. Stekhova, E.B. Shentsova, M.M. Anisimov & V.A. Stonik 1985.

167

Antimicrobial activity of glycosides from holothurians of the family stichopodidae. *Pharm. Chem. J.* 19(1): 54-56.

Maruyama, J., T. Noguchi, J.K. Jeon, T. Harada & K. Hashimoto 1984. Occurence of tetrodotoxin in the starfish *Astropecten latespinosus*. *Experientia* 40: 1395-1396.

Masuda, K., S. Funayama, K. Komiyama, I. Umezawa & K. Ito 1987. Antitumor acid polysaccharide NRP-1 isolated from starfish *Asterias amurensis* Lütken. *Kitasato Arch.of Exp. Med.* 60(3): 95-103.

Mc Kay, D., C.R. Jenkin & D. Rowley 1969. Immunity in the invertebrates. I. Studies on the naturally occuring haemagglutinins in the fluid from invertebrates. *Aust. J. exp. biol. med. Sci.* 47: 125-134.

Mendes, E.G., L. Abbud & S. Umiji 1963. Cholinergic action of homogenates of sea urchin pedicellariae. *Science* 139: 408-409.

Minale, L., C. Pizza, R. Riccio, O. Squille Greco, F. Zollo, J. Pusset & J.L. Menou 1984. New polyhydroxylated sterols from the starfish *Luidia maculata*. *J. nat. Prod.* 47: 784-789.

Minale, L., C. Pizza, F. Zollo & R. Riccio 1982. 5α-cholestane-3β,6β,15α,16β,26-pentol: a polyhydroxylated sterol from the starfish *Hacelia attenuata*. *Tetrahedron Lett.* 23: 1841-1844.

Minale, L., R. Riccio, C. Pizza & F. Zollo 1986. Steroidal oligoglycosides of marine origin. In H. Imura, T. Goto, T. Murachi & T. Nakajina (eds). *Natural Products and Biological Activities*, a NAITO Foundation Symposium. 59-73. New-York: University of Tokyo Press and Elsevier.

Miyazawa, K., M. Higashiyama, K. Hori, T. Noguchi, K. Ito & K. Hashimoto 1987. Distribution of tetrodotoxin in various organs of the starfish *Astropecten polyacanthus*. *Mar. Biol.* 96: 385-390.

Munro, M.H.E., R.T. Luibrand & J.W. Blunt 1987. The search for antivirae and anticancer compounds from marine organisms. In P.J. Scheuer (ed.). *Bioorganic marine chemistry.* 93-176. Berlin, Heidelberg: Springer-Verlag.

Nakamura, H., J. Kobayashi & Y. Hirata 1981. Isolation and structure of a 330 nm UV-absorbing substance, Asterina-330 from the starfish *Asterina pectinifera*. *Chem. Lett.* 1413-1414.

Nemanich, J.W., B.F. Theiler & L.P. Hager 1978. The occurence of cytotoxic compounds in marine organisms. In P.N. Kaul & C.J. Sindermann (eds). *Drugs and food from the sea, Myth or reality? Proceed. 1977.* 123-136. Univ. Oklahoma.

Nigrelli, R.F. 1952. The effects of Holothurin on fish, and mice with sarcome 180. *Zoologica N.Y.* 37: 89-90.

Nigrelli, R.F. & S. Jakowska 1960. Effect of Holothurin, a steroid saponin from the Bahamian sea cucumber *(Actinopyga agassizi)* on various biological systems. *Ann. N.Y. Acad. Sci.* 90: 884-892.

Nigrelli, R.F., M.F. Stempien Jr, G.O. Ruggieri, V.R. Liguori & J.T. Cecl 1967. Substances of potential biomedical importance from marine organisms. *Fed. Proc.* 26: 1197-1205.

Nigrelli, R.F. & P.L. Zahl 1952. Some biological characteristics of Holothurin. *Proc. Soc. exp. Biol. Med.* 81: 379-380.

Noguchi, T., H. Narita, J. Maryurama & K. Hashimoto 1982. *Bull. Jap. Soc. Scient. Fish.* 48: 1173 (in MARYUAMA et al., 1984).

Owellen, R.J., R.G. Owellen, M.A. Gorog & D. Klein 1973. Cytolytic saponin fraction from *Asterias vulgaris*. *Toxicon* 11: 319-323.

Parrinello, N., C. Canicatti & D. Rindone 1976. Naturally occuring hemagglutinins in the coelomic fluid of *Holothuria polii* Delle Chiaje and *Holothuria tubulosa* Gmelin. *Boll. Zoll.* 43: 259-271.

Parrinello, N., D. Rindone & C. Canicatti 1979. Naturally occurring hemolysins in the coelomic fluid of *Holothuria polii* Delle Chiaje (Echinodermata). *Dev. Comp. Immunol.* 3: 45-54.

Palumbo, A., M. D'Ischia, G. Misuraca & G. Prota 1982. Isolation and structure of a new sulfur-containing amino acid from sea urchin eggs. *Tetrahedron Lett.* 23: 3207-3208.

Pathirana, C. & R.J. Andersen 1986. Imbricatine, an unusual benzyltetrahydroquinoline alkaloid from the starfish *Dermasterias imbricata. J. Am. chem. Soc.* 108: 7981-7993.

Pettit, G.R., J.F. Day, J.L. Hatwell & H.B. Wood 1970. Antineoplastic components of marine animals. *Nature, Lond.* 227: 962-963.

Pettit, G.R., J.A. Hasler, K.D. Paull & C.L. Herald 1981a. Antineoplastic agents 76. The sea urchin *Strongylocentrotus* droebachiensis. *J. Nat. Prod.* 44: 701-704.

Pettit, G.R., C.L. Herald & D.L. Herald 1976. Antineoplastic agents XLV. Sea cucumber cytotoxic saponins. *J. Pharm. Sci.* 65: 1558-1559.

Pettit, G.R., C.L. Herald & L.D. Vanell 1979. Isolation and characterization of Strongylostatin 1. *J. Nat. Prod.* 42: 407-409.

Pettit, G.R., J.A. Rideout, J.A. Hasler, D.L. Doubek & P.R. Reucroft 1981b. Isolation and characterization of Lytechinastatin. *J. Nat. Prod.* 44: 713-716.

Pettit, G.R., R.B. Von Dreele, G. Bolliger, P.M. Traxler & P. Brown 1973. Isolation and structural elucidation of 3,6-dioxo-hexahydropyrrolo [1,2-a]-pyrazine from the echinoderm *Luidia clathrata. Experientia* 29: 521-522.

Pinkas, M., F. Trotin & L. Bezanger-Beauquesne 1972. Les saponosides. *Prod. Prob. Pharm.* 27(31): 187-198.

Pizza, C., P. Pezzullo, L. Minale, E. Brietmaier, J. Pusset & P. Tirard 1985. Starfish saponins. Part. 20. Two novel steroidal glycosides from the starfish *Acanthaster planci* (L.). *J. Chem. Research* (S) 76-77.

Poscidio, G.N. 1983a. The mutagenicity potential of holothurin of some Philippine holothurians. *Philipp. J. Sci.* 112: 1-12.

Poscidio, G.N. 1983b. Holothurin of some Philippine holothurians and its hemolytic activity. *Philipp. J. Sci.* 112: 13-28.

Prendergast, R.A. 1971. Macrophage activating and chemotactic factor from the sea star *Asterias forbesi. Fed. Proc.* 30: 647.

Prendergast, R.A., G.A. Cole & C.S. Henney 1974. Marine invertebrate origin of a reactanct to mammalian T cells. *Ann. N.Y. Acad. Sci.* 234: 7-17.

Prendergast, R.A. & M. Suzuki 1970. Invertebrate protein simulating mediators of delayed hypersensitivity. *Nature, Lond.* 227: 277-279.

Rao, G.S., J.E. Sinsheeimer & K.W. Cochran 1974. Antiviral activity of triterpenoid saponins containing acylated β-amyrin aglycones. *J. Pharm. Sci.* 63: 471-473.

Riccio, R., M.V. D'Auria & L. Minale 1986. Two new steroidal glycoside sulfates, longicaudoloside-A and -B, from the Mediterranean ophiuroid *Ophioderma longicaudum. J. Org. Chem.* 51: 533-536. R

Riccio, R., L. Minale, S. Pagonis, C. Pizza, R. Zollo & J. Pusset 1982b. A novel group of highly hydroxylated steroids from the starfish *Protoreaster nodosus. Tetrahedron* 38: 3615-3622.

Riccio, R., L. Minale, C. Pizza, F. Zollo & J. Pusset 1982a. Starfish saponins. Part. 8. Structure of nodoside, a novel type of steroidal glycoside from the starfish *Protoreaster nodosus. Tetrahedron Lett.* 23: 2899-2902.

169

Rinehart, L. R., Jr, P. D. Shaw & 25 other coll. 1981. Marine natural products as sources of antiviral, antimicrobial and antineoplastic agents. *Pure Appl. Chem.* 53: 795-817.

Rio, G. J., G. D. Ruggieri, S. J. Martin, M. F. Stempien Jr & R. F. Nigrelli 1963. Saponin-like toxin from the giant sunburst starfish, *Pycnopodia helianthoides* from the Pacific northwest. *Amer. Zool.* 3: 554-555.

Rio, J. R., M. F. Stempien Jr, R. F. Nigrelli & G. D. Ruggieri 1965. Echinoderm toxins. I. Some biological and physiological properties of toxins from several species of Asteroidea. *Toxicon* 3: 147-155.

Rogers, D. J., G. Blunden, J. A. Topliss & M. D. Guiry 1980. A survey of some marine organisms for hemagglutinins. *Bota. Mar.* 23: 569-577.

Rubtzov, B. V., A. O. Rugitsky, G. I. Klebanov, A. M. Sedov & Yu. A. Vladirimov 1980. The effect of triterpene glycosides of marine invertebrate animals on the permeability of natural and artificial membranes. *Izv. Akad. Nauk. SSSR (Biol.)* 402-407.

Ruggieri, G. D. 1965. Echinoderm toxins. II. Animalizing action in sea urchin development. *Toxicon* 3: 157-162.

Ruggieri, G. D. & R. F. Nigrelli 1960. The effect of Holothurin, a steroid saponin from the sea cucumber, on the development of the sea urchin. *Zoologica, N.Y.* 45: 1-16.

Ruggieri, G. D. & R. F. Nigrelli 1964. Effects of extracts of the red web starfish, *Patiria miniata*, on sea urchin eggs. *Ann. Zool.* 4: 131.

Ruggieri, G. D. & R. F. Nigrelli 1966a. Effects of extracts of the sea star, *Acanthaster planci*, on the developping sea urchin. *Am. Soc. Biol.* 6: 592-593.

Ruggieri, G. D. & R. F. Nigrelli 1966b. Animalization of *Arbacia punctulata* larvae by extracts of the sea cucumber, *Holothuria edulis. Am. Soc. Biol.* 6: 593.

Ruggieri, G. D. & R. F. Nigrelli 1974. Physiological active substances from echinoderms. In H. J. Humm & C. E. Lane (eds). *Bioactive compounds from the sea.* 183-195. New York: Dekker Inc.

Ryoyama, K. 1973. Studies on the biological properties of coelomic fluid of sea urchin. *Biochim. Biophys. Acta* 320: 157-165.

Ryoyama, K. 1974. Studies on the biological properties of coelomic fluid of sea urchin . II. Naturally occuring hemagglutinin in sea urchin. *Biol. Bull.* 146: 404-414.

Santhakumari, G. & J. Stephen 1988. Antimitotic effects of Holothurin. *Cytologia* 53: 163-168.

Sedov, A. M., I. B. Shepeleva, N. S. Zakharova, O. G. Sakandelidze, V. V. Sergeev & I. Ya. Moshiashvili 1984. Effect of Cucumarioside (triterpene glycoside from the sea cucumber *Cucumaria japonica)* on the development of immune response to corpuscular pertussis vaccine in mice. *Zh. Mikrobiol., Epidermiol. Immunobiol.* 9: 100-104.

Seeman, P., D. Cheng & G. H. Iles 1973. Structure of membrane holes in osmotic and saponin hemolyses. *J. Cell. Biol.* 56: 519-527.

Sen, D. K. & V. K. Lin 1975. Effect of Holothurin on *Trypanosoma duttoni* in Swiss Webster male mice. *J. Protozool.* 22: 25-26 A.

Sen, D. K. & V. K. Lin 1977. Effect of Holothurin on *Trypanosoma duttoni* in mice response of tripanosomes to biotoxin. *Virginia J. Sc.* 28: 9-12.

Service, M. & A. C. Wardlaw 1984. Echinochrome-A as a bactericidal substance in the coelomic fluid of *Echinus esculentus* (L.). *Comp. Biochem. Physiol.* 72B: 161-165.

Service, M. & A. C. Wardlaw 1985. Bactericidal activity of coelomic fluid of the sea urchin, *Echinus esculentus,* on different marine bacteria. *J. mar. biol. Assoc. U.K.* 65: 133-139.

Shaw, P. D., W. O. Mc Clure, G. V. Van Blaricom, J. Sims, W. Fenical & J. Rude 1976.

170

Antimicrobial activities from marine organisms. In H.H. Webber & G.D. Ruggieri (eds). *Food drugs from the sea, Proceed. 1974.* 429-433. Washington: Marine Technol. Soc.

Shibata, S. 1977. Saponins with biological and pharmacological activity. In H. Wagner & P. Wolff (eds). *New natural products and plant drugs with pharmacological, biological or therapeutical activity.* 177-196. Berlin: Springer Verlag.

Shimada, S. 1965. Verfahren zur Gewinnung von Holotoxin. *Germ. Patent* 1 199 923 (Cl. A 61 k), 2 Sept. 1965, Appl. 21 Nov. 1963.

Shimada, S. 1969. Antifungal steroid glycoside from sea cucumber. *Science* 163: 1462.

Shimizu, Y. 1971. Antiviral substances in starfish. *Experientia* 27: 1188-1189.

Shimizu, Y. 1973. Characterization of starfishtoxins. Identification of an aglycone as a 5-pregnane derivative. In L.R. Worthen (ed.). *Food drugs from the sea. Proceed. 1972.* 291-297. Washington: Marine Technol. Soc.

Shimizu, Y., M. Tanaka & T. Hayashi 1985. Antitumor marine biopolymers. In R.T. Colwell, E.R. Pariser & A.J. Sinskey (eds). *Biotechnology of marine polysaccharides.* 377-388. Washington: Hemisphere Publishing Corporation.

Sigel, M.M., L.L. Wellham, W. Lichter, L.E. Dudeck, J.L. Gargus & A.H. Lucas 1970. Anticellular and antitumor activity of extracts from tropical marine invertebrates. In H.W. Youngken Jr (ed.). *Food-drugs from the sea. Proceed. 1969.* 281-294. Washington: Mar. Technol. Soc.

Stonik, V.A. 1986. Some terpenoid and steroid derivatives from echinoderms and sponges. *Pure Appl. Chem.* 58: 423-436.

Stonik, V.A. & G.B. Elyakov 1988. Secondary metabolites from echinoderms as chemotaxonomic markers. In P. Scheuer (ed.). *Bioorganic Marine Chemistry.* 2: 43-82. Berlin: Springer Verlag.

Styles, T.J. 1970. Effect of Holothurin on *Trypanosoma lewisi* infections in rats. *J. Protozool.* 17: 196-198.

Suffness, M. & J.E. Thompson 1988. National Cancer Institute's role in the discovery of new antineoplastic agents. In D.G. FAUTIN (ed.).*Biomedical importance of marine organisms.* 151-157. San Francisco: California Acad. Sc.

Sullivan, T.D., K.T. Ladue & R.F. Nigrelli 1955. The effects of Holothurin, a steroid saponin of animal origin, on krebs-2 ascites tumors in swiss mice. *Zoologica, N.Y.* 40: 49-52.

Sullivan, T.D. & R.F. Nigrelli 1956. Antitumorous action of biologics of marine origin. I. Survival of swiss mice inoculated with krebs-2 ascites tumor and treated with holothurin, a steroid saponin from the sea cucumber, *Actinopyga agassizi. Proc. Am. Assoc. Cancer. Res.* 2: 151.

Suntory Ltd & Nippon Suisan Kaisha Ltd 1983. Antitumor glycoprotein extraction from sea urchin. *Jpn Kokai Tokkyo Koho* JP 58, 148, 825 [83, 148, 825] (Cl. A 61 k 35/36), 05 Sept. 1983, Appl. 82/30, 242, 26 Feb. 1982 ; 21 pp.

Suntory Ltd & Nippon Suissan Kaisha Ltd 1985. Antitumor glycoproteins from the urchin shells. *Jpn Kokai Tokkyo Koho* JP 60 58, 921 [85 58, 921] (Cl. A 61 k 35/36), 05 Apr. 1985, Appl. 83/166, 373, 09 Sept. 1983 ; 10 pp.

Thron, C.D. 1964. Hemolyse by Holothurin A, Digitonine and Quillaia saponin: estimates of the required cellular lysin uptakes and free lysin concentrations. *J. Pharm. Exp. Therap.* 145: 194-202.

Thron, C.D., R.C. Durant & S.L. Friess 1964. Neuromuscular and cytotoxic effects of Holothurin A and related saponins at low concentration levels III. *Toxicol. Appl. Pharmacol.* 6: 182-196.

171

Thron, C. D., R. N. Patterson & S. L. Friess 1963. Further biological properties of the sea cucumber toxin Holothurin A. *Toxicol. Appl. Pharmacol.* 5: 1-11.

Tschesche, R. 1971. Advances in the chemistry of antibiotic substances from higher plants. In H. Wagner & L. Horhammer (eds). *Pharmacognosy and Phytochemistry.* 274-289. Berlin: Springer Verlag.

Tschesche, R. & G. Wulff 1965. Uber die antimikrobielle Wirksamkeit von Saponinen. *Z. Naturforschung* 20b: 543-546.

Tsukagoshi, S. 1988. Lentinan, a new polysaccharide for the treatment of cancer. *Drugs of today* 24: 91-95.

Tyler, A. 1946. Natural heteroagglutinins in the body-fluids and seminal fluids of various invertebrates. *Biol. Bull.* 90: 213-219.

Uhlenbruck, G., U. Reifenberg & M. Heggen 1970. On the specificity of broad spectrum agglutinins. IV. Invertebrate agglutinins: current status, conceptions and further observations on the variation of the Hel receptor in pigs. *Z. Immun.-Forsch.* 139: 486-499.

Wakunaga Pharmaceutical Co. 1984. Antihypertensive steroid saponins of starfish. *Jpn Kokai Tokkyo Koho* JP 59, 231, 022 [84, 231, 022] (Cl. A 61 K 31/705), 25 Dec. 1984, Appl. 83/108, 375. 14 Jun. 1983; 10 pp.

Wardlaw, A.C. & S.E. Unkles 1978. Bactericidal activity of coelomic fluid from the sea urchin *Echinus esculentus*. *J. Invert. Pathol.* 32: 25-34.

Weinheimer, A.J., J.A. Matson, T.K.B. Karns, M.B. Hossain & D. Van der Helm 1978. Some new marine anticancer agents. In P.N. Kaul & C.J. Sindermann (eds). *Drugs and food from the sea, Myth or reality?* 117-121. Univ. Oklahoma.

Willenborg, D.O. & R.A. Prendergast 1974. The effect of sea star coelomocyte extract on cell-mediated resistance to *Listeria monocytogenes* in mice. *J. exp. Med.* 139: 820-833. Wilson, S. & S. Falkmer 1965. Starfish insulin. *Can. J. Biochem.* 43: 1615-1624.

World Health Organization 1984. Aquatic (marine and freshwater) biotoxins. *Environmental Health Criteria* 37. 95 p. Geneva: W.H.O.

Wu, C.H. & T. Narahashi 1988. Mechanism of action of novel marine neurotoxins on ion channels. *Ann. Rev. Pharmacol. Toxicol.* 28: 141-161.

Yamada, Y. & K. Aketa 1982. Purification and partial characterization of hemagglutinins in seminal plasma of the sea urchin, *Hemicentrotus pulcherrimus*. *Biochim. Biophys. Acta* 709: 220-226.

Yamanouchi, T. 1955. On the poisonous substance contained in holothurians. *Publ. Seto mar. biol. Lab.* 4: 184-203.

Yasumoto, T., T. Watanabe & Y. Hashimoto 1964. Physiological activities of starfish saponin. *Bull. Jap. Soc. scient. Fish.* 30: 357-364.

Zhang, P., S. Luo, G. Zhon & Q. Wang 1988. Anticoagulant effect of *Holothuria leucospilota* acid mucopolysaccharide. *Zhongguo Yaolixue Yu Dulixue Zazhi* 2: 98-101.

ANNEX: STRUCTURE OF PHARMACOLOGICAL SUBSTANCES FROM ECHINODERMS

A1 *Structure of some biological active holothurins and holothurigenins*

If unspecified, 1: structures are taken from Burnell and Apsimov's review (1983), 2: oses are on the pyranosyl form.

22,25-oxydoholothurinogenin   **R = H**

I   Holothurin A   R = β-D-quinovose $\overset{1\ 2}{\underline{\quad\quad}}$ β-D-xylose-4-0-sulfate[1]—

$\qquad\qquad\qquad\qquad\qquad\qquad\qquad\quad$ |4
$\qquad\qquad\qquad\qquad\qquad\qquad\qquad\quad$ |1
$\qquad\qquad\qquad\qquad\qquad\qquad\quad$ β-D-glucose
$\qquad\qquad\qquad\qquad\qquad\qquad\qquad\quad$ |3
$\qquad\qquad\qquad\qquad\qquad\qquad\qquad\quad$ |1
$\qquad\qquad\qquad\qquad\quad$ 3-0-methyl-β-D-glucose

II   Holothurin B   R = β-D-quinovose $\overset{1\ 2}{\underline{\quad\quad}}$ β-D-xylose-4-0-sulfate[1]—

III   Griseogenin

IV   Stichoposide A (MALTSEV *et al.* 1985)

$\qquad$ R = 6-desoxy-β-D-glucose $\overset{1\ 2}{\underline{\quad\quad}}$ β-D-xylose $\overset{1}{\underline{\quad\quad}}$

$\qquad\qquad$ Presence of a sulfate group

V   Stichoposide C (MALTSEV *et al.* 1985)

$\qquad$ R = 6-desoxy-β-D-glucose $\overset{1\ 2}{\underline{\quad\quad}}$ β-D-xylose $\overset{1}{\underline{\quad\quad}}$

$\qquad\qquad\quad$ |4 $\qquad\qquad\qquad\qquad\qquad\quad$ |4
$\qquad\qquad\quad$ |1 $\qquad\qquad\qquad\qquad\qquad\quad$ |1
$\qquad\quad$ β-D-xylose $\qquad\qquad\qquad\qquad$ β-D-glucose
$\qquad\qquad\quad$ |3 $\qquad\qquad\qquad\qquad\qquad\quad$ |3
$\qquad\qquad\quad$ |1 $\qquad\qquad\qquad\qquad\qquad\quad$ |1
$\quad$ 3-0-methyl-β-D-glucose $\qquad$ 3-0-methyl-β-D-glucose

$\qquad\qquad$ Presence a sulfate group

VI   Stichoposide D (MALTSEV et al. 1985)

$$R = \beta\text{-D-glucose} \xrightarrow{\quad 1 \qquad\qquad 2 \quad} \beta\text{-D-xylose} \xrightarrow{\quad 1 \quad}$$

```
R = β-D-glucose ──1────────2── β-D-xylose ──1──
        |4                      |4
        |1                      |1
    β-D-xylose              β-D-glucose
        |3                      |3
        |1                      |1
  3-O-methyl-β-D-glucose   3-O-methyl-β-D-glucose
```

VII   Stichoposide E (MALTSEV *et al.* 1985)

```
R = β-D-xylose ──1────────2── β-D-xylose ──1──
        |4                      |4·
        |1                      |1
    β-D-glucose             β-D-glucose
        |3                      |3
        |1                      |1
  3-O-methyl-β-D-glucose   3-O-methyl-β-D-glucose
```

VIII   Stichloroside C$_2$ or Astichoposide C (STONIK 1986)

```
R = β-D-quinovose ──1────────2── β-D-xylose ──1──
        |4                         |4
        |1                         |1
    β-D-xylose                 β-D-glucose
        |3                         |3
        |1                         |1
  3-O-methyl-β-D-glucose      3-O-methyl-β-D-glucose
```

IX   Thelenotoside A (MALTSEV *et al.* 1985)

```
R = 6-desoxy-β-D-glucose ──1 2── β-D-xylose ──1──
        |4
        |1
    β-D-xylose
        |3
        |1
  3-O-methyl-β-D-glucose
```

X   Thelenotoside B (MALTSEV *et al.* 1985)

```
R = β-D-glucose ──1 2── β-D-xylose ──1──
        |4
        |1
    β-D-xylose
        |3
        |1
  3-O-methyl-β-D-glucose
```

XI   Cucumarioside G   R = D-xylose, D-quinovose 3-O-methyl-
                                             D-xylose, D-glucose

XII Echinoside A (KITAGAWA *et al.* 1985) **R**: the same osidic chain as for I
XIII Echinoside B (KITAGAWA *et al.* 1985) **R**: the same osidic chain as for II

XIV 24-dehydroechinoside A   **R** = the same osidic chain as for I

Holotoxigenin   R = H

XV   Holotoxin A

R = 3-0-methyl-β-D-glucose $\xrightarrow{1\ 3}$ β-D-glucose $\xrightarrow{1\ 4}$ β-D-xylose $\xrightarrow{1}$

$\qquad\qquad\qquad\qquad\qquad\qquad\qquad\qquad\qquad\qquad\qquad$ $\Big|$ 2
$\qquad\qquad\qquad\qquad\qquad\qquad\qquad\qquad\qquad\qquad\qquad$ $\Big|$ 1

3-0-methyl-β-D-glucose $\xrightarrow{1\ 3}$ β-D-glucose $\xrightarrow{1\ 4}$ β-D-quinovose

XVI Holotoxin A$_1$ (STONIK 1986)

R = 3-0-methyl-β-D-glucose $\xrightarrow{1\ 3}$ β-D-glucose $\xrightarrow{1\ 4}$ β-D-xylose $\xrightarrow{1}$

$\qquad\qquad\qquad\qquad\qquad\qquad\qquad\qquad\qquad\qquad\qquad$ $\Big|$ 2
$\qquad\qquad\qquad\qquad\qquad\qquad\qquad\qquad\qquad\qquad\qquad$ $\Big|$ 1

3-0-methyl-β-D-glucose $\xrightarrow{1\ 3}$ β-D-xylose $\xrightarrow{1\ 4}$ β-D-quinovose

XVII Holotoxin B

R = 3-0-methyl-β-D-glucose $\xrightarrow{1\ 3}$ β-D-glucose $\xrightarrow{1\ 4}$ β-D-xylose $\xrightarrow{1}$

$\qquad\qquad\qquad\qquad\qquad\qquad\qquad\qquad\qquad\qquad\qquad$ $\Big|$ 2
$\qquad\qquad\qquad\qquad\qquad\qquad\qquad\qquad\qquad\qquad\qquad$ $\Big|$ 1

β-D-glucose $\xrightarrow{1\ 3}$ β-D-glucose $\xrightarrow{1\ 4}$ β-D-quinovose

175

XVIII Holotoxin C

    R = ß-D-quinovose, ß-D-xylose, ß-D-glucose and 3-0-methyl-ß-D-glucose

    Seicheyllogenin   R = H

XIX   Holothurin C

    R = quinovose, xylose, glucose and 3-0-methyl-glucose

A2 *Structure of some biological active asterosaponins and sterols from Asteroidea*

If unspecified, 1: structures are taken from Fusetani et al. (1984), 2: oses are on the pyranosyl form.

    Thornasterol (MINALE *et al.* 1986)
    R = SO$_3^-$
    R' = H

XX   Thornasteroside A
    R = SO$_3^-$

    R' = galactose $\xrightarrow{1\ 4}$ xylose $\xrightarrow{1\ 3}$ quinovose $\xrightarrow{1}$
                |2 |2
                |1 |1
           fucose   quinovose

XXI   R = H

176

$$R' = \text{galactose} \xrightarrow{1\ 4} \text{xylose} \xrightarrow{1\ 3} \text{quinovose} \xrightarrow{1}$$

```
R' =  galactose ―1 4→ xylose ―1 3→ quinovose ―1―
            |2                |2
            |1                |1
         fucose            quinovose
            |3
            |1
         galactose
```

XXII $R = SO_3^-$

R' = the same osidic chain as for **XXI**

XXIII   Marthasteroside A_1
$R = SO_3^-$

```
R' =  galactose ―1 4→ xylose ―1 3→ quinovose ―1―
            |2                |2
            |1                |1
         fucose            quinovose
            |3
            |1
         fucose
```

XXIV $R = SO_3^-$

```
R =  fucose ―1 4→ quinovose ―1―
        |2              |2
        |1              |1
     fucose          quinovose
```

XXV $R = SO_3^-$

R = fucose, quinovose

XXVI $R = SO_3^-$

```
R =  glucose ―1 4→ xylose ―1 3→ quinovose ―1―
        |2               |2
        |1               |1
     fucose           quinovose
```

XXVII $R = SO_3^-$

```
R =  quinovose ―1 4→ xylose ―1 3→ quinovose ―1―
         |2                |2
         |1                |1
      fucose            quinovose
```

XXVIII Marthasteroside A$_2$

$R = SO_3^-$

$R =$ quinovose $\xrightarrow{1\ 4}$ xylose $\xrightarrow{1\ 3}$ quinovose $\xrightarrow{1}$
            |2 |2
            |1 |1
        fucose   quinovose
            |3
            |1
        fucose

XXIX Pectinioside A (DUBOIS *et al.* 1988)

$R = SO_3^-$

$R =$ β-D-glucose $\xrightarrow{1\ 4}$ β-D-quinovose $\xrightarrow{1\ 3}$ β-D-quinovose $\xrightarrow{1}$
             |2 |2
             |1 |1
      β-D-fucose   β-D-quinovose

XXX Pectinioside E (DUBOIS *et al.* 1988)

$R = SO_3^-$

$R =$ β-D-fucose $\xrightarrow{1\ 4}$ β-D-glucose $\xrightarrow{1\ 4}$ β-D-quinovose $\xrightarrow{1\ 3}$ β-D-quinovose $\xrightarrow{1}$
                       |2 |2
                       |1 |1
                β-D-xylose   β-D-quinovose

XXXI Pectinioside F (DUBOIS *et al.* 1988)

$R = SO_3^-$

$R' =$ β-D-glucose $\xrightarrow{1\ 4}$ β-D-xylose $\xrightarrow{1\ 3}$ β-D-quinovose $\xrightarrow{1}$
             |2 |2
             |1 |1
      β-D-fucose   β-D-quinovose

XXXII Asterosaponin A (BURNELL and APSIMON, 1983)

$R = SO_3^-$

$R' =$ fucose $\xrightarrow{1\ 4}$ fucose $\xrightarrow{1\ 4}$ quinovose $\xrightarrow{1\ 4}$ quinovose $\xrightarrow{1}$

XXXIII $R = SO_3^-$

$R' =$ the same osidic chain as for **XXXVII**

178

XXXIV Sepositoside A (ANDERSON *et al.* 1989)

glucuronic acid
$|$ 2
$|$ 1
galactose $\xrightarrow{2 \quad 1}$ glucose

XXXV   R = SO_3^-

R' = the same osidic chain as for **XXVII**

XXXVI   R = H

R' = the same osidic chain as for **XXXIV**

XXXVII R = SO_3^-

R' = the same osidic chain as for **XXIII**

XXXVIII R = quinovose $\xrightarrow{1 \quad 4}$ quinovose $\xrightarrow{1 \quad 3}$ glucose $\xrightarrow{1}$
                    $|$ 2 $|$ 2
                    $|$ 1 $|$ 1
                  fucose   quinovose

179

XXXIX Marthasteroide B (ANDERSSON *et al.* 1989)

R = fucose $\xrightarrow{1\ \ 4}$ quinovose $\xrightarrow{1\ \ 3}$ glucose $\xrightarrow{1}$

fucose   quinovose

(with side chains: fucose $\begin{smallmatrix}2\\1\end{smallmatrix}$, quinovose $\begin{smallmatrix}2\\1\end{smallmatrix}$)

XL   R = the same osidic chain as for **XX**

XLI   R = the same osidic chain as for **XXII**

XLII   R = the same osidic chain as for **XXIII**

XLIII Nodoside (RICCIO *et al.* 1982a)

R =   2-O-methyl-β-D-xylopyranose $R_4 = R_7 = R_{16} = H$

$\qquad\qquad\qquad\qquad\qquad\qquad\qquad\quad$ $R_5 = R_{15} = OH\ (\alpha)$

α-L-arabinofuranose $\xrightarrow{1}$ $R_6 = OH\ (\beta)$

XLIV Halityloside D (ANDERSSON *et al.* 1989)

R = 2,4-di-O-methyle-β-D-xylose $R_5 = R_7 = H$

$\qquad\qquad\qquad\qquad\qquad\qquad\qquad\quad$ $R_6 = OH\ (\alpha)$

α-L-arabinose $\xrightarrow{1}$ $R_4 = R_{15} = R_{16} = OH\ (\beta)$

XLV (HIGUCHI *et al.* 1988)

R = H $R_4 = R_5 = R_{16} = H$

$\qquad\qquad\qquad\qquad\qquad\qquad\qquad\quad$ $R_6 = R_7 = R_{15} = OH\ (\alpha)$

180

XLVI   Crossasteroside A (ANDERSSON *et al.* 1989)

R = 4-O-methyl-β-D-xylose
    |1
    |2
3-O-methyl-β-D-xylose $\xrightarrow{\;1\;}$

$R_4 = R_5 = R_{16} = H$
$R_6 = R_7 = R_{15} = OH \; (\alpha)$

XLVII   Crossasteroside B (ANDERSSON *et al.* 1989)

R = the same osidic chain as for **LXI**
    $R_4 = R_5 = R_7 = R_6 = H$
    $R_6 = R_{15} = OH \; (\alpha)$

XLVIII Crossasteroside C (ANDERSSON *et al.* 1989)

R = β-D-xylose
    |1
    |2
3-O-methyl-β-D-xylose $\xrightarrow{\;1\;}$

$R_4 = R_5 = R_{16} = H$
$R_6 = R_7 = R_{15} = OH \; (\alpha)$

XLIX   (ANDERSSON *et al.* 1989)

L   (ANDERSSON *et al.* 1989)

LI   (RICCIO *et al.* 1982b)

R = $R_4 = R_7 = H$ $R_{15} = OH \; (\alpha)$

LII   (ANDERSSON *et al.* 1989)

R = 2-O-methyl-β-D-xylose $\xrightarrow{1}$

$R_4 = R_7 = H$
$R_{15} = OH \ (\alpha)$

LIII   (RICCIO *et al.* 1982b)

R = R_4 = H

$R_7 = R_{15} = OH \ (\alpha)$

LIV   (RICCIO *et al.* 1982b)

R = H R_4 OH (β)

$R_7 = R_5 = OH \ (\alpha)$

LV   (HIGUCHI *et al.* 1988)

R = H $R_7 = OH \ (\alpha)$

$R_4 = R_{15} = OH \ (\beta)$

LVI   Halityloside A (ANDERSSON *et al.* 1989)

R = 2-O-methyl-β-D-xylose $\xrightarrow{1 \ 2}$ β-D-xylose $\xrightarrow{1}$

LVII   (MINALE *et al.* 1982)

$R_4 = R_7 = R_8 = H$

$R_6 = OH \ (\beta)$

LVIII (ANDERSSON *et al.* 1989)

$R_4 = R_7 = H$

$R_6 = OH \ (\alpha)$

LIX   (ANDERSSON *et al.* 1989)

$R_4 = H$

$R_6 = R_7 = OH \ (\alpha)$

182

LX

R_4 = OH (β) R_6 = R_7 = OH (α)

LXI   Ophidianoside C (ANDERSSON *et al.* 1989)

$$R = \text{xylose} \xrightarrow{1\ 4} \text{xylose} \xrightarrow{1\ 3} \text{quinovose} \xrightarrow{1}$$

$$\begin{array}{cc} |2 & |2 \\ |1 & |1 \\ \text{fucose} & \text{quinovose} \end{array}$$

LXII   (ANDERSSON *et al.* 1989)

$$R = \text{glucose} \xrightarrow{1}$$

LXIII   (ANDERSSON *et al.* 1989)

R_{12} = OH (β)

LXIV   (ANDERSSON *et al.* 1989)

R_{12} = H

183

LXV (ANDERSSON *et al.* 1989)

$R_{24} = CH_2$

LXVI (ANDERSSON *et al.* 1989)

$R_{24} = H$

Asterone   R = R' = H

LXVII R = SO_3^-

R' = the same osidic chain as for **XX**

LXVIII Asterosaponin L (BURNELL and APSIMOV 1983)

R ou R' = D-fucose, D-quinovose, D-xylose

LXIX Pectinioside D (DUBOIS *et al.* 1988)

R = SO_3^-

R' = the same osidic chain as for **XIX**

LXX Regularoside A (ANDERSSON *et al.* 1989)

R = the same osidic chain as for **XXXIII**

184

LXXI   Pectinioside B (DUBOIS *et al.* 1988)

R' = the same osidic chain as for **XXXI**

LXXII   Pectinioside C (DUBOIS *et al.* 1988)

R = β-D-glucose $\xrightarrow{1\ 4}$ β-D-xylose $\xrightarrow{1\ 3}$ β-D-quinovose $\xrightarrow{1}$

|2
|1
β-D-fucose   β-D-quinovose

|3
|1
β-D-fucose

LXXIII   Laeviusculoside A (D'AURIA *et al.* in press)

R = β-D-glucose $\xrightarrow{1\ 2}$ β-D-glucose $\xrightarrow{1}$

LXXIV   Echinochrome A (SERVICE and WARDLAW 1984)

LXXV   Imbricatine (PATHIRANA and ANDERSEN 1986)

An index of names of recent Asteroidea – Part 2: Valvatida

AILSA M. CLARK
Department of Zoology, Natural History Museum [formerly British Museum (Natural History)], London, UK
Present address: Gyllyngdune, Wivelsfield Green, Sussex, RH17 7QS, UK

Final manuscript acceptance: October 1990.

INTRODUCTION

Explanation of the procedure followed in this index was given in the Introduction to part 1. However, the type conventions followed may be briefly repeated here: Valid names for genera and species are given in bold type when in their definitive position but in italics in cross references where either genus-group names have been altered in rank or species have been transferred to other genera; names in ordinary type are synonyms or otherwise invalid. Asterisks before names signify doubtful or threatened ones needing further attention, while asterisks under Range indicate the type localities (however, these have not been checked for some species).

This part includes the rather heterogeneous order Valvatida Perrier, 1884, the current scope and sequence of which largely follows the proposals of Blake (1981 and 1987) rather than those of Spencer & Wright (1966, 'Treatise of Invertebrate Paleontology') who revived the order for the non-paxillosidan Phanerozonia of Sladen (1889) followed by Fisher (1911c). Blake (1981) transferred to the Valvatida a number of families formerly included in the Spinulosida.

In recent years there has been some controversy about the ordinal position of certain families, notably the Caymanostellidae Belyaev, 1974. Aziz & Jangoux (1984b) believed that its affinities are closest to the Asterinidae, which was one of the families transferred to the Valvatida by Blake (1981). However, Blake himself (1987) aligned the Caymanostellidae close to the Korethrasteridae in the Velatida, being followed in this by Smith (1988). Rowe et al. (1988) and Rowe (1989b) thought that all three families – the Asterinidae, Korethrasteridae and Caymanostellidae – are related and should be included in the order Valvatida. Since the classification used by A.M. Clark & Downey in 'Starfishes of the Atlantic' (finally published 1992 while this index was in press), generally follows that of Blake, the Caymanostellidae accordingly follows the Korethrasteridae in the Velatida. The same alignment is being adopted here so both families are deferred to part 3 of this index.

The limit between the Goniasteridae and the Oreasteridae has also been a subject of controversy with the views of Döderlein (1935) differing markedly from those of Fisher (1911c and 1919a), the latter postulating several subfamilies of the Goniasteridae, certain genera of which were abstracted for inclusion in the Oreasteridae (sensu ext.) by Döderlein. For want of a detailed revision, the compromise treatment used by Spencer & Wright (1966), when they listed many genera of Goniasteridae as incertae sedis, is followed here.

Since submission of this typescript, 'Starfishes of the Atlantic' by Clark and Downey (Natural History Museum, London & Chapman and Hall, 1992) has been published, including a number of charges in synonymy in addition to those mentioned in preliminary papers and included here.

Order VALVATIDA Perrier

Family CHAETASTERIDAE Sladen

Linckiidae (Chaetasterinae) Sladen, 1889: 398; Perrier 1893: 852; 1894: 328.
Chaetasteridae: Ludwig 1897: 134, 156; Tortonese 1965: 153; Spencer & Wright 1966: U56.
Ophidiasteridae (pt) Downey 1973: 62.
  Genus-group names: [Astropus], **Chaetaster**.

[ASTROPUS Gray, 1840: 182. As subgenus of *Astropecten*. Type species: *Astropecten (Astropus) longipes* Gray. Type locality 'Isle of France' (Mauritius). Assumed by Sladen (1889: 398) followed by Döderlein (1917) to be conspecific with *Asterias longipes* Retzius, 1805 so making *Astropus* a synonym of *Chaetaster* Müller & Troschel, 1840. However, three factors suggest otherwise: The non-atlantic locality, the recognition by Gray of *Chaetaster longipes* (though under the name of *Nepanthia tessellata)* isolated from *Astropecten* in his classification, and finally his description, which accords better with *Archaster* (Archasteridae). The holotype could not be traced in the British Museum collections.]
CHAETASTER Müller & Troschel, 1840
  Müller & Troschel, 1840a: 103; 1840b: 321; 1842: 27; Viguier 1878: 147; Sladen 1889: 398; Perrier 1896: 43; Ludwig 1897: 155; Fisher 1911c: 18,20,21; Verrill 1914a: 115; 1915: 115; Tortonese, 1965: 154; Downey 1973: 62.
  Type species: *Asterias subulata* Lamarck, 1816, a synonym of *Asterias longipes* Retzius, 1805.

borealis Düben, 1845: 113; a NOMEN NUDUM, undescribed

*californicus Grube, 1865

Grube 1865: 52.

Type locality presumably California but not stated.

Possibly a synonym of *Henricia leviuscula* (Stimpson, 1857) (Echinasteridae) according to Fisher (1911).

cylindratus Möbius 1859: 1. Probably a synonym of *Nepanthia maculata* Gray, 1840 (Asterinidae) according to Sladen (1889).

hermanni Müller & Troschel, 1842: 27. A synonym of *Stichastrella rosea* (O. F. Müller, 1776) (Asteriidae) according to Döderlein in Jangoux (1986b).

longipes (Retzius, 1805) (with synonyms *Asterias subulata* Lamarck, 1816 and *Nepanthia tessellata* Gray, 1840)

Retzius 1805: 20 (as *Asterias*).

M. Sars 1857: 51; Perrier 1875: 329 [1876: 249]; Perrier 1894: 329; Ludwig 1897: 134; Verrill 1915: 117; Koehler 1924: 140; Mortensen 1927: 95; Nobre 1930: 44; Madsen 1950: 219; Tortonese 1965: 154; Pawson 1978: 9.

Range: Mediterranean, Atlantic Spain and Portugal, S to Liberia and W to Azores, Ascension and St Helena, 30-1140 m.

longipes: Sladen, 1889: 398, non *Chaetaster longipes* (Retzius, 1805), = *C. nodosus* Perrier, 1875 according to A.M. Clark (1951).

maculatus: Müller & Troschel, 1842: 28 *(?Chaetaster)*, see *Nepanthia* (Asterinidae).

moorei Bell, 1894

Bell 1894: 404; A.M. Clark 1951: 1256; Jangoux 1984: 283; 1986a: 142.

Range: Macclesfield Bank, S China Sea*, New Caledonia, 22-73 m.

munitus Möbius, 1859: 3. A synonym of *Nectria ocellifera* (Lamarck, 1816) according to Sladen (1889) or possibly of *N. multispina* H.L. Clark, 1928 according to A.M. Clark (1966).

nodosus Perrier, 1875

Perrier 1875: 330 [1876: 250]; Verrill 1915: 116; A.M. Clark 1951: 1260; Downey 1973: 63; Jangoux 1978a: 96.

Range: Lesser Antilles*, N and W to Bermuda, South Carolina and the Gulf of Mexico, S to Guyana, 34-110 m.

subulata (Lamarck, 1816) Müller & Troschel, 1840: 103; 1842: 127. A synonym of *Chaetaster longipes* (Retzius, 1805) according to Perrier (1875).

tessellata (Gray, 1840) Müller & Troschel, 1842: 28. A synonym of *Chaetaster longipes* (Retzius, 1805) according to Perrier (1875).

*troscheli Müller & Troschel, 1842

Müller & Troschel, 1842: 28.

Type locality unknown; probably not *Chaetaster;* 'type lost and name should be discarded' according to Sladen (1889).

189

vanzolinicus Tommasi, 1972
 Tommasi 1972: 4; 1974: 9.
 Range: N Brazil.
vestitus Koehler, 1910
 Koehler 1910a: 136.
 Range: Andaman Is, Bay of Bengal.

Family ODONTASTERIDAE Verrill

Archasteridae (Gnathasterinae) Perrier 1893: 851; 1894: 244, 251, 323.
Odontasteridae Verrill 1899: 201; Fisher 1911c: 153; Verrill 1914a: 302-303;
 Tortonese 1936: 53; Fisher 1940: 97; Bernasconi 1962b: 27-51; A.M.
 Clark 1962a: 14-15; Spencer & Wright 1966: U55; A.M. Clark &
 Courtman-Stock 1976: 58.
'Gnathasteridées' Koehler 1920: 179-195.
Gnathasteridae: Koehler 1924: 183; Mortensen 1927: 76-77.
 Nomenclatural note. The earliest family-group name – Gnathasteridae – is
 invalidated by the synonymy of *Gnathaster* Sladen, 1889 with *Odontaster*
 Verrill, 1880.

 Genus-group names: **Acodontaster**, Asterodon, **Diplodontias**, Epidontaster, **Eurygonias**, Gnathaster, Gnathodon, Goniodon, Gymnognathaster, Heuresaster, **Hoplaster**, Metadontaster, **Odontaster**, Peridontaster, Pseudontaster, Tridontaster.

ACODONTASTER Verrill, 1899 (with synonyms *Heuresaster* Bell, 1908,
 Pseudontaster Koehler, 1912, *Metadontaster* Koehler, 1920 and *Tridontaster* Koehler 1920).
 Verrill 1899: 204; Koehler 1920: 192; Fisher 1940: 109; A.M. Clark
 1962a: 16; Bernasconi 1962b: 42; H.E.S. Clark 1963b: 38.
 Type species: *Gnathaster elongatus* Sladen, 1889.
abbreviatus Koehler, 1923: 81 (as variety of *Acodontaster elongatus*). A
 synonym of *Acodontaster conspicuus* (Koehler, 1920) according to Fisher
 (1940).
capitatus (Koehler, 1912)
 Koehler 1912b: 82 (as *Odontaster*).
 Koehler 1920: 195; A.M. Clark 1962a: 19; H.E.S. Clark 1963: 41 (as
 Acodontaster).
 Fisher 1940: 99, 110 (as forma of *Acodontaster elongatus*).
 Range: Graham Land (Antarctic Peninsula)*, Bellingshausen Sea, Ross
 Sea, Enderby Land, 193-647 m.
ceramoideus Fisher, 1940. A forma of *Acodontaster elongatus* (Sladen,
 1889).

190

conspicuus (Koehler, 1920) (with synonym *Acodontaster abbreviatus* Koehler, 1923)

Koehler 1920: 202 (as *Pseudontaster*).

Fisher 1940: 113; H. E. S. Clark 1963b: 39 (as *Acodontaster*).

Range: Adelie Land, Ross Sea, Graham Land, South Georgia, 24-460 m.

cremeus (Ludwig, 1903)

Ludwig 1903: 21 (as *Odontaster*).

Koehler 1920: 199 (as *Acodontaster*).

Range: Bellingshausen Sea, 450 m.

A synonym of *Acodontaster elongatus* according to Fisher (1940) but possibly a valid species according to A. M. Clark (1962). Type lost according to Jangoux & Massin (1986).

elongatus (Sladen, 1889) (with synonyms *Odontaster propinquus* A. H. Clark, 1917 and *Odontaster glaber* Barattini, 1938, also subspecies *granuliferus* (Koehler, 1912 (as *Odontaster granuliferus*), forma *capitatus* (Koehler, 1912) (as *Odontaster capitatus*) and forma *ceramoideus* Fisher, 1940).

elongatus elongatus (Sladen, 1889)

Sladen 1889: 288 (as *Gnathaster*).

Verrill 1899: 204; Fisher 1940: 109; Madsen 1955: 12; A. M. Clark 1962a: 19.

Range: Kerguelen*, Heard Id., Marion Id., Palmer Archipelago, 91-600 m.

elongatus granuliferus (Koehler, 1912)

Koehler 1912b: 77 (as *Odontaster*).

Koehler 1920: 192; Fisher 1940: 111; Bernasconi 1962b: 45; 1964b: 252; 1973: 294.

Range: E of Tierra del Fuego*, the Falkland Is and N to Uruguay, 40-840 m.

granuliferus: Koehler, 1920. A subspecies of *Acodontaster elongatus* (Sladen, 1889) according to Fisher (1940).

hodgsoni (Bell, 1908) (with synonyms *Tridontaster laseroni* Koehler, 1920, *Pseudontaster moderatus* Koehler, 1923 and forma *stellatus* (Koehler, 1920) (as *Pseudontaster*).

Bell 1908: 8 (as *Heuresaster*).

Fisher 1940: 115; A. M. Clark 1962a: 20; H. E. S. Clark 1963b: 38 (as *Acodontaster*).

Range: circum-south polar N to South Georgia (forma *stellatus* from Enderby Land E to Queen Mary Land), 4-540 m.

marginatus (Koehler, 1912)

Koehler 1912b: 86 (as *Pseudontaster*).

Fisher 1940: 99 (as *Acodontaster*).

Range: Bellingshausen Sea, 250 m.

miliaris: Verrill, 1899, see *Diplodontias*

waitei (Koehler, 1920)

Koehler 1920: 219 (as *Metadontaster*).

Fisher 1940: 114; A. M. Clark 1962a: 21 (as *Acodontaster*).

Range: Adelie Land* W to MacRobertson Land and the Palmer Archipelago, 160-647 m.

***ASTERODON** Perrier, 1891 *(Asterodon* Perrier, 1888: 765 a nomen nudum, no species named)

Perrier 1891a: K129; Fell 1953: 75.

Type species: *Goniodiscus singularis* Müller & Troschel, 1843, by subsequent designation by Koehler (1920).

Invalidated as a junior homonym of *Asterodon* Münster, 1841 (Pisces) but replaceable by *Diplodontias* Fisher, 1908, formerly a junior synonym according to Fell (1953). [New observation included in Clark & Downey (1992)]

dilatatus: Fell, 1953, see *Diplodontias*

granulosus Perrier, 1891a: 132; also Bernasconi 1962: 32; Guzman 1979: 114. A forma of *Asterodon* [i.e. *Diplodontias] singularis* (Müller & Troschel, 1843) according to Fisher (1940) but a subspecies according to Bernasconi (1964), though meantime treated as a species by Bernasconi (1962b) followed by Guzman (1979).

grayi: Perrier, 1891a: 138, non *Odontaster grayi:* Bell 1893, a synonym of *O. penicillatus* (Philippi, 1870), = *Acodontaster elongatus granuliferus* according to Fisher (1940).

miliaris: Benham, 1909, see *Diplodontias*

pedicellaris Perrier, 1891a: 135. A synonym of *Odontaster penicillatus* (Philippi, 1870) according to Fisher (1940).

robustus Fell, 1953, see *Diplodontias*

singularis: Perrier, 1891, see *Diplodontias*

DIPLODONTIAS Fisher, 1908 (restored as valid generic name, including both *Asterodon* Perrier 1891 and *Goniodon* Perrier, 1894, invalid homonyms) Fisher 1908a: 89.

Type species: *Pentagonaster dilatatus* Perrier, 1894; type species of *Goniodon.*

[Now proposed as replacement name for *Goniodon* Perrier, 1894, an invalid junior homonym, and itself available as a replacement name for *Asterodon* Perrier, 1891, found to be invalid.]

dilatatus (Perrier, 1875)

Perrier 1875: 217 [1876: 32] (as *Pentagonaster*).

Sladen 1889: 286 (as *Gnathaster*).

Perrier 1894: 244; Farquhar 1907: 126 (as *Goniodon*).

Fisher 1908a: 89; Mortensen 1925a: 286 (as *Diplodontias*).

Fell 1953: 77; Pawson 1965: 254 (as *Asterodon*).

192

Range: Cook Strait S to Stewart Id and the Snares, New Zealand, [0]-90 m.

miliaris (Gray, 1847) **new comb.** (with synonym *Astrogonium rugosum* Hutton, 1872)

Gray 1847: 80 (as *Astrogonium*).
Perrier 1875: 220 [1876: 35] (as *Pentagonaster*).
Sladen 1889: 286; Farquhar 1907: 126 (as *Gnathaster*).
Verrill 1899: 204 (as *Acodontaster*).
Benham 1909b: 90 [8]; Mortensen 1925a: 287; Fell 1962: 24; McKnight 1967a: 299 (as *Asterodon*).
Range: New Zealand from Cook Strait southwards.

*paxillosus (Gray, 1847) **new comb.**

Gray 1847: 79 (as *Astrogonium*).
Perrier 1875: 221 [1876: 37] (as *Pentagonaster*).
Sladen 1889: 286 (as *Gnathaster*).
Bell 1893b: 262 (as *?Odontaster*).
Range: 'Port Essington' [Northern Territory, Australia!].
[Holotype extant but tropical type locality probably a mistake.]

robustus (Fell, 1953) **new comb.**

Fell 1953: 80; H.E.S. Clark 1970b: 2 (restricted).
Range: Auckland Is, S of New Zealand.

singularis (Müller & Troschel, 1843) **new comb.** (with subspecies *granulosus* (Perrier, 1891))

singularis singularis (Müller & Troschel, 1843)

Müller & Troschel 1843: 116 (as *Goniodiscus*).
Perrier 1875: 222 [1876: 37]; Bell 1881a: 95 (as *Pentagonaster*).
Sladen 1889: 750 (as *Gnathaster*).
Perrier 1891a: K134; Fisher 1940: 116; Bernasconi 1962b: 32; 1964b: 251 (as *Asterodon*).
Bell 1893b: 262; Meissner 1896: 92 (as *Odontaster*).
Range: southern Chile*, Magellan area and Argentina S of the Rio de la Plata, 0-84 m.

singularis granulosus Perrier, 1891

Perrier 1891a: K132; Bernasconi 1964b: 251; Guzman 1979: 114 (as *Asterodon granulosus*).
Fisher 1940: 116 (as forma of *Asterodon singularis*).
Bernasconi 1962b: 32 (as subspecies).
Range: Magellan Strait*, N in Chile to c.41.5°S and southern Argentina.

EPIDONTASTER Koehler, 1920

Koehler 1920: 191, 195.
Type species: *Epidontaster pentagonalis* Koehler, 1920.
A synonym of *Odontaster* Verrill, 1880 according to Fisher (1940).

pentagonalis Koehler, 1920: 235. A synonym of *Odontaster meridionalis* (E.

A. Smith, 1876) according to Fisher (1940).

EURYGONIAS Farquhar, 1913

Farquhar 1913: 213; Fell 1959: 134.

Type species: *Eurygonias hylacanthus* Farquhar, 1913.

hylacanthus Farquhar, 1913

Farquhar 1913: 213; Fell 1959: 13; H.E.S. Clark 1970b: 3.

Range: Cook Strait and North and South Islands, New Zealand, 3-37 m.

GNATHASTER Sladen, 1889

Sladen 1889: 285.

Type species: *Astrogonium meridionalis* E. A. Smith, 1876.

A synonym of *Odontaster* Verrill, 1880 according to Fisher (1911).

dilatatus Sladen, 1889, see *Diplodontias*

elegans (Koehler, 1912) Koehler, 1920: 297. A synonym of *Odontaster meridionalis* (E. A. Smith, 1876) according to Fisher (1911).

elongatus Sladen, 1889, see *Acodontaster*

grayi (Bell, 1881a) Sladen, 1889: 286. A forma of *Odontaster penicillatus* (Philippi, 1870) according to Fisher (1940).

mediterraneus Marenzeller, 1893, see *Odontaster*

meridionalis: Sladen, 1889, see *Odontaster*

miliaris: Sladen, 1889, see *Diplodontias*

miliaris: Hutton, 1872: 7, non *Astrogonium miliare* Gray, 1847, = *Goniodon* [i.e. *Diplodontias*] *dilatatus* (Perrier, 1875) according to Farquhar (1907).

paxillosum: Sladen, 1889, see *Odontaster*

pedicellaris (Perrier, 1891) Perrier, 1894: 244. A synonym of *Odontaster penicillatus* (Philippi, 1870) according to Fisher (1940).

penicillatus: Koehler, 1920, see *Odontaster*

pilulatus Sladen, 1889: 292. A synonym of *Odontaster penicillatus* (Philippi, 1870) according to Fisher (1940).

rugosus (Hutton, 1872)(as *Pentaceros)* Farquhar, 1897: 194; 1898: 311. A synonym of *Asterodon* [i.e. *Diplodontias*] *miliaris* (Gray, 1847) according to Farquhar (1907).

singularis: Sladen, 1889, see *Diplodontias*

tenuis: Koehler, 1920: 232. A forma of *Odontaster validus* Koehler, 1906 according to Fisher (1940).

validus: Koehler, 1920, see *Odontaster.* (But *G. validus:* Koehler, 1923 probably another species according to Bernasconi (1956).

verrucosus: Sladen, 1889, see *Cycethra* (Ganeriidae)

GNATHODON Verrill, 1899: 203, lapsus for *Odontaster* according to Fisher (1940)

GONIODON Perrier, 1894

Perrier 1894: 244.

Type species: *Pentagonaster dilatatus* Perrier, 1875.

Invalidated as a junior homonym of *Goniodon* Herrick, 1888 (Mollusca) but replaceable by *Diplodontias* Fisher, 1908.

angustus Koehler, 1911b: 9. A synonym of *Diplodontias dilatatus* (Perrier, 1875) according to Mortensen (1925).

dilatatus: Perrier, 1894, see *Diplodontias*

GYMNOGNATHASTER Döderlein, 1928

Döderlein 1928: 297.

Type species: *Gymnognathaster gaussae* Döderlein, 1928.

A synonym of *Odontaster* Verrill, 1880 according to Fisher (1940).

gaussae Döderlein, 1928: 297. Probably a synonym of *Odontaster meridionalis* (E. A. Smith, 1876) according to Fisher (1940); certainly an *Odontaster*.

HEURESASTER Bell, 1908

Bell 1908: 8.

Type species: *Heuresaster hodgsoni* Bell, 1908.

A synonym of *Acodontaster* Verrill, 1899 according to Fisher (1940).

hodgsoni Bell, 1908, see *Acodontaster*

HOPLASTER Perrier, 1882

Perrier in Milne Edwards 1882: 48; 1894: 323; Verrill 1899: 197; Mortensen 1927: 77, 439; McKnight 1973b: 227.

Type species: *Hoplaster spinosus* Perrier in Milne Edwards, 1882.

kupe McKnight, 1973

McKnight 1973b: 227.

Range: NW of New Zealand (c.37°S, 170.5°E), 1995-2010 m.

lepidus (Sladen, 1889) Verrill, 1899: 198. Probably a synonym of *Hoplaster spinosus* Perrier, 1882 according to Verrill (1899), confirmed by A.M.C. in Gage et al. (1983).

spinosus Perrier, 1882 (with synonym *Pentagonaster lepidus* Sladen, 1889)

Perrier in Milne Edwards 1882: 48; 1894: 324; Verrill 1899: 197; Mortensen 1927: 77, 439; Gage et al. 1983: 278.

Range: Bay of Biscay* N to the Rockall Trough and W to the Azores (also unpublished records from the Porcupine Seabight, SW of Ireland, Morocco and W of Cape Province, South Africa, identified by A.M.C.), 1795-3310 m.

METADONTASTER Koehler, 1920

Koehler 1920: 191.

Type species: *Metadontaster waitei* Koehler, 1920.

A synonym of *Acodontaster* Verrill, 1899 according to Fisher (1940).

waitei Koehler, 1920, see *Acodontaster*

ODONTASTER Verrill, 1880 (with synonyms *Gnathaster* Sladen, 1889, *Gnathodon* Verrill, 1899, *Peridontaster* Koehler, 1920, *Epidontaster* Koehler, 1920 and *Gymnognathaster* Döderlein, 1927).

Verrill 1880b: 402; Bell 1893c: 260; 1899: 204; Fisher 1911c: 154; Verrill

1914a: 303; 1915: 118; Fisher 1940: 99; Bernasconi 1962b: 33; Downey 1973: 44.

Type species: *Odontaster hispidus* Verrill, 1880.

aucklandensis McKnight, 1973

McKnight 1973b: 225.

Range: Auckland Is, S of New Zealand, 216 m.

australis H.L. Clark, 1926

H.L. Clark 1926a: 16; Mortensen 1933a: 246; A.M. Clark & Courtman-Stock 1976: 58.

Range: W coast of South Africa, 243-366 m.

belli (Studer, 1884) Bell, 1893c: 260. A synonym of *Odontaster penicillatus* (Philippi, 1870) according to Madsen (1956).

benhami (Mortensen, 1925)

Mortensen 1925a: 288 (nom nov. for *O. grayi:* Benham, 1909); Fell 1952: 7; 1958: 7 (as *Peridontaster*).

McKnight 1967a: 299; H.E.S. Clark 1970b: 18 (as Odontaster).

Range: Cook Strait, New Zealand S to Auckland Is and E to Chatham Is, 37-555 m.

capitatus Koehler, 1912a: 156. A forma of *Acodontaster elongatus* (Sladen, 1889) according to Fisher (1940).

crassus Fisher, 1905

Fisher 1905: 302; 1911c: 154; Ziesenhenne 1937: 214.

Range: mid- to Lower California, 168-444 m.

cremeus Ludwig, 1903: 21. A synonym of *Acodontaster elongatus* (Sladen, 1889) according to Fisher (1940) but possibly valid according to A.M. Clark (1962a).

elegans Koehler, 1912a: 157. A synonym of *Odontaster meridionalis* (E. A. Smith, 1876) according to Fisher (1940).

glaber Barattini, 1938: 22. A synonym of *Acodontaster elongatus granuliferus* (Koehler, 1912) according to Fisher (1940).

granuliferus Koehler, 1912b: 77. A subspecies of *Acodontaster elongatus* (Sladen, 1889) according to Bernasconi (1964).

grayi: Bell, 1893c: 261. A forma of *Odontaster penicillatus* (Philippi, 1870) according to Fisher (1940).

grayi: Benham, 1909: 7, non *Odontaster grayi* (Bell, 1881), = *Peridontaster* [i.e. *Odontaster] benhami* Mortensen, 1925 nom. nov.

hispidus Verrill, 1880

Verrill 1880b: 402; 1894: 263; Marenzeller 1895: 131; Verrill 1899: 206; 1915: 119; Fisher 1928b: 489; H.L. Clark 1941: 30; Gray, Downey & Cerame-Vivas 1968: 148; Downey 1973: 45.

Range: George's Bank, NE of Cape Cod to the Florida Strait, 50-1160 m.

mediterraneus (Marenzeller, 1893) *(mediterraneus* Marenzeller, 1891: 443 a nomen nudum)

Marenzeller 1893a: 68; 1893b: 6 (as *Gnathaster*).

Marenzeller 1895: 7; Ludwig 1897: 125; Koehler 1909a: 83; 1924: 183; Mortensen 1927: 77; Koehler 1930 (13) pl.; Tortonese 1956: 189; Gautier-Michaz 1957: 80; Tortonese 1965: 152; Sibuet 1974: 791; Tortonese 1983: 25 (as *Odontaster*).

Range: Mediterranean, ? also W of Ireland, 100-1800 m.

meridionalis (E. A. Smith, 1876) (with synonyms *Odontaster elegans* Koehler, 1912, *Epidontaster pentagonalis* Koehler, 1920 and *?Gymnognathaster gaussae* Döderlein, 1928).

E. A. Smith 1876: 109 (as *Astrogonium*).

E. A. Smith 1879: 276 (as *Pentagonaster*).

Sladen 1889: 287 (as *Gnathaster*).

Fisher 1940: 99; A.M. Clark 1962a: 15; H.E.S. Clark 1963b: 34; Guille 1974: 32 (as *Odontaster*).

Range: antarctic Southern Ocean, N to South Georgia, Kerguelen and Marion Id., 0-646 m.

meridionalis: Leipoldt, 1895: 620, also Meissner, 1896: 93 non *Odontaster meridionalis* (E. A. Smith, 1876), = *Odontaster penicillatus* (Philippi, 1870) according to Fisher (1940) and Madsen (1956).

*paxillosus (Gray, 1847), see under *Diplodontias*

pedicellaris (Perrier, 1891a) Bell, 1893c: 262. A synonym of *Odontaster penicillatus* (Philippi, 1870) according to Fisher (1940).

penicillatus (Philippi, 1870) (with synonyms *Pentagonaster belli* Studer, 1884, *Gnathaster pilulatus* Sladen, 1889 and *Asterodon pedicellaris* Perrier, 1891, also forma *grayi* (Bell, 1881)).

Philippi 1870: 268 (as *Goniodiscus*).

Meissner 1898: 394; Ludwig 1905: 42; Fisher 1940: 105; Madsen 1956: 19; Bernasconi 1962b: 34; 1966: 162; 1973: 295; Leeling 1985: 529 (as *Odontaster*).

Range: S Chile*, Falkland-Magellan region and N to Rio de la Plata, 8-350 m. (forma *grayi* from Magellan Strait).

propinquus A.H. Clark, 1917a: 7. A synonym of *Acodontaster elongatus granuliferus* (Koehler, 1912) according to Fisher (1940).

*pusillus Koehler, 1907

Koehler 1907a: 143; 1908a: 540; Fisher 1940: 99; A.M. Clark 1962a: 14. Koehler 1920: 190 (as *Peridontaster*).

Range: Southern Ocean WNW from Bouvet Id, 3120 m. Holotype immature, identity uncertain according to Koehler (1920).

robustus Verrill, 1899

Verrill 1899: 209.

Range: E from New Jersey, U.S.A., 600-673 m.

setosus Verrill, 1899

Verrill 1899: 207; 1915: 120; Downey 1973: 45.

Range: S of Cape Cod to Florida and the northern Gulf of Mexico, 102-730 m.

singularis: Bell, 1893c: 262, see *Diplodontias*

tenuis Koehler, 1906c: 8; 1920: 232. A forma of *Odontaster validus* Koehler, 1906 according to Fisher (1940).

validus Koehler, 1906 (with forma *tenuis* Koehler, 1906)

Koehler 1906c: 6; 1908a: 540; 1912b: 68; Fisher 1940: 101; A.M. Clark 1962a: 15; H.E.S. Clark 1963b: 35; Pearse 1965: 39; Bernasconi 1967b: 7.

Koehler 1920: 228 (as *Gnathaster*).

Range: circumpolar antarctic, N to South Georgia and Bouvet Id., 0-653 m.

PERIDONTASTER Koehler, 1920

Koehler 1920: 192,194.

Type species: *Calliderma grayi* Bell, 1881.

A synonym of *Odontaster* Verrill, 1880 according to Fisher (1940).

benhami Mortensen, 1925, see *Odontaster*

crassus: Koehler, 1920, see *Odontaster*

grayi: Koehler, 1920: 194; 1923: 86. A forma of *Odontaster penicillatus* (Philippi, 1870) according to Fisher (1940).

pusillus: Koehler, 1920, see *Odontaster*

PSEUDONTASTER Koehler, 1912

Koehler 1912a: 157; 1912b: 85.

Type species: *Pseudontaster marginatus* Koehler, 1912.

A synonym of *Acodontaster* Verrill, 1899 according to Fisher (1940).

conspicuus Koehler, 1920, see *Acodontaster*

marginatus Koehler, 1912, see *Acodontaster*

moderatus Koehler, 1923: 89. A synonym of *Acodontaster hodgsoni* (Bell, 1908) according to A.M. Clark (1962).

stellatus Koehler, 1920: 210. A forma of *Acodontaster hodgsoni* (Bell, 1908) according to A.M. Clark (1962).

TRIDONTASTER Koehler, 1920

Koehler 1920: 191.

Type species: *Tridontaster laseroni* Koehler, 1920.

A synonym of *Acodontaster* Verrill, 1899 according to Fisher (1940).

laseroni Koehler, 1920: 214. A synonym of *Acodontaster hodgsoni* (Bell, 1908) according to Fisher (1940).

Family GANERIIDAE Sladen

Asterinidae (Ganeriinae) Sladen 1889: xxxiv, 375.

Ganeriidae: Perrier 1893: 849; 1894: 171; Fisher 1911c: 251; Verrill 1914a: 365; A.H. Clark 1939b: 497; Fisher 1940: 127; A.M. Clark 1962a: 23;

Bernasconi 1964a: 59; Spencer & Wright 1966: U69; Blake 1981: 380; A.M. Clark 1983: 359-364 (reviewed and rediagnosed).

Genus-group names: **Aleutiaster**, Cribellopsis, Cribraster, Cryaster, **Cycethra, Ganeria, Hyalinothrix, Knightaster**, Lebrunaster, [*Leilaster* see Leilasteridae], [Magdalenaster], **Perknaster, Scotiaster, Tarachaster, Vemaster**.

ALEUTIASTER A.H. Clark 1939
A.H. Clark 1939b: 497.
Type species: *Aleutiaster schefferi* A.H. Clark, 1939.
schefferi A.H. Clark, 1939
A.H. Clark 1939b: 498; Baranova 1957: 166.
Range: Aleutian Is, 2-12 m.
CRIBELLOPSIS Koehler 1917
Koehler 1917: 36.
Type species: *Cribellopsis rallieri* Koehler, 1917.
A synonym of *Perknaster* Sladen 1889 according to Fisher (1940).
rallieri Koehler, 1917: 36. A synonym of *Perknaster fuscus* Sladen, 1889 according to A.M. Clark (1962).
CRIBRASTER Perrier, 1891 *(Cribraster* Perrier, 1888: 675 a nomen nudum).
Perrier 1891a: K104.
Type species: *Cribraster sladeni* Perrier, 1891.
A synonym of *Perknaster* Sladen, 1889 according to Fisher (1940).
sladeni Perrier, 1891, see *Perknaster*
CRYASTER Koehler, 1906
Koehler 1906c: 24; 1920: 131.
Type species: *Cryaster antarcticus* Koehler, 1906. A synonym of *Perknaster* Sladen, 1889 according to Fisher (1940).
antarcticus Koehler, 1906c, see *Perknaster*
aurorae Koehler, 1920, see *Perknaster*
brachyactis H.L. Clark, 1923, see *Spoladaster* (Poraniidae)
charcoti Koehler, 1912a, see *Perknaster*
CYCETHRA Bell, 1881 (with synonym *Lebrunaster* Perrier, 1891)
Bell, 1881a: 96; Sladen 1889: 376; Perrier 1891a: K170; Fisher 1940: 128; A. M. Clark 1962a: 24; Bernasconi 1964a: 65; 1970: 239.
Type species: *Cycethra simplex* Bell, 1881, a synonym of *Goniodiscus verrucosus* Philippi, 1857.
A monotypic genus according to A.M. Clark (1983), though without specifying full synonymy of *Cycethra verrucosa* (Philippi).
cingulata Koehler, 1923: 68; also Madsen, 1956: 21; Bernasconi 1964a: 67; 1973: 303. Conspecific with *Cycethra verrucosa* (Philippi, 1857) according to A. M. Clark (1983) [by inference], though thought to be distinct by

Madsen (1956) [and probably recognisable as at least a forma].

electilis Sladen, 1889: 377. A forma of *Cycethra verrucosa* (Philippi, 1857) according to Fisher (1940) but not used by Bernasconi (1964) [so better treated as a synonym].

lahillei de Loriol, 1904: 73; also Koehler 1923: 68; Madsen 1956: 21; Bernasconi 1964a: 77. Conspecific with *Cycethra verrucosa* (Philippi, 1857) according to A. M. Clark (1962) and [by inference] 1983, though meantime treated as a valid species by Bernasconi (1964) [and probably recognisable as at least a forma].

*macquariensis Koehler, 1920

Koehler 1920: 139; A. M. Clark 1962a: 24.

Range: Macquarie Id., Southern Ocean.

Affinities with *Asterina hamiltoni* Koehler, 1920 and *A. frigida* Koehler, 1917 (Asterinidae) according to A. M. Clark (1962a).

nitida Sladen, 1889: 379; also Bernasconi 1964a: 75; 1973: 305. A forma of *Cycethra verrucosa* (Philippi, 1857) according to Fisher (1940) and [by inference] A. M. Clark (1983), though meantime treated as a valid species by Bernasconi (1964, 1973) [and probably recognisable as at least a forma].

pinguis Sladen, 1889: 380. A forma of *Cycethra verrucosa* (Philippi, 1857) according to Fisher (1940) and [by inference] A. M. Clark (1983), though meantime treated as a valid species by Bernasconi (1964) [and probably recognisable as at least a forma].

simplex Bell, 1881a: 96; also Perrier 1891a: K122, 170 (with names of additional forms not subsequently recognised); Meissner 1896: 95; 1904: 14; de Loriol 1904: 21. A synonym of *Cycethra verrucosa* (Philippi, 1857) according to Meissner (1904).

verrucosa (Philippi, 1857) (with synonyms *Cycethra simplex* Bell, 1881, *C. electilis* Sladen, 1889, *C. simplex* forma *subelectilis* Perrier, 1891, *Lebrunaster paxillosus* Perrier, 1891 and *Parasterina obesa* H.L. Clark, 1910, also subspecies *Cycethra verrucosa mawsoni* A.M. Clark (1962) and formae *nitida* Sladen, 1889, *pinguis* Sladen, 1889, *lahillei* de Loriol, 1904 and *cingulata* Koehler, 1923).

verrucosa verrucosa (Philippi, 1857)

Philippi 1857: 132 (as *Goniodiscus*).

Perrier 1878: 84 (as *Pentagonaster*).

Sladen 1889: 750 (as *?Gnathaster*).

Meissner 1898: 394; 1904: 14; Ludwig 1905: 53; Koehler 1912b: 64; 1923: 60; Fisher 1940: 129; A. M. Clark 1962a: 27; Bernasconi 1964a: 71; 1966: 164; A. M. Clark 1983: 364 (as *Cycethra*).

H.L. Clark 1910: 331 (as *Tosia*).

Range: southern Chile, c.33°S*, the Falkland-Magellan area and N to southern Brazil (Bernasconi, 1973, no details), 0-279 m.

200

verrucosa mawsoni A. M. Clark, 1962

A. M. Clark 1962a: 25; Bernasconi 1967b: 8.

Range: off MacRobertson Land, Antarctica (c.66°S, 62°E)*, Enderby and Weddell Quadrants of Southern Ocean N to South Georgia, 60-540 m.

verrucosa: Bell, 1908: 10, non *Cycethra verrucosa* (Philippi, 1857), = *Odontaster validus* Koehler, 1906 (Odontasteridae) according to Fisher (1940).

verrucosa: Döderlein, 1928: 296, non *Cycethra verrucosa* (Philippi, 1857), probably = *Asterina frigida* Koehler, 1917 (Asterinidae) according to A. M. Clark (1962a).

verrucosa: Grieg, 1929: 6, non *Cycethra verrucosa* (Philippi, 1857), = *Kampylaster incurvatus* Koehler, 1920 (Asterinidae) according to Madsen (1955).

GANERIA Gray, 1847

Gray 1847: 83; Perrier 1875: 327 [1876: 247]; Sladen 1889: 382; Perrier 1891a: K118; Verrill 1914a: 365; Bernasconi 1964a: 69.

Type species: *Ganeria falklandica* Gray, 1847.

attenuata Koehler, 1907

Koehler 1907a: 144; 1908a: 547.

Range: E of South Orkney Is, Weddell Quadrant* and Bouvet Id, Southern Ocean, 3250 and 200 m.

falklandica Gray, 1847 (with synonyms *Ganeria robusta* Perrier, 1891 and *G. papillosa* Perrier, 1891)

Gray 1847: 83; Perrier 1875: 327 [1876: 247]; Sladen 1889: 383; Meissner 1896: 94; Fisher 1940: 127; Bernasconi 1964a: 61; 1973: 305.

Range: Falkland Is*, N to Uruguay and W to southern Chile, 0-137 m.

hahni Perrier, 1891

Perrier 1891a: K118; Bernasconi 1964a: 64; 1973: 306; Hernandez & Tablado, 1985: 5.

Range: Magellan area*, S Chile and Argentina N to c.45°S, 0[?]-135 m.

A forma of *Ganeria falklandica* Gray, 1847 according to Fisher (1940) but a valid species according to Bernasconi (1964).

papillosa Perrier, 1891a: K121. A synonym of *Ganeria falklandica* Gray, 1847 according to Fisher (1940).

robusta Perrier, 1891a: K119. A synonym of *Ganeria falklandica* Gray, 1847 according to Fisher (1940).

HYALINOTHRIX Fisher, 1911

Fisher 1911b: 659.

Type species: *Hyalinothrix millespina* Fisher, 1911.

millespina Fisher, 1911

Fisher 1911b: 661.

Range: Hawaiian Is, 232-281 m.

KNIGHTASTER H. E. S. Clark, 1972

H. E. S. Clark 1972: 147.

Type species: *Knightaster bakeri* H. E. S. Clark, 1972.

bakeri H. E. S. Clark, 1972

H. E. S. Clark 1972: 147.

Range: northern New Zealand, 30-54 m.

LEBRUNASTER Perrier, 1891 *(Lebrunaster* Perrier, 1888: 765 a nomen nudum, no species named)

Perrier, 1891a: K116.

Type species: *Lebrunaster paxillosus* Perrier, 1891.

A synonym of *Cycethra* Bell, 1881 according to Fisher (1940).

paxillosus Perrier, 1891a: K116. A synonym of *Cycethra verrucosa* (Philippi, 1857) according to Fisher (1940).

LEILASTER A. H. Clark, 1938, see Leilasteridae

MAGDALENASTER Koehler, 1907

Koehler 1907b: 18; 1909a: 104.

Type species: *Magdalenaster arcticus* Koehler, 1907.

A synonym of *Henricia* Gray, 1840 (Echinasteridae) according to Mortensen (1927).

arcticus Koehler, 1907b: 19; 1909a: 105. A synonym of *Henricia sanguinolenta* (O. F. Müller, 1776) (Echinasteridae) according to Mortensen (1927).

PERKNASTER Sladen, 1889 (with synonyms *Cribraster* Perrier, 1891, *Cryaster* Koehler, 1906 and *Cribellopsis* Koehler, 1917)

Sladen 1889: 550; Perrier 1891a: K161(pt); Koehler 1912b: 39; Fisher 1940: 133; A. M. Clark 1962a: 28; H. E. S. Clark 1963b: 50; Bernasconi 1964a: 79.

Type species: *Perknaster fuscus* Sladen, 1889.

antarcticus Koehler, 1906

Koehler 1906c: 24; 1920: 126 (as *Cryaster*).

Fisher 1940: 136; H. E. S. Clark 1963b: 50; Bernasconi 1967: 9 (as *Perknaster fuscus antarcticus*).

A. M. Clark 1962a: 29 (as *Perknaster antarcticus*).

Range: Ross, Victoria and Weddell Quadrants of Southern Ocean, 11-457 m.

Maintained as subspecies of *Perknaster fuscus* Sladen, 1889 by Bernasconi (1967) but without reference to A. M. C. (1962).

aurantiacus Koehler, 1912

Koehler 1912a: 153; 1912b: 36; 1923: 73; Fisher 1940: 142; A. M. Clark 1962a: 28.

Range: Graham Land* and South Georgia area, 75-254 m.

aurorae (Koehler, 1920)

Koehler 1920: 120 (as *Cryaster*).

Fisher 1940: 146; A. M. Clark 1962a: 31 (as *Perknaster*).

Range: Adelie and Queen Mary Lands*, Enderby Land and Weddell Quadrant N to South Georgia and Shag Rocks, 45-310 m.

charcoti (Koehler, 1912)

Koehler 1912a: 152; 1912b: 33 (as *Cryaster*).

Fisher 1940: 144 (as *Perknaster*).

Range: Antarctic Peninsula* to South Georgia, 0-235 m.

densus Sladen, 1889 (with subspecies *patagonicus* Bernasconi, 1962)

densus densus Sladen, 1889

Sladen 1889: 552; Fisher 1940: 138; A.M. Clark 1962a: 32; H.E.S. Clark 1963b: 52.

Range: Kerguelen*, Marion Id, Ross Sea, 101-232 m.

densus patagonicus Bernasconi, 1962

Bernasconi 1962a: 257; 1964a: 81.

Range: Santa Cruz Province, southern Argentina.

fuscus Sladen, 1889 (with synonym *Cribellopsis rallieri* Koehler, 1917)

Sladen 1889: 551; Fisher 1940: 136; A.M. Clark 1962a: 30; Guille 1974: 343.

Range: Kerguelen* and Heard Id, 45-137 m.

georgianus Fisher, 1940. A subspecies of *Perknaster sladeni* (Perrier, 1891).

patagonicus Bernasconi, 1972. A subspecies of *Perknaster densus* Sladen, 1889.

sladeni (Perrier, 1891) (with subspecies *georgianus* Fisher, 1940)

sladeni sladeni (Perrier, 1891)

Perrier 1891a: K104; Koehler 1912b: 39 (as *Cribraster*).

Fisher 1940: 140; ?A.M. Clark 1962a: 31; H.E.S. Clark 1963b: 53 (as *Perknaster*).

Range: See below.

sladeni georgianus Fisher, 1940

Fisher 1940: 143; ?Bernasconi 1964a: 84.

Range: Falkland Is*, and probably southern Argentina for *P. sladeni sladeni* [but antarctic records including Palmer Archipelago, Ross Sea and MacRobertson Land with South Georgia probably apply to *P. sladeni georgianus*], 199-500 m.

SCOTIASTER Koehler, 1907

Koehler 1907a: 144; 1908a: 548.

Type species: *Scotiaster inornatus* Koehler, 1907.

inornatus Koehler, 1907

Koehler 1907a: 144; 1908a: 549.

Range: Southern Ocean between Bouvet and Gough Islands, 3850 m.

TARACHASTER Fisher, 1913

Fisher 1913c: 216; 1919a: 401.

Type species: *Tarachaster tenuis* Fisher, 1913.

tenuis Fisher, 1913

Fisher 1913c: 216; 1919a: 402.
Range: Philippine Is, 296 m.
VEMASTER Bernasconi, 1965
Bernasconi 1965: 333; A.M. Clark 1983: 362.
Type species: *Vemaster sudatlanticus* Bernasconi, 1965.
sudatlanticus Bernasconi, 1965
Bernasconi 1965: 334; 1966: 163.
Range: Argentine Basin, E from southern Argentina (also unpublished record from off Rio de la Plata, identified by A.M.C.), 5055-5208 m.

Family ASTERINIDAE Gray

Asterinidae Gray 1840: 228; Perrier 1875: 27 [292]; Viguier 1878: 205; Sladen 1889: xxxiii, 374; Fisher 1911c: 253; Verrill 1913: 477; 1914a: 262, 364; 1915: 56; H. L. Clark 1923a: 279; 1946: 128; Madsen 1956: 22; Bernasconi 1964b: 241; Spencer & Wright 1966: U68; Bernasconi 1973: 335; A.M. Clark & Courtman-Stock 1976: 75; Blake 1981: 380-391(pt); A.M. Clark 1983: 359-378 (rediagnosed and reviewed).

Genus-group names: **Allopatiria, Anseropoda, Asterina,** Asterinides, **Asterinopsis,** Asteriscus, **Callopatiria,** Carna, Ctenaster, **Desmopatiria, Disasterina,** Enoplopatiria, Habroporina, Henricides, **Kampylaster, Manasterina,** [*Mirastrella* see Leilasteridae], **Nepanthia,** Palmipes, **Paranepanthia,** Parasterina, Patiria, **Patiriella, Paxillasterina, Pseudasterina, Pseudonepanthia,** Socomia, **Stegnaster, Tegulaster, Tremaster.**

ALLOPATIRIA Verrill, 1913
Verrill 1913: 480; Tortonese 1965: 174; A.M. Clark 1983: 372.
Type species: *Patiria ocellifera* Gray.
ocellifera (Gray, 1847) (with synonyms *Patiria pulla* Koehler & Vaney, 1906, *P. rosea* Koehler & Vaney, 1906 and *Parasterina africana* Engel & Croes, 1960).
Gray 1847: 82; Perrier 1875: 324 [1876: 244]; Koehler & Vaney 1906: 62; A. M. Clark 1963: 1 (as *Patiria)* [See Zeidler & Rowe 1986: 118].
Verrill 1913: 480; Tortonese 1963: 289; 1965: 174; 1983: 28; A.M. Clark 1983: 372; Jangoux & Massin 1986: 89 (as *Allopatiria).*
Range: Mauritania and Sahara, NW Africa and S of Spain and Italy, Mediterranean, 6-55 m.
ANSEROPODA Nardo, 1834 (with synonyms *Palmipes* L. Agassiz, 1835, *Asteriscus* Müller & Troschel, 1840 [following new type designation] and *Carna* Gistel, 1848).
Nardo 1834: 716; Bell 1891a: 245; Fisher 1906a: 1088; 1908a: 87; Koehler 1921: 33; 1924: 134; H.L. Clark 1946: 144; Tortonese 1965: 176;

A. M. Clark & Courtman-Stock 1976: 75.

Type species: *Asterias placenta* Pennant, 1777, by subsequent designation by Fisher (1906).

antarctica Fisher, 1940

Fisher 1940: 149.

Range: Clarence Id., off Antarctic Peninsula, 342 m.

aotearoa McKnight, 1973

McKnight 1973c: 12.

Range: W of South Island, New Zealand, 366 m.

diaphana (Sladen)

Sladen 1889: 395 (as *Palmipes*).

Fisher 1919a: 425; Rowe 1974: 214 (as *Anseropoda*).

Range: Admiralty Is, N of New Guinea, 272 m.

fisheri Aziz & Jangoux, 1985

Aziz & Jangoux 1985: 286.

Range: Philippine Is, 130-137 m.

grandis Mortensen, 1933

Mortensen 1933a: 251 (as *Anseropoda (Palmipes)*).

A. M. Clark & Courtman-Stock 1976: 75.

Range: W and S coasts of South Africa, 275-315 m.

habracantha H. L. Clark, 1923

H. L. Clark 1923a: 287; Mortensen 1933a: 251; A. M. Clark & Courtman-Stock 1976: 76.

Range: S coast of South Africa and Natal, 125-155 m.

insignis Fisher 1906

Fisher 1906a: 1088.

Koehler 1909b: 105 (as *Palmipes*).

Range: Hawaiian Is, 222-332 m.

lobianci: Koehler, 1924: 139; Tortonese 1965: 179, lapsus for

lobiancoi (Ludwig, 1897)(as *lobianci,* corrected Cherbonnier, 1970)

Ludwig 1897: 267 (as *Palmipes*).

Cherbonnier 1970b: 946 (as *Anseropoda*).

Range: Isle of Capri, SE Italy, 90-100 m (?S France [from aquarium]).

***ludovici** (Koehler, 1909) **new comb.**

Koehler 1909a: 100 (as *Palmipes*).

Range: S India, 186 m.

macropora Fisher 1913c

Fisher 1913c: 219; 1919a: 424; Aziz & Jangoux 1985: 287.

Range: Philippine Is, 45 m.

membranacea (Retzius, 1783) Nardo, 1834: 716. A synonym of *Anseropoda placenta* (Pennant, 1777) according to Bell (1891).

novemradiata (Bell, 1905)

Bell 1905: 248 (as *Palmipes*).

H.L. Clark 1923a: 287; 1926a: 18; A.M. Clark & Courtman-Stock 1976: 76 (as *Anseropoda*).

Range: Natal, South Africa, 46-88 m.

petaloides (Goto, 1914)

Goto 1914: 659 (as *Palmipes*).

Hayashi 1973b: 73 (as *Anseropoda*).

Range: SE Japan, 50-60 m.

***pellucida** (Alcock, 1893) **new comb.**

Alcock 1893a: 100 (as *Palmipes*).

Range: Andaman Sea, c.200 m.

placenta (Pennant, 1777)(with synonyms *Asterias membranacea* Retzius, 1783, *A. palmipes* Olivi, 1792 and *Asterias cartilaginea* Fleming, 1828)

Pennant 1777: 62 (as *Asterias*).

Lütken 1865: 143 (as *Asteriscus*).

Bell 1891a: 235; Fisher 1906a: 1088; Tortonese 1952a: 194; Zavodnik 1961: 50; Tortonese 1965: 176; Harvey et al. 1988: 161 (as *Anseropoda*).

Range: Dorset, S England*, Shetland Is, Scotland, W England and Ireland, Belgium, N France, S to Mediterranean and Sierra Leone, 10-600 m.

***rosacea** (Lamarck, 1816) (with synonym *Palmipes stokesi* Gray, 1840)

Lamarck 1816: 558 (as *Asterias*).

Müller & Troschel, 1842: 40 (as *Asteriscus*).

Fisher 1906a: 1089; H.L. Clark 1916: 59; 1946: 144; Endean 1956: 125; Chang & Liao 1964: 373 (as *Anseropoda*).

Koehler 1910a: 127 (as *Palmipes*).

Range: Bay of Bengal, East Indies and northern Australia.

A synonym of *Asterias luna* Linnaeus, 1758 'without any reasonable doubt' according to Verrill (1914b) but overlooked or ignored by other authors. If Verrill could be correct, then the long-forgotten name *Asterias luna* needs to be formally suppressed by the ICZN.

tenuis (Goto, 1914)

Goto 1914: 656 (as *Palmipes*).

Hayashi 1973b: 9, 14 (as *Anseropoda*).

Range: SE Japan.

ASTERINA Nardo, 1834 (with synonyms *Patiria* Gray, 1840, *Asterinides* Verrill, 1913 and *Enoplopatiria* Verrill, 1913, possibly also *Patiriella* Verrill, 1913).

Nardo 1834: 716; L. Agassiz 1836: 192; Forbes 1839: 7; Perrier 1875: 294 [1876: 214]; Viguier 1878: 207; Fisher 1911c: 254; Verrill 1913: 481; Koehler 1924: 130; Mortensen 1927: 98; 1933a: 258; H.L. Clark 1946: 129; Tortonese 1965: 168; A.M. Clark 1983: 365, 378.

Type species: *Asterias minuta:* Nardo, 1834 = *Asterias gibbosa* Pennant,

1777, by subsequent designation by Fisher, 1906 (see also A.M.C., 1983: 378).

agustincasoi Caso, 1978
Caso 1978: 213.
Range: Ixtapa Id, W Mexico.

alba H.L. Clark, 1938
H.L. Clark 1938: 150; 1946: 132.
Range: Lord Howe Id, E Australia.

alutaceus Philippi in Quijada, 1911; a NOMEN NUDUM according to Madsen (1956).

anomala H.L. Clark, 1921
H.L. Clark 1921: 95; 1938: 143; Ely 1942: 25; A.H. Clark 1952: 289; McKnight 1972: 38; Marsh 1974: 92; 1977: 274; Oguro 1983: 222.
Range: Lord Howe Id, E Australia*, N and E to Caroline Is, Hawaiian Is and Polynesia.

atyphoida H.L. Clark, 1916
H.L. Clark 1916: 57; 1928: 389; A.M. Clark 1966: 324; Shepherd 1968: 745; Dartnall 1970a: 73; 1970b: 19.
Range: Victoria, N Tasmania and South Australia, 30-40 m.

aucklandensis Koehler, 1920
Koehler 1920: 135; Fell 1953: 88; 1959: 138.
Mortensen 1925a: 300 (as *Asterina (Asterinopsis)*).
Range: southern New Zealand, Auckland and Campbell Is.

batheri Goto, 1914
Goto 1914: 651; Hayashi 1938a: 116; 1940: 119; 1973b: 71; Chang & Liao 1964: 63.
Range: mid- and S Japan, E China Sea.

belcheri Perrier, 1875, see *Nepanthia*

bispinosa Perrier, 1891a. A variety of *Asterina fimbriata* Perrier, 1875.

borealis Verrill, 1878: 213. A synonym of *Poraniomorpha hispida* (M. Sars, 1872) according to Koehler (1909a), possibly of *P. hispida rosea* Danielssen & Koren, 1881 (Poraniidae) according to A.M. Clark (1984).

brevis Perrier, 1875: 321 [1876: 241]. A synonym of *Nepanthia belcheri* (Perrier, 1875) according to Rowe & Marsh (1982).

burtoni Gray, 1840 (with synonym *Asteriscus wega* Perrier, 1869, also subspecies *cepheus* (Müller & Troschel, 1842) and variety *iranica* Mortensen, 1941).

burtoni burtoni Gray, 1840
Gray 1840: 289; G.A. Smith 1927: 641; H.L. Clark 1946: 133; Tortonese 1966: 3; A.M. Clark 1967c: 146; Achituv 1969: 341; A.M.C. in Clark & Rowe 1971: 68; James & Pearse 1971: 84; Jangoux 1973a: 35; Sloan et al. 1979: 98; Price 1983: 47.
Verrill 1913: 482 (as *Asterinides*).

Range: Red Sea*, Indian Ocean, also easternmost Mediterranean; var. *iranica* from Arabian (Iranian) Gulf.

burtoni cepheus (Müller & Troschel, 1842)

Müller & Troschel, 1842: 41 (as *Asteriscus cepheus*).

Martens 1866: 85 (as *Asterina*).

A.M.C. in Clark & Rowe 1971: 69; Liao 1980: 161 (as *Asterina burtoni cepheus*).

Range: Java, Indonesia* to South China Sea, Wake Id, northern Australia.

cabbalistica Lütken, 1871: 242. A synonym of *Patiriella regularis* Verrill, 1870 according to Perrier (1875).

calcar: Gray, 1840, see *Patiriella*

calcarata: Perrier, 1875, see *Patiriella*

calcarata: Quijada, 1911: 162, non *Asteriscus calcarata* Perrier, 1869, = *Patiriella chilensis* (Lütken, 1859) according to Madsen (1956).

calcarata: Koehler, 1908: 296, non *Asteriscus calcarata* Perrier, 1869, = *Patiriella exigua* (Lamarck, 1816) according to Mortensen (1933).

cepheus: Martens, 1866: 85. A subspecies of *Asterina burtoni* Gray according to A.M.C. in Clark & Rowe (1971).

chilensis Lütken, 1859 (with synonym *Patiria gayi* Perrier, 1875)

Lütken 1859: 61; 1871: 302; Perrier 1875: 302 [1876: 222]; A.M. Clark 1983: 367.

Verrill 1870: 334; 1913: 482; Madsen 1956: 23 (as *Patiria*).

Range: Chile to Peru.

coccinea (Gray, 1840) Perrier, 1875: 314. A synonym of *Patiria* [i.e. *Asterina*] *miniata* (Brandt, 1835) according to Mortensen (1933a).

coccinea: Bell, 1905: 248, non *Patiria coccinea* Gray, 1840, = *Patiria bellula* Sladen, 1889 [i.e. *Callopatiria granifera* (Gray, 1847)] according to Mortensen (1933a).

corallicola Marsh, 1977

Marsh 1977: 271; Oguro 1983: 224.

Range: Caroline Is.

coronata Martens, 1866 (with subspecies *cristata* Fisher 1916, *euerces* Fisher, 1917 and *fascicularis* Fisher, 1918, also forma *japonica* Hayashi 1940).

coronata coronata Martens, 1866

Martens 1866: 73; Fisher 1918: 110; 1919a: 414; H.L. Clark 1946: 132; A.H. Clark 1949a: 76.

Range: Amboina*, W to Sri Lanka, N to the S China Sea, E to Polynesia and S to northern Australia; forma *japonica* from S Japan.

coronata cristata Fisher, 1916

Fisher 1916b: 27 (as species).

Fisher 1918: 111; 1919a: 411; A.H. Clark 1952: 289 (as subspecies).

Range: Caroline Is.
coronata euerces Fisher, 1917
Fisher 1917c: 91; 1918: 110; 1919a: 414.
Range: Palawan, Philippine Is.
coronata fascicularis Fisher, 1918
Fisher 1918: 110; H.L. Clark 1928: 390.
Range: Philippine Is, 'Northern Territory', Australia.
coronata: Jangoux, 1973a: 38, non *Asterina coronata* Martens, 1866, ? = *A. burtoni* Gray, 1840 according to A.M.C. in Clark & Courtman-Stock (1976).
crassispina H.L. Clark, 1928
H.L. Clark 1928: 390.
Range: Australia, ?N coast.
cristata Fisher, 1916. Reduced to subspecies of *Asterina coronata* Martens, 1866 by Fisher (1918).
dyscrita H.L. Clark, 1923, see *Patiriella*
euerces Fisher, 1917c. A subspecies of *Asterina coronata* Martens, 1866.
exigua: Perrier, 1875, see *Patiriella*
fascicularis Fisher, 1918. A subspecies of *Asterina coronata* Martens, 1866.
fimbriata Perrier, 1875 (with synonym *Asterina perrieri* de Loriol, 1904 and variety *bispinosa* Perrier, 1891).
Perrier 1875: 307 [1876: 227]; Ludwig 1905: 59; Koehler 1923: 55; Jangoux 1985b: 29 (as *Asterina*).
Verrill 1913: 484; Fisher 1931: 5; 1940: 148; Madsen 1956: 26; Hernandez & Tablado 1985: 3 (as *Patiriella*).
Range: S Chile*, S to Falkland-Magellan area and S Argentina, 0-250 m; ?Graham Land, Antarctic Peninsula, 74-95 m.
fimbriata: Benham, 1909: 295, non *Asterina fimbriata* Perrier, 1875, = *Asterina aucklandensis* Koehler, 1920 according to Mortensen (1925a).
folium (Lütken, 1859) (including '*Asterina minuta* var. 1' Gray, 1840)
Lütken 1859: 60 (as *Asteriscus*).
A. Agassiz 1877: 106; H.L. Clark 1898: 6; 1899: 130; Verrill 1907: 281; H.L. Clark 1933: 26; Ummels 1963: 89; A.M. Clark 1983: 364 (as *Asterina*).
Verrill 1913: 479; 1915: 58; Brito 1968: 17 (as *Asterinides*).
Range: Virgin Is*, Bermuda, the Bahamas and Florida S and E to South Trinidad Id, E of Brazil, 0-15 m.
frigida Koehler, 1917
Koehler 1917: 46; A.M. Clark 1962a: 33; Guille 1974: 34.
Range: Kerguelen, 0-20 m.
gayi Perrier, 1875: 305 [1876: 225]. A synonym of *Asterina chilensis* Lütken, 1859 according to Koehler (1920).
gibbosa (Pennant, 1777) (with synonyms *Asterias verruculata* Retzius,

1805, *Asteriscus ciliatus* Lorenz, 1860, *A. pulchellus* Perrier, 1869 and *A. arrecifensis* Greeff, 1872).

Pennant 1777: 62; Fleming 1828: 486 (as *Asterias*).

Forbes 1839: 119; 1841: 119; Norman 1865: 121; Perrier 1875: 295 [1876: 215]; Ludwig 1897: 207; Fisher 1906a: 1087; Koehler 1924: 131; Mortensen 1927: 98; Emson & Crump 1979: 81; A.M. Clark 1983: 365 (as *Asterina*).

Range: western Scotland S to the Mediterranean, Canary Is and Azores, intertidal – 125 m.

gibbosa: Martens, 1866: 72, non *Asterina gibbosa* (Pennant, 1777), = *A. cepheus* [i.e. *A. burtoni cepheus*] (Müller & Troschel, 1842) according to Perrier (1875).

gracilispina H.L. Clark, 1923

H.L. Clark 1923a: 286; Mortensen 1933a: 255; A.M. Clark 1974: 437; A.M. Clark & Courtman-Stock 1976: 77.

Range: East London, S coast of South Africa*, W to False Bay, 25-86 m.

granifera: H.L. Clark, 1923a, also Mortensen, 1933a, see *Callopatiria*

granulatus Philippi in Quijada, 1911. A NOMEN NUDUM according to Madsen (1956).

granulosa Perrier 1875: 312 [1876: 232]. A synonym of *Patiria miniata* (Brandt, 1835) according to A.M. Clark (1983).

gunni Gray, 1840, see *Patiriella*

hamiltoni Koehler, 1920

Koehler 1920: 133.

Range: Macquarie Id, Southern Ocean.

hartmeyeri Döderlein, 1910 (including '*Asterina minuta* var. 2' Gray, 1840).

Döderlein in Döderlein & Hartmeyer 1910: 154; H.L. Clark 1933: 27; Fontaine 1953: 183; Ummels 1963: 92.

Range: Barbados* to N coast of Jamaica and islands of southern Caribbean, 0-1m.

heteractis H.L. Clark, 1938

H.L. Clark 1938: 152; Marsh 1977: 274.

Range: Lord Howe Id, E Australia.

*hexabrachia crispa and hexabrachia tripora Ping, 1930: 133. NOMINA NUDA. [Provisional non-binomial names with only the briefest 'indications' in lieu of descriptions.]

inopinata Livingstone, 1933

Livingstone 1933: 3; Dartnall 1970a: 73.

Range: N.S.W. to Tasmania, Australia.

iranica Mortensen, 1941, a variety of *Asterina cepheus* (Müller & Troschel, 1842) [i.e. *A. burtoni* Gray].

japonica Hayashi, 1940. A forma of *Asterina coronata* Martens, 1866.

kraussi Gray, 1840: 289. A synonym of *Asterina* [i.e. *Patiriella] exigua* (Lamarck, 1816) according to Perrier (1875).

*laevigatus Philippi in Quijada, 1911. A NOMEN NUDUM according to Madsen (1956).

*laeviusculus Philippi in Quijada, 1911. A NOMEN NUDUM according to Madsen (1956).

leptalacantha H. L. Clark, 1916, see *Disasterina*

limboonkengi G. A. Smith, 1927

 G. A. Smith 1927: 273; Chang & Liao 1964: 63; Liao 1980: 161, 171; A. M. Clark 1982b: 490.

 Range: Amoy, SE China* to Hong Kong.

lorioli Koehler, 1910

 Koehler 1910a: 129; A. M. Clark & Rowe 1971: 38.

 Range: Karachi, Pakistan*, S India.

luederitziana Döderlein, 1908: 298; 1910: 250. A synonym of *Asterina stellifera* (Möbius, 1859) according to Madsen (1950).

lutea H. L. Clark, 1938

 H. L. Clark 1938: 153; A. M.C. in Clark & Rowe 1971: 68.

 Range: NW Australia.

lymani Perrier, 1881, see *Asterinopsis*

maculata: Perrier, 1875, see *Nepanthia*

*marginata: Koehler, 1911c: 9 (from *Asteriscus marginatus* Valenciennes in Hupé, 1857). A NOMEN NUDUM according to Bell (1893b) and = *Asterina stellifera* (Möbius, 1859).

miniata (Brandt, 1835) (with synonyms *Patiria coccinea* Gray, 1840 and *P. granulosa* Perrier, 1875).

 Brandt 1835: 774 [68] (as *Asterias*).

 Stimpson 1857: 90 (as *Asteriscus*).

 Sladen 1889: 774; Fisher 1911c: 254; Caso 1961a: 81; A. M. Clark 1983: 367 (as *Asterina*).

 Verrill 1913: 482; H. L. Clark 1940: 333; Caso 1978: 210 (as *Patiria*).

 Range: Alaska* to Lower California, 0-300 m.

minor Hayashi, 1974

 Hayashi 1974: 41.

 Range: Kushimoto, Japan.

minuta: Nardo, 1834: 716, non *Asterias minuta* Linnaeus, 1761, = *Asterina gibbosa* (Pennant, 1777) according to Perrier (1875).

minuta var. 1, Gray, 1840, non *Asterina minuta:* Nardo, 1834, = *Asterina folium* (Lütken, 1859) according to H. L. Clark (1933).

minuta var. 2, Gray, 1840, non *Asterina minuta:* Nardo, 1834, = *Asterina hartmeyeri* Döderlein, 1910 according to H. L. Clark (1933).

modesta Verrill, 1870

 Verrill 1870: 277 (as *Asterina (Asteriscus)*).

211

Fisher 1919a: 410 (as *Asterina*).
Verrill 1913: 482; 1915: 61 (as *Asterinides*).
Range: Panama, pacific coast and Pearl Is.
neozelanica Farquhar, 1909: 126, lapsus for *Asterina novaezelandiae*, Perrier, 1875.

*novaezelandiae Perrier, 1875
Perrier 1875: 308 [1876: 228]; Koehler 1920: 135; Fell 1959: 138.
Verrill 1913: 482 (as *Patiria*).
Mortensen 1925a: 299 (as *Asterina (Patiria) novaezelandiae*).
Range: 'New Zealand'.
Locality mistaken according to Mortensen (1925a) supported by Farquhar (1927). Possibly a variant of *Asterina regularis* (Verrill, 1870) according to Fell (1959). [Holotype originally with specimens of *Patiriella gunni* supposedly from New Zealand but almost certainly from southern Australia.]
novaezelandiae: Goto, 1914, non *Asterina novaezelandiae* Perrier, 1875, = *A. coronata* Martens, 1866 according to Fisher (1919a).

nuda H.L. Clark, 1921
H.L. Clark 1921: 98; Endean 1956: 125; A.M.C. in Clark & Rowe 1971: 67, 68.
Range: Torres Strait*, Great Barrier Reef.
obtusa: Perrier, 1875, see under *Patiria*

oliveri Benham, 1911
Benham 1911: 147.
Range: Kermadec Is, N Tasman Sea.

orthodon Fisher, 1922
Fisher 1922: 415; Döderlein 1926: 20; Mortensen 1934: 8; A.M.C. in Clark & Rowe 1971: 68; A.M. Clark 1982b: 491.
Range: Hong Kong*, ?W Australia.
pacifica Hayashi, 1977. A subspecies of *Asterina* [i.e. *Patiriella]* *pseudoexigua* Dartnall, 1971.

panceri Gasco, 1870
Gasco 1870: 86 (as *Asteriscus*).
Tortonese 1952: 163; 1960b: 1 (neotype); 1965: 172; Lopez-Ibor et al., 1982: 10 (as *Asterina*).
Koehler 1924: 133 (as *Asterina gibbosa* var. *panceri*).
Range: western Mediterranean.

pectinifera (Müller & Troschel, 1842)
Müller & Troschel, 1842: 40 (as *Asteriscus*).
Martens 1865: 352; Döderlein 1902: 330; Goto 1914: 634; Uchida 1928: 788; Hayashi 1973b: 70; A.M. Clark 1983: 367; Jangoux & de Ridder 1987: 90 (as *Asterina*).
Djakonov 1950a: 62 [1968: 53]; Baranova 1971: 255 (as *Patiria*).

Range: N Japan Sea to SE Japan, 0-300 m.

penicillaris: Martens, 1866, see *Asterinopsis*

penicillata: Whitelegge 1889: 40, lapsus for *penicillaris* according to Livingstone (1933) but = *Asterinopsis praetermissa* Livingstone, 1933.

penicillatus: A.M. Clark, 1983: 364, lapsus for *penicillaris.*

pentagona (Müller & Troschel, 1842) Martens, 1866: 74. A synonym of *Asterina* [i.e. *Patiriella*] *exigua* (Lamarck, 1816) according to Perrier (1875).

perplexa H.L. Clark, 1938

H.L. Clark 1938: 155.

Range: Lord Howe Id, E Australia.

perrieri de Loriol, 1904: 27. A synonym of *Asterina* [i.e. *Patiriella*] *fimbriata* Perrier, 1875 according to Koehler (1923).

phylactica Emson & Crump, 1979

Emson & Crump 1979: 77; Crump & Emson 1983: 867.

Range: S Wales*, SW England, S Ireland, S France and N Adriatic in Mediterranean, near HW to c.2 m.

pilosa Perrier, 1881, see under *Asterinopsis*

pseudoexigua: Hayashi, 1977 (with subspecies *pacifica* nov.), see *Patiriella*

pusilla Perrier, 1875: 306. Probably a synonym of *Patiriella calcarata* (Perrier, 1869) according to Madsen (1956).

pygmaea Verrill, 1878, see under *Poranisca* (Poraniidae)

regularis Verrill, 1870, see *Patiriella*

sarasini (de Loriol, 1897)

de Loriol 1897: 11; Koehler 1910a: 127 (as *Palmipes*).

Livingstone 1933: 4; A.M.C. in Clark & Rowe 1971: 67; A.M. Clark 1983: 364; Jangoux 1985: 31 (lectotype)(as *Asterina*).

Range: Sri Lanka*, Andaman and Nicobar Is.

scobinata Livingstone 1933

Livingstone 1933: 1; Dartnall 1970a: 73; 1970b: 19.

Range: Tasmania* and Victoria, SE Australia.

selkirki Meissner 1896: 97. A variety of *Asterina* [i.e. *Patiriella*] *calcarata* (Perrier, 1869) but not distinct according to Madsen (1956).

*****setacea** (Müller & Troschel, 1842)

Müller & Troschel 1842: 43 (as *Asteriscus*).

Perrier 1875: 318 [1876: 238] (as *Asterina*).

Range: Type locality unknown.

[Validity uncertain, no subsequent records.]

spinigera Koehler 1911

Koehler 1911b: 20.

Range: Singapore.

sporacantha H.L. Clark, 1923. A variety of *Asterina* [i.e. *Callopatiria*] *granifera* (Gray, 1847).

*squamata (Perrier, 1869)
    Perrier 1869: 101 (as *Asteriscus*).
    Perrier 1875: 312 [1876: 232] (as *Asterina*).
    Range: type locality ?Senegal. [Validity uncertain. Type large, possibly *Asterina stellifera* (Möbius, 1859), or *A. miniata* (Brandt, 1835).]
*stellaris Perrier, 1875
    Perrier 1875: 313 [1876: 232].
    Range: Type locality unknown.
    Validity uncertain.
stellifera (Möbius, 1859) (with synonyms *Asteriscus brasiliensis* Lütken, 1859 and *Asterina luderitziana* Döderlein, 1908; also includes *Asteriscus marginatus* Valenciennes in Hupé, 1857, [nomen nudum] and *A. marginatus* Perrier, 1869).
    Möbius 1859: 4 (as *Asteriscus*).
    Lütken 1871: 301; Bell 1893b: 26; Madsen 1950: 213; A.M. Clark 1955: 33; Nataf & Cherbonnier 1975: 824; A.M. Clark 1983: 367 (as *Asterina*).
    Fisher 1919a: 410(inferred); Tortonese 1956: 197; A.M. Clark & Courtman-Stock 1976: 80 (as *Patiria*).
    Range: Rio de Janeiro, Brazil*, S to northern Argentina, also Senegal to Namibia, 0-50 m.
stellifera: Tortonese, 1962: 1, non *Asterina stellifera* (Möbius, 1859), = *Patiria* [i.e. *Allopatiria*] *ocellifera* Gray, 1847 according to A.M. Clark (1963)
*trochiscus (Retzius, 1805)
    Retzius 1805: 10 (as *Asterias*).
    Müller & Troschel 1842: 44 (as *Asteriscus*).
    Sladen 1889: 392 (as *Asterina*).
    Range: Type locality 'in mare indico'.
    Validity doubtful according to Sladen (1889).
tuberculata Danielssen & Koren, 1881, as variety of
tumida: Danielssen & Koren, 1881, (as *Asterina (Solaster) tumida*), see *Poraniomorpha* (Poraniidae).
wega (Perrier, 1869) Perrier, 1875: 318 [1876: 238]. A synonym of *Asterina burtoni* Gray, 1840 according to A.M. Clark (1952 and 1967b) also A.M.C. in Clark & Rowe (1971), though meantime treated as a valid species by Achituv (1969).
wesseli Perrier, 1875, see *Stegnaster*
ASTERINIDES Verrill, 1913
    Verrill 1913: 479; Brito 1968: 17; Tommasi 1970: 15.
    Type species: *Asteriscus folium* Lütken, 1859.
    A synonym of *Asterina* Nardo, 1834 according to Fisher (1919a) but possibly of subgeneric rank according to A.M. Clark (1983).
burtoni: Verrill, 1913, see *Asterina*.

214

folium: Verrill 1913, see *Asterina.*

minuta (Gray, 1840 var. 1) Verrill, 1913 = *Asterina folium* (Lütken, 1859) according to H.L. Clark (1933).

modesta: Verrill, 1913, see *Asterina.*

wega (Perrier, 1869) Verrill, 1913: 482. A synonym of *Asterina burtoni* Gray, 1840 according to A.M. Clark (1952). etc., see under *Asterina.*

*****ASTERINOPSIS** Verrill, 1913 (probably congeneric with *Paranepanthia* Fisher, 1917)

Verrill 1913: 480; Fisher 1919a: 417; Mortensen 1933a: 258; Livingstone 1933: 14; H.L. Clark 1938: 137.

Type species: *Asterias penicillaris* Lamarck, 1816.

Invalid and of uncertain relationship with *Paranepanthia* Fisher, 1917, owing to uncertainty about identity of type species according to H.L. Clark (1938). [However, Jangoux (pers. comm.) believes *Asterinopsis penicillaris* to be congeneric with *Paranepanthia platydisca* (Fisher, 1913), type species of *Paranepanthia,* on the basis of a drawing by Lesueur of Lamarck's lost holotype. Pending publication and reassessment of the species involved, only those referred to *Asterinopsis* by Verrill or Fisher are included under this generic heading here.]

grandis: Livingstone, 1933, see *Paranepanthia*

lymani (Perrier, 1881)

Perrier 1881: 15; 1884: 219 (as *Asterina*).

Verrill 1913: 480; 1915: 64 (as *Asterinopsis*).

A.M. Clark 1983: 372 (*?Asterinopsis*).

Range: Barbados. 220-256 m.

pedicellaris (Fisher, 1913)

Fisher 1913c: 217 (as *Nepanthia*).

Fisher 1919a: 417 (as *Asterinopsis*).

Range: Philippine Is, 123 m.

*****penicillaris** (Lamarck, 1816)

Lamarck 1816: 555 (as *Asterias*).

Müller & Troschel, 1842: 42 (as *Asteriscus*).

Martens 1866: 74; Sladen 1889: 393; Mortensen 1933a: 258 (as *Asterina*).

Verrill 1913: 480; H.L. Clark 1946: 137 (as *Asterinopsis*).

Range: Indo-West Pacific. Identity and validity uncertain, but see under genus.

pilosa (Perrier, 1881)

Perrier 1881: 16 (as *Asterina*).

Verrill 1913: 480; 1915: 64 (as *Asterinopsis*).

A.M. Clark 1983: 372 (*?Asterinopsis*).

Range: Dominica, Leeward Is, 216 m.

praetermissa Livingstone, 1933, see *Paranepanthia*

ASTERISCUS Müller & Troschel, 1840

Müller & Troschel, 1840a: 140; 1842: 39.

Type species: none of 4 named species designated, so name invalid; genus originally a composite of *Anseropoda* Nardo, 1834 and *Asterina* Nardo, 1834. By designation of the first-named, *Asterias membranacea* Lamarck, as type species, *Asteriscus* can be disposed of in the synonymy of *Anseropoda* Nardo, 1834, since this is a synonym of *Anseropoda placenta* (Pennant, 1777).]

arrecifensis Greeff, 1872: 105. A synonym of *Asterina gibbosa* (Pennant, 1777) according to Greeff (1882).

australis Müller & Troschel, 1842: 43. Partly = *Asterina* [i.e. *Patiriella*] *gunni* Gray, 1840 according to Sladen (1889) and partly *Asterina* [i.e. *Patiriella*] *calcar* (Lamarck, 1816). [By nomination of six-armed lectotype, restriction to a synonym of *Patiriella gunni* can be achieved.]

brasiliensis Lütken 1859: 57. A synonym of *Asterina stellifera* (Möbius, 1859) according to Lütken (1871).

calcar: Dujardin & Hupé, 1862: 377, non *Asterias calcar* Lamarck, 1816 as already restricted by Gray, 1840, = *Asterina* [i.e. *Patiriella*] *gunni* Gray, 1840 according to Sladen (1889).

calcaratus Valenciennes in Gay, 1854: 427, also in Quijada, 1911: 162. A NOMEN NUDUM, undescribed and = *Patiria* [i.e. *Asterina*] *chilensis* (Lütken, 1859) according to Madsen (1956).

calcaratus Perrier, 1869: 100 [297], non *Asteriscus calcaratus* Gay, 1854 [but invalidated as primary homonym], see *Patiriella*.

cepheus Müller & Troschel, 1842. A subspecies of *Asterina burtoni* Gray, 1840 according to A.M.C. in Clark & Rowe (1971).

chilensis Lütken, 1859, see *Asterina*

ciliatus Lorenz, 1860: 658. A synonym of *Asterina gibbosa* (Pennant, 1777) according to Lütken (1865).

coccineus (Gray, 1840) Müller & Troschel, 1842: 43. A synonym of *Patiria* [i.e. *Asterina*] *miniata* (Brandt, 1835) according to Mortensen (1933a).

diesingi Müller & Troschel, 1842: 43. A synonym of *Asterina* [i.e. *Patiriella*] *gunni* Gray, 1840 according to Sladen (1889).

exiguus: Perrier, 1869: 100 [297], non *Asterias exigua* Lamarck, 1816, = *Asterina* [i.e. *Patiriella*] *gunni* Gray, 1840 according to Perrier (1875).

folium Lütken, 1859, see *Asterina*

gibbosa: Fischer, 1869, see *Asterina*

kraussi (Gray, 1840) Müller & Troschel, 1842: 42. A synonym of *Asterina* [i.e. *Patiriella*] *exigua* (Lamarck, 1816) according to Sladen (1889).

marginatus Hupé, 1857: 100. A NOMEN NUDUM according to Bell (1893b) and = *Asterina stellifera* (Möbius, 1859), invalidating

marginatus Perrier, 1869: 92 [289].

membranaceus (Retzius, 1783) Müller & Troschel, 1840a: 184. A synonym of *Anseropoda placenta* (Pennant, 1777) according to Fisher (1940).

militaris: Müller & Troschel, 1842, see *Pteraster* (Pterasteridae)

miniatus: Stimpson, 1857, see *Asterina*

minutus: Müller & Troschel, 1842: 41, non *Asterias minuta* Gmelin 1791 = *Asterina marginata* Hupé, 1857 [i.e. *A. stellifera* (Möbius, 1859)] according to Perrier (1875).

palmipes (Olivi, 1792) Müller & Troschel, 1842: 39. A synonym of *Anseropoda placenta* (Pennant, 1777) according to Fisher (1940).

panceri Gasco, 1870, see *Asterina*

pectinifera Müller & Troschel, 1842, see *Asterina*

penicillaris: Müller & Troschel, 1842, see *Asterinopsis*

*pentagonus Müller & Troschel, 1842: 42. A synonym of *Asterina* [i.e. *Patiriella*] *exigua* (Lamarck, 1816) according to Sladen (1889) and Jangoux & de Ridder (1987) who found types still extant. However, type locality Java indicates synonymy with *Patiriella pseudoexigua* Dartnall, 1971, which would therefore need action by the ICZN to remain valid.

placenta: Lütken, 1864, see *Anseropoda*

pulchellus Perrier, 1869: 99 [296]. A synonym of *Asterina gibbosa* (Pennant, 1777) according to Ludwig (1897).

rosaceus: Müller & Troschel, 1842, see *Anseropoda*

setaceus Müller & Troschel, 1842, see *Asterina*

squamata Perrier, 1869, see *Asterina*

stellifer Möbius, 1859, see *Asterina*

trochiscus: Müller & Troschel, 1842, see *Asterina*

verruculatus (Retzius, 1805) Müller & Troschel, 1842: 41. A synonym of *Asterina gibbosa* (Pennant, 1777) according to Lütken (1865).

verruculatus: Peters, 1852: 178, non *Asterias verruculata* Retzius, 1805, probably = *Asterina cephea* (Müller & Troschel, 1842) [i.e. *Asterina burtoni cepheus*] according to Perrier (1875).

wega Perrier, 1869: 102. A synonym of *Asterina burtoni* Gray, 1840 according to A.M.C. in Clark & Rowe (1971).

CALLOPATIRIA Verrill, 1915

Verrill 1913: 480; A.M. Clark 1983: 367.

Type species: *Patiria bellula* Sladen, 1889, a synonym of *Patiria* [i.e. *Callopatiria*] *granifera* Gray, 1847.

A synonym of *Patiria* Gray according to Fisher (1941) but restored by A. M. Clark (1983).

bellula (Sladen, 1889) Verrill, 1913: 480. A synonym of *Patiria* [i.e. *Callopatiria*] *granifera* Gray, 1847 according to A.M. Clark (1956).

formosa (Mortensen, 1933)

Mortensen 1933a: 261 (as *Parasterina*).

A.M. Clark 1956: 382; A.M. Clark & Courtman-Stock 1976: 78-79 (as *Patiria*).

A.M. Clark 1983: 369 (as *Callopatiria*).

Range: False Bay, South Africa, 12-55 m.

granifera (Gray, 1847) (with synonym *Patiria bellula* Sladen, 1889 and variety *sporacantha* H.L. Clark, 1923).

Gray 1847: 82; Fisher 1940: 271; A.M. Clark 1956: 380; A.M. Clark & Courtman-Stock 1976: 79 (as *Patiria*).

H.L. Clark 1923a: 281 (and variety *sporacantha); Mortensen 1933a: 256 (as Asterina*).

A.M. Clark 1983: 367 (as *Callopatiria*).

Range: southern Namibia to Natal, intertidal – 82 m.

obtusa: Verrill, 1913, see under *Patiria*

sporacantha H.L. Clark, 1923: 282. A variety of *Callopatiria granifera* (Gray, 1847).

CARNA Gistel, 1848

Gistel 1848: 176.

Type species: *Asterias membranacea* Retzius, 1783.

A synonym of *Anseropoda* Nardo, 1834 according to Fisher (1919a).

membranacea (Retzius, 1783) Gistel, 1848: 176. A synonym of *Anseropoda placenta* (Pennant, 1777) according to Bell (1891).

CTENASTER L. Agassiz, 1836: 192. A NOMEN NUDUM, cited only as a synonym of *Asterina*.

***DESMOPATIRIA** Verrill, 1913

Verrill 1913: 481, 484.

Type species: *Desmopatiria flexilis* Verrill, 1913.

[Validity doubtful since identity of type species uncertain.]

***flexilis** Verrill, 1913

Verrill 1913: 481, 484.

Range: supposedly Chile; doubted by Madsen (1956) to account for lack of subsequent records.

DISASTERINA Perrier, 1875 (with synonym *Habroporina* H.L. Clark, 1921)

Perrier 1875: 289 [1876: 209]; Livingstone 1933: 5.

Type species: *Disasterina abnormalis* Perrier, 1875.

abnormalis Perrier, 1875 (with synonym *Habroporina pulchella* H.L. Clark, 1921).

Perrier 1875: 289 [1876: 209]; Livingstone 1933: 7; Jangoux 1978b: 296; Liao 1980: 164, 170; Jangoux 1986a: 144.

Range: northern Queensland, New Caledonia, Celebes, Indonesia.

africana Mortensen, 1933a. A subspecies of *Disasterina leptalacantha* (H.L. Clark, 1916).

ceylanica Döderlein, 1888, see *Tegulaster*

leptacantha Livingstone 1933: 8, lapsus for *leptalacantha*

218

leptalacantha (H.L. Clark, 1916) (with subspecies *africana* Mortensen, 1933)
leptalacantha leptalacantha (H.L. Clark, 1916)
H.L. Clark 1916: 57 (as *Asterina*).
Livingstone 1933: 8; Endean 1953: 54 (as *Disasterina*).
Range: Queensland.
leptalacantha africana Mortensen, 1933
Mortensen 1933a: 259 (as *variety*).
A.M. Clark & Courtman-Stock 1976: 78 (as subspecies).
Range: Port Elizabeth* to S Natal, South Africa, 0-366 m.
odontacantha Liao, 1980
Liao 1980: 163, 169.
Range: Paracel (Xisha) Is, S China Sea.
praesignis Livingstone, 1933
Livingstone 1933: 10.
Range: Queensland, Australia, 5-7 m.
spinosa Koehler, 1910
Koehler 1910a: 131; Livingstone 1933: 6.
Range: Andaman Is, Bay of Bengal.
spinulifera H.L. Clark, 1938
H.L. Clark 1938: 156; Julka & Sumita Das 1978: 349.
Range: NW Australia*, Andaman Is.
ENOPLOPATIRIA Verrill, 1913
Verrill 1913: 480; Tommasi 1970: 15.
Type species: *Asterina marginata* Hupé, 1857, a nomen nudum, = *Asterina stellifera* (Möbius, 1859).
A synonym of *Patiria* Gray, 1840 according to Fisher (1919a) and of *Asterina* according to A.M. Clark (1983), though meantime treated as valid by Bernasconi (1955) and Tommasi (1970).
marginata (Hupé, 1857) Verrill, 1913: 480; also 1915: 62; Bernasconi 1955: 70. A NOMEN NUDUM and = *Asterina stellifera* (Möbius, 1859) according to Madsen (1950).
*siderea Verrill, 1913
Verrill 1913: 480; 1914a: 365; 1915: 63.
A.M. Clark 1983: 378 (as '*Enoplopatiria*', in need of reassessment).
Range: pacific side of Panama.
stellifera: Tommasi, 1970: 15, see *Asterina*
HABROPORINA H.L. Clark, 1921
H.L. Clark 1921: 34.
Type species: *Habroporina pulchella* H.L. Clark, 1921.
A synonym of *Disasterina* Perrier, 1875 according to Livingstone (1933).
pulchella H.L. Clark, 1921: 34. A synonym of *Disasterina abnormalis* Perrier, 1875 according to Livingstone (1933).

HENRICIDES Verrill, 1914
Verrill 1914a: 210.
Type species: *Manasterina longispina* H.L. Clark, 1938.
thia belcheri (Perrier, 1875) according to H.L. Clark (1938), [so *Henricides* a synonym of *Nepanthia* Gray, 1840.]
heteractis (H.L. Clark, 1909) Verrill, 1914a: 210. A synonym of *Nepanthia belcheri* (Perrier, 1875) according to H.L. Clark (1938).
KAMPYLASTER Koehler, 1920
Koehler 1920: 136.
Type species: *Kampylaster incurvatus* Koehler, 1920.
ganulatus Koehler, 1920: 8. A NOMEN NUDUM according to Fisher (1940) and =
incurvatus Koehler, 1920
Koehler 1920: 138; Fisher 1940: 150; Madsen 1955: 13; A.M. Clark 1962a: 33; H.E.S. Clark 1963b: 48.
Range: Adelie Land*, Enderby Land, Ross Sea, Palmer Archipelago, South Shetland Is, 93-750 m.
MANASTERINA H.L. Clark, 1938
H.L. Clark 1938: 157.
Type species: *Manasterina longispina* H.L. Clark, 1938.
longispina H.L. Clark, 1938
H.L. Clark 1938: 157.
Range: Western Australia.
[MIRASTRELLA Fisher, 1940, incertae sedis according to A.M. Clark (1983).]
NEPANTHIA Gray, 1840 (with synonyms *Parasterina* Fisher, 1908 and *Henricides* Verrill, 1914).
Gray 1840: 287; Fisher 1941: 452; Rowe & Marsh 1982: 89 (reviewed).
Type species: *Nepanthia maculata* Gray, 1840.
belcheri (Perrier, 1875) (with synonyms *Asterina (Nepanthia) brevis* Perrier, 1875, *N. suffarcinata* Sladen, 1889, *Nepanthia joubini* Koehler, 1908, *Henricia heteractis* H.L. Clark, 1909, *N. polyplax* Döderlein, 1926, *N. variabilis* H.L. Clark, 1938 and *N. magnispina* H.L. Clark, 1938).
Perrier 1875: 320 [1876: 240] (as *Asterina (Nepanthia)*).
H.L. Clark 1938: 169; Rowe & Marsh 1982: 99; Aziz & Jangoux 1984a: 136 (as *Nepanthia*).
Range: Lord Howe Id, E Australia*, N Australia, W Indonesia, Vietnam, Burma, Mergui Archipelago, 0-46 m.
brachiata Koehler, 1910, see *Paranepanthia*
brevis (Perrier, 1875) Verrill, 1913: 480. A synonym of *Nepanthia belcheri* (Perrier, 1875) according to Rowe & Marsh (1982).
briareus (Bell, 1894)
Bell 1894: 404 (as *Patiria*).

220

A. M. Clark 1956: 374; Jangoux 1978b: 297; Rowe & Marsh 1982: 109 (as *Nepanthia*).

Range: Macclesfield Bank, S China Sea*, S Philippine Is, Moluccas, 27-83 m.

crassa (Gray, 1847)

Gray 1847: 83 (as *?Patiria*).

Perrier 1875: 326 [1876: 246] (as *Patiria*).

Fisher 1908a: 90; H.L. Clark 1923b: 243; 1946: 143 (as *Parasterina*).

Fisher 1941: 453 (as *?Nepanthia*).

Rowe & Marsh 1982: 97; A.M. Clark 1983: 370 (as *Nepanthia*).

Range: Western Australia, Victoria, 0-38 m.

fisheri Rowe & Marsh, 1982

Rowe & Marsh 1982: 103.

Range: Philippines, Sabah, northern Borneo and Timor Sea, 29-84 m.

gracilis Rowe & Marsh, 1982

Rowe & Marsh 1982: 107.

Range: Philippine Is*, ?N.S.W., Australia, 100-124 m.

grandis H.L. Clark, 1928, see *Paranepanthia*

hadracantha A.M. Clark, 1966: 320. A synonym of *Nepanthia troughtoni* (Livingstone, 1934) according to Shepherd (1968).

joubini Koehler 1908b: 232; also Fisher 1919a: 423. A synonym of *Nepanthia belcheri* (Perrier, 1875) according to Rowe & Marsh (1982).

maculata Gray, 1840 (with synonym *Nepanthia tenuis* H.L. Clark, 1938, possibly also *Chaetaster cylindratus* Möbius, 1859).

Gray 1840: 287; Fisher 1919a: 422; A.M. Clark 1956: 377; Rowe & Marsh 1982: 106; Aziz & Jangoux 1984a: 136.

Müller & Troschel, 1842: 28 (as *?Chaetaster*).

Perrier 1875: 322 [1876: 242]; Studer 1884: 42 (as *Asterina (Nepanthia)*).

Range: Philippine Is*, Indonesia, New Guinea, N Australia, 0-84 + m.

magnispina H.L. Clark, 1938: 174. A synonym of *Nepanthia belcheri* (Perrier, 1875) according to Rowe & Marsh (1982).

nigrobrunnea Rowe & Marsh, 1982

Rowe & Marsh 1982: 95.

Range: N.S.W., Australia, 10-30 m.

polyplax Döderlein, 1926: 20. A synonym of *Nepanthia belcheri* (Perrier, 1875) according to H.L. Clark (1938).

suffarcinata Sladen, 1889: 328; also Koehler 1910a: 133. A synonym of *Nepanthia belcheri* (Perrier, 1875) according to Rowe & Marsh (1982).

tenuis H.L. Clark, 1938: 175. A synonym of *Nepanthia maculata* Gray, 1840 according to Rowe & Marsh (1982).

tessellata Gray, 1840: 287. A synonym of *Chaetaster longipes* (Retzius, 1805) (Chaetasteridae) according to Perrier (1875).

troughtoni (Livingstone, 1934) (with synonyms *Nepanthia hadracantha* A. M. Clark, 1966 and *Parasterina occidentalis* H.L. Clark, 1938).
Livingstone 1934: 179 (as *Parasterina*).
Shepherd 1968: 748; Rowe & Marsh 1982: 94 (as *Nepanthia*).
Range: W Australia to Victoria, 0-73 m.
variabilis H.L. Clark, 1938: 176, also Fisher 1941: 454. A synonym of *Nepanthia belcheri* (Perrier, 1875) according to Rowe & Marsh (1982).
PALMIPES L. Agassiz, 1836 (from Linck, 1733)
L. Agassiz 1836: 192; also Forbes 1839: 114; Gray 1840: 288; Norman 1865: 120; Perrier 1875: 290 [1876: 210]; Viguier 1878: 212; Sladen 1889: 394; Norman 1891: 384; Bell 1891a: 234; Mortensen 1927: 99.
Type species: *Asterias placenta* Pennant, 1777, by subsequent designation by Fisher (1906).
A synonym of *Anseropoda* Nardo, 1834 according to Bell (1891a).
diaphanus Sladen, 1889, see *Anseropoda*
inflatus: Perrier, 1875, see *Stegnaster*
insignis: Koehler, 1909, see *Anseropoda*
lobianc[o]i Ludwig, 1897, see *Anseropoda*
ludovici Koehler, 1909, see *Anseropoda*
membranaceus (Retzius, 1783) L. Agassiz, 1836: 192. A synonym of *Anseropoda placenta* (Pennant, 1777) according to Bell (1891a).
novemradiatus Bell, 1905, see *Anseropoda*
pellucidus Alcock, 1893, see *Anseropoda*
placenta: Norman, 1865, see *Anseropoda*
rosacea: Dujardin & Hupé, 1862, see *Anseropoda*
sarasini de Loriol, 1897, see *Asterina*
stokesi, Gray, 1840: 288. A synonym of *Asteriscus* [i.e. *Anseropoda*] *rosacea* (Lamarck, 1816) according to Müller & Troschel, 1842.
tenuis Goto, 1914, see *Anseropoda*
*****PARANEPANTHIA** Fisher, 1917 (possibly a synonym of *Asterinopsis* Verrill, 1913)
Fisher 1917e: 172; 1919a: 419; H.L. Clark 1946: 136.
Type species: *Nepanthia platydisca* Fisher, 1913.
Maintained as valid by H.L. Clark (1938) pending information about the identity of the type species of *Asterinopsis* (q.v.).
brachiata (Koehler, 1910)
Fisher 1917e: 173 (as *?Paranepathia*).
Koehler 1910a: 133 (as *Nepanthia*).
Fisher 1919a: 420; Macan 1938: 411 (as *Paranepanthia*).
Range: Andaman Is, SE Arabia, 38-200 m.
grandis (H.L. Clark, 1928)
H.L. Clark 1928: 393 (as *Nepanthia*).
Livingstone 1933: 15 (as *Asterinopsis*).

H. L. Clark 1938: 159; 1946: 137; Shepherd 1968: 748; Dartnall 1970a: 76
(as *Paranepanthia*).
Range: southern half of Australia, 2-40 m.

platydisca (Fisher, 1913)
Fisher 1913c: 218 (as *Nepanthia*).
Fisher 1919a: 420 (as *Paranepanthia*).
Range: Sulawesi (Celebes), Indonesia, 377 m.

praetermissa (Livingstone, 1933)
Livingstone 1933: 14 (as *Asterinopsis*).
H. L. Clark 1938: 161; 1946: 137 (as *Paranepanthia*).
Range: N.S.W., Australia.

rosea H. L. Clark, 1938
H. L. Clark 1938: 161; 1946: 137.
Range: Western Australia.

PARASTERINA Fisher, 1908
Fisher 1908a: 90; Fisher 1940: 270; A. M. Clark 1956: 380.
Type species: *Patiria crassa* Gray, 1840.
Probably a synonym of *Nepanthia* according to Fisher (1941), confirmed
by Rowe and Marsh (1982).

africana Engel & Croes, 1960: 13. A synonym of *Allopatiria ocellifera* (Gray,
1847) according to Tortonese (1963).

bellula (Sladen, 1889) H. L. Clark, 1923a: 280. A synonym of *Callopatiria*
granifera (Gray, 1840) according to A. M. Clark (1983).

crassa: Fisher, 1908, see *Nepanthia*

formosa Mortensen, 1933a, see *Callopatiria*

obesa H. L. Clark, 1910: 334. A synonym of *Cycethra verrucosa* (Philippi,
1857) (Ganeriidae) according to A. M. Clark (1983).

occidentalis H. L. Clark, 1938: 180. A synonym of *Nepanthia troughtoni*
(Livingstone, 1934) according to A. M. Clark (1966).

troughtoni Livingstone, 1934, see *Nepanthia*

PATIRIA Gray, 1840
Gray 1840: 290; Verrill 1913: 482; Fisher 1941: 451; A. M. Clark &
Courtman-Stock 1976: 78.
Type species: *Patiria coccinea* Gray, 1840, a synonym of *Asterina miniata*
(Brandt, 1835).
A synonym of *Asterina* Nardo, 1834 according to Mortensen (1933a),
Hayashi (1940) and A. M. Clark (1983), though meantime treated as a
valid genus by [e.g] Fisher (1940), Tortonese (1956) and A. M.C. in Clark
& Courtman-Stock (1976).

PATIRIA: Perrier, 1875: 323 [1876: 243], non *Patiria* Gray, 1840 = *Allopati-*
ria Verrill, 1913

bellula Sladen, 1889: 385. A synonym of *Callopatiria granifera* (Gray, 1840)
according to A. M. Clark (1983).

briareus Bell, 1894, see *Nepanthia*

chilensis: Verrill, 1870, see *Asterina*

coccinea Gray, 1840: 290. A synonym of *Asterina miniata* (Brandt, 1835) according to Mortensen (1933).

crassa Gray, 1847, see *Nepanthia*

crassa: Bell, 1884a: 131, non *Patiria crassa* Gray, 1847, = *Nepanthia belcheri* (Perrier, 1875) according to H.L. Clark (1938).

crassa: Whitelegge, 1889, non *Patiria crassa* Gray, 1847, = *Henricia heteractis* H. L. Clark, 1909, a synonym of *Nepanthia belcheri* (Perrier, 1875), according to H. L. Clark (1938).

formosa: A.M. Clark, 1956, see *Callopatiria*

gayi Perrier, 1875: 305 [1876: 235]. A synonym of *Asterina chilensis* (Lütken, 1859) according to Koehler (1920).

granifera Gray, 1847, see *Callopatiria*

granulosa (Perrier, 1875) Verrill, 1913: 482. A synonym of *Asterina miniata* (Brandt, 1835) according to A.M. Clark (1983).

miniata: Verrill, 1913, see *Asterina*

novaezelandiae: Verrill, 1913, see *Asterina*

obesa (H.L. Clark, 1910) Madsen, 1956: 24. A synonym of *Cycethra verrucosa* (Philippi, 1857) according to A.M. Clark (1983).

*obtusa Gray, 1847 *[' Patiria']*

Gray 1847: 82; [?]Verrill 1870: 276.

Perrier 1875: 318 [1876: 238](as *Asterina).*

Verrill 1913: 480 (as *Callopatiria).*

A.M. Clark 1983: 372 (position uncertain, in need of reassessment).

Range: Panama, probably pacific side, 11-18 m.

ocellifera Gray, 1847, see *Allopatiria*

pectinifera: Djakonov, 1950a, see *Asterina*

pulla Koehler & Vaney, 1906: 60. A synonym of *Allopatiria ocellifera* (Gray, 1847) according to A.M. Clark (1983).

rosea Koehler & Vaney, 1906: 58. A synonym of *Allopatiria ocellifera* (Gray, 1847) according to A.M. Clark (1983).

stellifera: A.M. Clark & Courtman-Stock, 1976, see *Asterina*

*PATIRIELLA Verrill, 1913

Verrill 1913: 483; H.L. Clark 1946: 134; Dartnall 1971a: 39; A.M. Clark 1983: 365.

Type species: *Asterina (Asteriscus) regularis* Verrill, 1870.

Thought to be a synonym of *Asterina* Nardo, 1834 by Hayashi (1940) but this conclusion overlooked or unadopted by non-Japanese authors. Possibly a synonym according to A.M. Clark (1983).

brevispina H.L. Clark, 1938

H.L. Clark 1938: 1656; Shepherd 1968: 747; Dartnall 1970a: 75.

Range: southern half of Australia, 0-10 m.

calcar (Lamarck, 1816)

Lamarck 1816: 557 (only var. *c. octagona*, as *Asterias*).

Gray 1840: 290 (restricted, as *Asterina*).

Verrill 1913: 484; H.L. Clark 1946: 134; Shepherd 1968: 746; Dartnall 1971a: 46 (as *Patirella*).

Range: S and E coasts of Australia, N to southern Queensland.

*calcarata (Perrier, 1869) (with synonym *selkirki* Meissner, 1896 [as variety])

Perrier 1869: 100 [298] (as *Asteriscus*).

Perrier 1875: 302 [1876: 222]; Meissner 1896: 97 (var. *selkirki*); H. L. Clark 1910: 33; Lieberkind 1924: 383 (as *Asterina*).

Madsen 1956: 25 (as *Patiriella*; var. *selkirki* (Meissner, 1896) not distinct).

Range: Juan Fernandez Id, SE Pacific.

Invalidated as junior primary homonym of *Asteriscus calcarata* Gay, 1854, even though that is a nomen nudum, so needs submission to the ICZN.

carcar Shepherd 1968: 746, lapsus for *calcar*

dyscrita (H.L. Clark, 1923)

H.L. Clark 1923a: 284 (as *Asterina*).

Mortensen 1933a: 252 (as *Asterina (Patiriella)*).

A.M. Clark 1974: 438; A.M. Clark & Courtman-Stock 1976: 80 (as *Patiriella*).

Range: S coast of South Africa.

exigua (Lamarck, 1816) (with synonyms *Asterina kraussi* Gray, 1840 and *Asteriscus pentagonus* Müller & Troschel, 1842).

Lamarck 1816: 554 (as *Asterias*).

Perrier 1875: 302 [1876: 222]; Döderlein 1910: 250; Hayashi 1973b: 72 (as *Asterina*).

Mortensen 1933a: 252; 1933b: 432; Engel 1938a: 20; 1938b: 2 (as *Asterina (Patiriella)*).

Verrill 1913: 484; Fisher 1919a: 416; Fisher 1940: 272; Shepherd 1968: 746; A.M.C. in Clark & Rowe 1971: 42; Dartnall 1971a: 40 (neotype); A.M. Clark 1976: 250 (as *Patiriella*).

Range: False Bay, South Africa*(neotype), NW to Namibia and St Helena, E to Mozambique, Amsterdam and St Paul Is, S Indian Ocean and southern Australia, 0-3 m.

exigua [tropical records] e.g. Fisher 1919a: 416, Djakonov, 1930a: 251, non *Patiriella exigua* (Lamarck, 1816), = *Patiriella pseudoexigua* Dartnall, 1971 (nom. nov.)

fimbriata: Verrill, 1913, see *Asterina*

gunni (Gray, 1840) (with synonyms *Asteriscus australis* Müller & Troschel, 1842 and *A. diesingi* Müller & Troschel, 1842)

Gray 1840: 289 (as *Asterina*).

Verrill 1913: 484; H.L. Clark 1946: 135; Shepherd 1968: 747; Dartnall 1970a: 74 (as *Patirella*).
Range: Tasmania*, SW to SE Australia, 0-30 m.
inornata Livingstone, 1933
Livingstone 1933: 17.
Range: Western Australia.
mimica Livingstone, 1933: 16. A synonym of *Patiriella regularis* (Verrill, 1870) according to Dartnall (1969a).
nigra H.L. Clark, 1938
H.L. Clark 1938: 167.
Range: Lord Howe Id, E of Australia.
obscura Dartnall, 1971
Dartnall 1971a: 45.
Range: Bowen, Queensland*, Philippine Is.
[octagona Lamarck, 1816: 557 (as variety of *Asterias calcar*). Effectively a synonym of *Asterina* [i.e. *Patiriella*] *calcar* (Lamarck, 1816) as restricted by Gray (1840).]
parvivipara Keough & Dartnall, 1978
Keough & Dartnall 1978: 407.
Range: South Australia.
*****pseudoexigua** Dartnall, 1971 (with subspecies *pacifica* (Hayashi, 1977)) but see under *Asteriscus pentagonus*.
pseudoexigua pseudoexigua Dartnall, 1971
Dartnall 1971a: 43; Marsh 1977: 275.
Range: Queensland*, Solomon Is to Caroline and Ryu Kyu Is, S Japan.
pseudoexigua pacifica (Hayashi, 1977) **new comb.**
Hayashi 1977: 89 (as *Asterina*).
Range: Japan Sea.
*****regularis** (Verrill, 1870) (with synonym *Patiriella mimica* Livingstone, 1933).
Verrill 1870: 250 (as *Asterina (Asteriscus) regularis*).
Hutton 1872: 9; Koehler 1920: 136; Farquhar 1927: 237 (as *Asterina*).
Verrill 1913: 480, 483; Dartnall 1969a: 53 (as *Patiriella*).
Range: New Zealand, Tasmania, N.S.W., 0-15 m.
[Conspecific with var. *a, quinqueangula* of *Asterias calcar* Lamarck, 1816 according to Perrier (1875), of which it would be a junior synonym if *quinqueangula* Lamarck is not rejected by the ICZN.]
tangribensis Domantay & Acosta, 1970
Domantay & Acosta 1970: 60.
Range: Philippine Is.
vivipara Dartnall, 1969
Dartnall 1969b: 294.
Range: Tasmania.

PAXILLASTERINA A.M. Clark, 1983
    A.M. Clark 1983: 373.
    Type species: *Paxillasterina pompom* A.M. Clark, 1983.
pompom A.M. Clark, 1983
    A.M. Clark 1983: 374.
    Range: atlantic coast of Panama, 3-6 m.
PSEUDASTERINA Aziz & Jangoux, 1985
    Aziz & Jangoux 1985: 283.
    Type species: *Pseudasterina delicata* Aziz & Jangoux, 1985.
delicata Aziz & Jangoux, 1985
    Aziz & Jangoux 1985: 284.
    Range: Philippine Is, 192-220 m.
granulosa Aziz & Jangoux, 1985
    Aziz & Jangoux 1985: 285.
    Range: Philippine Is, 130-137 m.
PSEUDONEPANTHIA A.H. Clark, 1916
    A.H. Clark 1916c: 118.
    Type species: *Pseudonepanthia gotoi* A.H. Clark, 1916.
gotoi A.H. Clark, 1916
    A.H. Clark 1916c: 118.
    Range: Sagami Bay, SE Japan, 90 m.
*SOCOMIA Gray, 1840
    Gray 1840: 290.
    Type species: *Socomia paradoxa* Gray, 1840.
    [Identity uncertain.]
*paradoxa Gray, 1840
    Gray 1840: 290.
    [Brief description under generic name only; type lost; locality unknown.]
STEGNASTER Sladen, 1889
    Sladen 1889: 375.
    Type species: *Pteraster inflatus* Hutton, 1872.
inflatus (Hutton, 1872)
    Hutton 1872: 812 (as *Pteraster*).
    Perrier 1875: 291[1876: 211] (as *Palmipes*).
    Sladen 1889: 778; Farquhar 1894: 199; Mortensen 1925a: 303; Fell 1959:
    138; McKnight 1967: 300 (as *Stegnaster*).
    Range: New Zealand and Chatham Is.
wesselli (Perrier, 1875)
    Perrier 1875: 311 [1876: 231]; 1884: 220 (as *Asterina*).
    Sladen 1889: 375, 778; H.L. Clark 1898: 6; Verrill 1915: 66; H.L. Clark
    1933: 28 (as *Stegnaster*).
    Range: Barbados*, around Caribbean to Gulf of Mexico and Bahamas,
    0-183 m.

TEGULASTER Livingstone, 1933

Livingstone 1933: 11.

Type species: *Tegulaster emburyi* Livingstone, 1933.

ceylanica (Döderlein, 1888)

Döderlein 1888: 825 (as *Disasterina*).

Livingstone 1933: 6; James 1989: 105 (as *Tegulaster*).

Range: Sri Lanka (Ceylon) and Lakshadweep (Laccadive) Is.

emburyi Livingstone, 1933

Livingstone 1933: 11.

Range: Queensland.

TREMASTER Verrill, 1880

Verrill 1880: 201; Sladen 1889: xxxiv, 375, 394; Jangoux 1982: 155.

Type species: *Tremaster mirabilis* Verrill, 1880.

laevis H. L. Clark, 1941: 53. A synonym of *Tremaster mirabilis* Verrill, 1880 according to Jangoux (1982).

mirabilis Verrill, 1880 (with synonym *Tremaster laevis* H. L. Clark, 1941)

Verrill 1880: 201; 1885: 638; Dons 1929: 98; Jangoux 1982: 158.

Range: New England Fishing Banks*, N and E to Greenland, Iceland, the Barents Sea and Norway, also Cuba and Kerguelen, Southern Ocean, 230-485 m.

novaecaledoniae Jangoux, 1982

Jangoux 1982: 159.

Range: New Caledonia, 480-505 m.

Family PORANIIDAE Perrier

Poraniidae Perrier 1893: 849; 1894: 163-164; Verrill 1914a: 17; 1915: 68; Fisher 1919a: 407; Mortensen 1927: 89-90; Fisher 1940: 154; Spencer & Wright 1966 (pt): U69; Hotchkiss & A.M. Clark 1976: 263-266; Blake 1981: 380-381; Tablado 1982: 88; A. M. Clark 1984: 20-47 (reviewed).

Gymnasteriidae: Bell 1893a: 21, 78; Ludwig 1900: 459; Farran 1913: 16. [Gymnasteriadae (pt only) Perrier 1884: 165, 168, 229.]

Asteropidae: Koehler 1921: 40-41; 1924: 151. [Asteropidae (pt) Fisher 1911c: 247-248.]

The confusion between the Poraniidae and what is now known as the Asteropseidae was mostly resolved by Hotchkiss & Clark (1976), followed up by A.M.C. (1984).

Genus-group names: Alexandraster, Cheilaster, **Chondraster**, **Culcitopsis**, Glabraster, Habroporina, Lahillea, Lasiaster, **Marginaster**, Ortmannia, **Porania**, **Poraniomorpha**, **Poraniopsis**, Poranisca, **Pseudoporania**, Rhegaster, Sphaeraster, Sphaeriaster, **Spoladaster**, **Tylaster**.

ALEXANDRASTER Ludwig, 1905

Ludwig 1905: 210.

228

Type species: *Alexandraster mirus* Ludwig, 1905.

A synonym of *Poraniopsis* Perrier, 1891 according to Fisher (1911c).

inflatus Fisher, 1906, see *Poraniopsis*

mirus Ludwig, 1905, non *Lahillea* [i.e. *Poraniopsis*] *mira* de Loriol, 1904, probably = *Poraniopsis inflata flexilis* Fisher, 1910 according to Fisher (1940).

*CHEILASTER Bell, 1893

Bell 1893a: 78, 81.

Type species: *Marginaster fimbriatus* Sladen, 1889.

Thought to be a synonym of *Marginaster* Perrier, 1881 by Mortensen (1927) since Ludwig (1897) synonymized the type species with *Marginaster capreensis* (Gasco, 1876). [However, *M. fimbriatus* is more likely a synonym of *Chondraster grandis* (Verrill, 1878) according to A. M. Clark (1984) so *Cheilaster* synonymous with *Chondraster* Verrill, 1895, which it antedates, accordingly needing formal suppression by the ICZN.]

fimbriatus (Sladen, 1889) Bell, 1893a: 81. Probably a synonym of *Chondraster grandis* (Verrill, 1878) according to A. M. Clark (1984).

CHONDRASTER Verrill, 1895 (see under *Cheilaster* for threat to validity)

Verrill 1895: 137 (as *Porania (Chondraster)*).

A. M. Clark 1984: 27 (as genus).

Type species: *Porania grandis* Verrill, 1878.

elattiosus: A. M. Clark 1952: 196, lapsus for *elattosis*

elattosis H. L. Clark, 1923

H. L. Clark 1923a: 274; 1926a: 17; Madsen 1959: 156; A. M. Clark & Courtman-Stock 1976: 73.

Range: S and W coasts of South Africa, 402-1025 m.

grandis (Verrill, 1878) (with probable synonym *Marginaster fimbriatus* Sladen, 1889)

Verrill 1878: 371; 1885b: 542 (pt) (as *Porania*).

Verrill 1895: 137 (as *Porania (Chondraster)*).

A. M. Clark 1984: 27; Harvey et al. 1988: 163 (as *Chondraster*).

Range: E of Cape Cod*, S to c.38°N off New Jersey, also SW of the Faeroe Is to the southern Bay of Biscay, 300-2490 m.

hermanni Madsen, 1959a, see *Porania*

CULCITOPSIS Verrill, 1914. A subgenus of *Poraniomorpha* Danielssen & Koren, 1881 according to A. M. Clark (1984).

borealis: Verrill, 1914b, see *Poraniomorpha (Culcitopsis)*

GLABRASTER A. H. Clark, 1916

A. H. Clark 1916c: 122.

Type species: *Porania magellanica* Studer, 1876.

A synonym of *Porania* Gray, 1840 according to Fisher (1940).

antarctica: A. H. Clark, 1916c, see *Porania*

magellanica: A. H. Clark, 1916c. A subspecies of *Porania antarctica* E. A.

Smith, 1876 according to Madsen (1956).

HABROPORINA H.L. Clark, 1921

H.L. Clark 1921: 34.

Type species: *Habroporina pulchella* H.L. Clark, 1921.

A synonym of *Disasterina* Perrier, 1875 (Asterinidae) according to Livingstone (1933).

pulchella H.L. Clark 1921: 34. A synonym of *Disasterina abnormalis* Perrier, 1875 (Asterinidae) according to Livingstone (1933).

LAHILLEA de Loriol, 1904

de Loriol 1904: 32.

Type species: *Lahillea mira* de Loriol, 1904.

A synonym of *Poraniopsis* Perrier, 1891 according to Fisher (1911c).

mira de Loriol, 1904, see *Poraniopsis*

LASIASTER Sladen, 1889

Sladen 1889: 371.

Type species: *Lasiaster villosus* Sladen, 1889.

A synonym of *Poraniomorpha* Danielssen & Koren, 1881 according to Grieg (1907).

hispidus Sladen, 1889, see *Poraniomorpha*

villosus Sladen, 1889: 372. A synonym of *Poraniomorpha hispida* (M. Sars, 1872) according to Grieg (1907).

MARGINASTER Perrier, 1881 *(Marginaster* Perrier 1881: 16 a nomen nudum, undiagnosed and neither of the two species designated as type; synonym *Poranisca* Verrill, 1914b)

Perrier, 1884: 229; 1894: 164; Verrill 1914b: 18; 1915: 75; Downey 1973: 82; A.M. Clark 1984: 25.

Type species: *Marginaster pectinatus* Perrier, 1881 by subsequent designation by Sladen (1889).

Possibly a synonym of *Porania* Gray, 1840 according to A.M. Clark (1984).

*austerus Verrill, 1899

Verrill 1899: 221; A.M. Clark 1984: 25.

Verrill 1914b: 20; 1915: 78 (as *?Porania).*

Range: 'West Indies.'

Possibly referable to *Porania* Gray, 1840 according to Verrill (1914).

Possibly a synonym of *Marginaster pectinatus* Perrier, 1881 according to A.M. Clark (1984).

capreensis (Gasco, 1876)

Gasco 1876: 9[38] (as *Asteropsis).*

Sladen 1889: 366, 768; Ludwig 1897: 189 (as *?Marginaster).*

Marenzeller 1893b: 68 [6]; Fisher 1919a: 409; Mortensen 1927: 93; Tortonese 1965: 166; A.M. Clark 1984: 27.

Range: Mediterranean, 50-600 m.

echinulata Perrier, 1881, see *Poraniella* (Asteropseidae)

fimbriatus Sladen, 1889: 365. Probably a synonym of *Chondraster grandis* (Verrill, 1878) according to A. M. Clark (1984).

littoralis Dartnall, 1970
  Dartnall 1970: 207.
  Range: Tasmania.

paucispinus Fisher, 1913
  Fisher 1913c: 216; 1919a: 407; A. M. Clark 1962a: 99.
  Range: S China Sea near Hong Kong, c.183 m.

pectinatus Perrier, 1881 (with synonym *Poranisca lepidus* Verrill, 1914)
  Perrier 1881: 16; 1884: 229; 1894: 167; Verrill 1914b: 18; 1915: 76; H. L. Clark 1941: 53; Downey 1973: 82; Tommasi & Oliveira 1976: 89; [?]Carrera-Rodriguez & Tommasi 1977: 102; A. M. Clark 1984: 25.
  Range: W of Cuba*, Yucatan Channel to Brazil [but ?southernmost Brazil], 166-230 m.

*pentagonus Perrier, 1882
  Perrier in Milne-Edwards 1882: 47; 1894: 165; A. M. Clark 1984: 33.
  Range: NW Spain, 400 m.
  Possibly conspecific with *Poraniomorpha (Culcitopsis) borealis* (Süssbach & Breckner, 1911), which it antedates and threatens. As noted in 1984, the name *pentagonus* needs to be formally suppressed by the ICZN.

ORTMANNIA de Loriol, 1906
  de Loriol 1906: 78 (nom. nov. for *Lahillea* de Loriol, 1904. [Though *Lahillea* not invalidated by *Lahillia* Cossmann, 1899.] Type species: *Lahillea mira* de Loriol, 1894.
  A synonym of *Poraniopsis* Perrier, 1891 according to Fisher (1911c).

mira: de Loriol, 1906, see *Poraniopsis*

PORANIA Gray, 1840 (with synonym *Glabraster* A.H. Clark, 1916, possibly also *Marginaster* Perrier, 1881, and subgenus *Pseudoporania* Dons, 1936)

PORANIA (PORANIA) Gray, 1840
  Gray 1840: 288; Perrier 1875: 280 [1876: 96]; Viguier 1878: 220; Sladen 1889: 358; Bell 1893a: 79; Perrier 1894: 163; Fisher 1911c: 248; Koehler 1921: 41; 1924: 152; Mortensen 1927: 90; Fisher 1940: 154; Madsen 1956: 160; Bernasconi 1964b: 263; Tablado 1982: 88; A. M. Clark 1984: 42.
  Type species: 'Porania gibbosa' Leach (an MS name derived from *Asterias gibbosa),* a nomen nudum and = *Goniaster templetoni* Forbes, 1839, a synonym of *Asterias pulvillus* O. F. Müller, 1776.

*antarctica E. A. Smith, 1876 (with synonyms *Porania spiculata* Sladen, 1889 and *P. armata* Koehler, 1917, also subspecies *magellanica* Studer, 1876 and forma *glabra* Sladen, 1889).

As noted in 1984, the name *Astrogonium fonki* Philippi, 1858 needs to be formally rejected by the ICZN in order to preserve the much better known name *Porania antarctica* Smith.

antarctica antarctica E. A. Smith, 1876

E. A. Smith 1876: 108; 1879: 275; Sladen 1889: 360; Ludwig 1903: 22; 1905: 51; Koehler 1906c: 10; 1911d: 27; 1912b: 66; 1917: 42; 1920: 178; Fisher 1940: 154; A. M. Clark 1962a: 34; H. E. S. Clark 1963b: 45; Guille 1974: 35; Tablado 1982: 92.

[Subantarctic records of *antarctica*, (e.g. Perrier 1891a: K107), represent subspecies *magellanica*].

A. H. Clark 1916c: 122 (as *Glabraster*).

Range: Kerguelen area* and circumpolar antarctic, 4-2930 m.

antarctica magellanica Studer, 1876

Studer, 1876: 459; Sladen 1889: 363; Barattini 1938: 20; Tablado 1982: 96 (as species).

Madsen 1956: 28; A. M. Clark 1984: 22, in caption (as subspecies).

Range: Magellan Strait*, S Chile, Falkland Is, Argentina, Uruguay and [?]southernmost Brazil, 74-405 m.

armata Koehler, 1917: 43. A synonym of *Porania antarctica* Smith, 1876 according to Fisher (1940).

austera: Verrill, 1914b (?*Porania*), see under *Marginaster*

borealis (Verrill, 1878) Verrill, 1882: 218; 1884: 659. A synonym of *Poraniomorpha hispida* (M. Sars, 1872) according to Koehler (1909a).

gibbosa Leach in Gray, 1840: 288. A NOMEN NUDUM, undescribed, = *Porania pusillus* (O. F. Müller, 1776) according to Sladen (1889).

glaber Sladen, 1889: 360. A synonym, or forma, of *Porania antarctica* E. A. Smith, 1876 according to A. M. Clark (1962) followed by Guille (1974), though treated as a subspecies by McKnight 1976.

grandis Verrill, 1878, see *Chondraster*

hermanni Madsen, 1959

Madsen 1959a: 153 (as *Porania (Chondraster)*).

Madsen in A. M. Clark 1984: 33 (as *Porania (Porania)*).

Range: East Greenland.

insignis Verrill, 1895. A subspecies of *Porania pulvillus* (O. F. Müller, 1776) according to A. M. Clark (1984).

magelhaenica Studer, 1884: 42, lapsus for *magellanica*.

magellanica Studer, 1876. A subspecies of *Porania antarctica* E. A. Smith, 1876 according to Madsen (1956) and A. M. Clark (1984), though meantime treated as a distinct species by Tablado (1982).

patagonica Perrier, 1878: 27, 50, 85. A NOMEN NUDUM, undescribed, probably = *Porania magellanica* [i.e. *P. antarctica magellanica*] Studer, 1876 according to Sladen (1889).

pulvillus (O. F. Müller, 1776) (with synonyms *Goniaster templetoni* Forbes,

1839 and *Asteropsis ctenacantha* Müller & Troschel, 1842, possibly also *Asterina pygmaea* Verrill, 1878; subspecies *insignis* Verrill, 1895)

pulvillus pulvillus (O. F. Müller, 1776)

O. F. Müller 1776: 234 (as *Asterias*).

Müller & Troschel 1842: 63 (as *Asteropsis*).

Norman 1865: 122; Perrier 1875: 280 [1876: 96]; Bell 1893a: 79; Farran 1913: 16; Koehler 1921: 41; Mortensen 1927: 90; Madsen 1959a: 169; Jangoux 1982: 149, 151; Gage et al. 1983: 281; A.M. Clark 1984: 42.

Range: SW Sweden*, W Norway to the Rockall Plateau and S to N Spain, 5-c.500 m.

pulvillus *insignis Verrill, 1895

Verrill 1895: 138; 1915: 70; Gray, Downey & Cerame-Vivas 1968: 152; Franz, Worley & Merrill 1981: 397 (as species).

A.M. Clark 1984: 47 (as subspecies).

Range: Gulf of Maine to Cape Hatteras, North Carolina, 35-680 m.

As noted in 1984, *insignis* is probably conspecific with and threatened by *Asterina* [subsequently *Poranisca*] *pygmaea* Verrill, 1878, which needs to be formally rejected by the ICZN.

spiculata Sladen, 1889: 362. A synonym of *Porania antarctica* Smith, 1876 according to Fisher (1940), though treated as a subspecies of *Porania magellanica* Studer by Tablado (1982).

spinulosa Verrill, 1880a: 202; see also A.M. Clark 1984: 34. A synonym of *Poraniomorpha hispida* (M. Sars, 1872) according to Koehler (1909a).

PORANIA (CHONDRASTER) Verrill, 1895: 138. Raised to generic rank by Verrill (1914b).

Porania (Chondraster): Madsen, 1959a: 160, non *Porania (Chondraster)* Verrill, 1895, = *Porania (Porania)* Gray, 1840 according to Madsen in A.M. Clark (1984).

hermanni Madsen, 1959a, see *Porania (Porania)*

insignis Verrill, 1895. A subspecies of *Porania pulvillus* (O. F. Müller, 1776) according to A.M. Clark (1984).

PORANIA (PSEUDOPORANIA) Dons, 1936

Dons 1936: 17 (as genus).

A.M. Clark 1984: 44 (as subgenus).

Type species: *Pseudoporania stormi* Dons, 1936.

stormi Dons, 1936

Dons 1936: 19; 1938: 167 (as *Pseudoporania*).

A.M. Clark 1984: 44 (as *Porania (Pseudoporania)*).

Range: Trondheim-Lofoten Is area, Norway*, to SW of Faeroe Is and SW of Ireland, 300-770 m.

PORANIOMORPHA Danielssen & Koren, 1881 (with synonyms *Rhegaster* Sladen, 1883, *Lasiaster* Sladen, 1889, *Sphaeraster* Dons, 1938 [non †*Sphaeraster* Schöndorff, 1906] and *Sphaeriaster* Dons, 1939, also subgenus *Culcitopsis* Verrill, 1914).

PORANIOMORPHA (PORANIOMORPHA) Danielssen & Koren, 1881
  Danielssen & Koren, 1881: 189; 1884b: 67; Grieg 1907: 41; Koehler 1924: 157; Mortensen 1927: 92; Gallo 1937: 1664; Djakonov 1950a: 58 [1968: 48]; A.M. Clark 1984: 33.
  Type species: *Poraniomorpha rosea* Danielssen & Koren, 1881, a subspecies of *Poraniomorpha hispida* (M. Sars, 1872).

abyssicola (Verrill, 1895)
  Verrill 1895: 140 (as *Rhegaster*).
  Grieg 1907: 42 (as *Poraniomorpha*).
  Range: W of North American Basin (c.37°N), 3740 m.

bidens Mortensen, 1932
  Mortensen 1932: 9; Djakonov 1950a: 60 [1968: 50]; Grainger 1966: 5, 28; A. M. Clark 1984: 41.
  Range: W of Greenland*, Faeroe Channel, Kara Sea, 53-1200 m.

borealis (Verrill, 1878) Verrill, 1882: 218. A synonym of *Poraniomorpha (Poraniomorpha) hispida* (M. Sars, 1872) according to Grieg (1907).

hispida (M. Sars, 1872) (with synonyms *Asterina borealis* Verrill, 1878, *Porania spinulosa* Verrill, 1880, *Rhegaster murrayi* Sladen, 1883 and *Lasiaster villosus* Sladen, 1889, also subspecies *rosea* Danielssen & Koren, 1881).

hispida hispida (M. Sars, 1872)
  M. Sars in G. O. Sars, 1872: 28; 1877: 72; Storm 1878: 253 (as *Goniaster*).
  Danielssen & Koren 1884b: 58 (as *Pentagonaster*).
  Sladen 1889: 374; Grieg 1902: 22; Süssbach & Breckner 1911: 219 (as *Lasiaster*).
  Ostergren 1904: 615; Koehler 1909a: 100; Koehler 1924: 157; Mortensen 1927: 92; Djakonov 1946: 163 (pt); 1950a: 59 [1968: 50]; A.M. Clark 1984: 33.
  Range: northern Norway*, north boreal Atlantic S to Cape Cod area [?to Cape Hatteras] and to the Skagerrak c.100-1000 m.

hispida rosea Danielssen & Koren, 1881
  Danielssen & Koren 1881: 189; 1884b: 67; Ludwig 1900: 459; Grieg 1902: 21 (as species).
  Ostergren 1904: 615; Grieg 1907: 42; Mortensen 1927: 92; Djakonov 1950a: 59 [1968: 50] (as variety).
  A.M. Clark 1984: 34; Harvey et al. 1988: 163 (as subspecies).
  Range: W of Norway*, Rockall Trough to southern Bay of Biscay, c.250-1500 m.

rosea Danielssen & Koren, 1881. A subspecies of *Poraniomorpha hispida* (M. Sars, 1872).

spinulosa (Verrill, 1880) Verrill 1885b: 542. A synonym of *Poraniomorpha hispida* (M. Sars, 1872).

tuberculata: Grainger, 1964. A variety of *Poraniomorpha tumida* (Stuxberg, 1878).

tumida (Stuxberg, 1878) (with probable synonym *Sphaeraster bjoerlykkei* Dons, 1938 and variety *tuberculata* (Danielssen & Koren, 1881))
Stuxberg, 1878: 31 (as *Solaster*).
Sladen 1883b: 156; Ludwig 1900: 459 (as *Rhegaster*).
Danielssen & Koren 1881: 182 (variety *tuberculata,* as *Asterina (Solaster)*).
Danielssen & Koren 1884b: 60 (as *Asterina*).
Grieg 1907: 34 (as *Poraniomorpha (Rhegaster)*).
Mortensen 1932: 9; Djakonov 1933: 44; 1946: 163 (pt); 1950a: 58 [1968: 49]; Grainger 1964: 36 (var. *tuberculata); A.M.* Clark 1984: 33.
Range: Barents Sea, Spitzbergen, Greenland, 9-1200 m.

PORANIOMORPHA (CULCITOPSIS) (Verrill, 1914)
Verrill 1914b: 21; Koehler 1924: 160 (as genus).
A.M. Clark 1984: 33 (as subgenus).
Type species: *Culcita borealis* Süssbach & Breckner, 1911.

*borealis (Süssbach & Breckner, 1911) (with synonym *Sphaeraster berthae* Dons, 1938)
Süssbach & Breckner, 1911: 217; Farran 1913: 15 (pt) (as *Culcita*).
Verrill 1914b: 21; Koehler 1924: 160; Cherbonnier & Sibuet 1973: 1348; A. M. Clark 1976: 257 (as *Culcitopsis*).
Grieg 1927: 131 (as variety of *Poraniomorpha hispida*).
A.M. Clark 1984: 34 (as species of *Poraniomorpha (Culcitopsis)*).
Range: NE of Shetland Is*, southern Norwegian Sea, W of British Is to NW Spain, 110-1170 m.

PORANIOPSIS Perrier, 1891 *(Poraniopsis* Perrier, 1888: 765 a nomen nudum, no species named; synonyms *Lahillea* de Loriol, 1904, *Alexandraster* Ludwig, 1905 and *Ortmannia* de Loriol, 1906).
Perrier 1891a: K105; Fisher 1910a: 568; 1911c: 260; 1940: 158; Bernasconi 1980: 253; A.M. Clark 1984: 21, 24.
Type species: *Poraniopsis echinaster* Perrier, 1891.

capensis H.L. Clark, 1923a: 289; also 1926a: 20; A.M. Clark 1952a: 208. A synonym of *Poraniopsis echinaster* according to Madsen (1956).

echinaster Perrier, 1891 (with synonym *Poraniopsis capensis* H.L. Clark, 1923).
Perrier 1891a: K106; Fisher 1940: 158; Madsen 1956: 29; A.M. Clark & Courtman-Stock 1976: 90; Bernasconi 1980: 255; A.M. Clark 1984: 23, 24 (captions).
Range: Magellan area*, S Chile, Gough Id, W coast of South Africa, 30-420 m.

echinasteroides Perrier, 1891a: 197 (caption), lapsus for *echinaster*.

flexilis Fisher, 1910. A subspecies of *Poraniopsis inflatus* (Fisher, 1906).

inflatus (Fisher, 1906) (with probable synonym *Alexandraster mirus* Ludwig, 1905 and subspecies *flexilis* Fisher, 1910).

inflatus inflatus (Fisher, 1906)

Fisher 1906b: 300 (as *Alexandraster*).

Fisher 1910a: 569; 1911c: 261; Djakonov 1950a: 85 [1968: 73].

Range: Oregon to Southern California, also N Honshu, Japan according to Djakonov, 1950 [but ? = *P. japonica* Fisher], 48-290 m.

inflatus flexilis Fisher, 1910

Fisher 1910: 568.

Range: California, 610-1100 m.

japonica Fisher, 1939

Fisher 1939: 470.

Range: NE Honshu, Japan, 333 m.

mira (de Loriol, 1904)

de Loriol 1904: 33 (as *Lahillea*).

de Loriol 1906: 78 (as *Ortmannia*).

Barattini 1938: 20; Fisher 1939: 472; 1940: 159; Bernasconi 1973: 307; 1980: 254; Jangoux 1985b: 27 (lectotype) (as *Poraniopsis*).

Range: southern Argentina*, to southern Brazil (c.33°S), 0(?)-500 m.

PORANISCA Verrill, 1914

Verrill 1914b: 19.

Type species: *Poranisca lepidus* Verrill, 1914.

A synonym of *Marginaster* Perrier, 1881 according to A.M. Clark (1984).

lepidus Verrill, 1914b: 19. A synonym of *Marginaster pectinatus* Perrier, 1881 according to A.M. Clark (1984).

*pygmaea (Verrill, 1878)

Verrill 1878: 372 (as *Asterina*).

Verrill 1914b: 19 (as *Poranisca*).

Range: Gulf of Maine, U.S.A.

Probably conspecific with *Porania pulvillus insignis* Verrill, 1895 and, as noted in 1984, needs formally suppressing by the ICZN.

PSEUDOPORANIA Dons, 1936. A subgenus of *Porania* Gray, 1840 according to A. M. Clark (1984).

stormi Dons, 1936, see *Porania (Pseudoporania)*)

RHEGASTER Sladen, 1883

Sladen 1883b: 155.

Type species: *Solaster tumidus* Stuxberg, 1878.

A synonym of *Poraniomorpha* Danielssen & Koren, 1881 according to Grieg (1907).

abyssicola Verrill, 1895, see *Poraniomorpha*

borealis (Verrill, 1878) Verrill, 1914b: 17. A synonym of *Poraniomorpha hispida* (M. Sars, 1872) according to Koehler (1909a) [indirectly].

murrayi Sladen, 1883b: 156. A synonym of *Poraniomorpha hispida* (M. Sars, 1872) according to Grieg (1907).

spinulosa (Verrill, 1880) Verrill, 1914b: 18. A synonym of *Poraniomorpha hispida* (M. Sars, 1872) according to Koehler (1909a) [indirectly].

tumidus: Ludwig, 1900, see *Poraniomorpha*

SPHAERASTER Dons, 1938

Dons 1938: 161.

Type species: *Sphaeraster berthae* Dons, 1938.

A homonym of †*Sphaeraster* Schöndorff, 1906, as noted by Dons, 1939, = *Poraniomorpha (Culcitopsis)* Verrill, 1914 according to A.M. Clark (1984).

berthae Dons, 1938: 163. A synonym of *Poraniomorpha (Culcitopsis) borealis* (Süssbach & Breckner, 1911) according to A.M. Clark (1984).

bjoerlykkei Dons, 1938: 165. A probable synonym of *Poraniomorpha tumida* (Stuxberg, 1878) according to Madsen in A.M. Clark (1984).

SPHAERIASTER Dons, 1939

Dons, 1939: 37 (nom. nov. for *Sphaeraster* Dons, 1938, non *Sphaeraster* Schöndorf, 1906).

Type species: *Sphaeraster berthae* Dons, 1938.

A synonym of *Poraniomorpha (Culcitopsis)* Verrill, 1914 according to A. M. Clark (1984).

berthae (Dons, 1938) Dons, 1939: 37. A synonym of *Poraniomorpha (Culcitopsis) borealis* (Süssbach & Breckner, 1911) according to A.M. Clark (1984).

bjoerlykkei (Dons, 1938) Dons, 1939: 37. A probable synonym of *Poraniomorpha tumida* (Stuxberg, 1878) according to Madsen in A.M. Clark (1984).

SPOLADASTER Fisher, 1940

Fisher 1940: 136; A.M. Clark 1976: 254; 1984: 41.

Type species: *Cryaster brachyactis* H.L. Clark, 1923.

brachyactis (H.L. Clark, 1923) (with synonym *Tylaster meridionalis* Mortensen, 1933)

H.L. Clark 1923a: 293; Mortensen 1933a: 249 (as *Cryaster*).

Fisher 1940: 136; A.M. Clark 1952a: 208; 1976: 255; 1984: 41.

Range: S and W coasts of South Africa, 47-205 m.

veneris (Perrier, 1879)

Perrier 1879: 48 (as *Culcita*).

A.M. Clark 1952a: 208; 1976: 251 (as *Spoladaster*).

Range: St Paul and Amsterdam Islands, S Indian Ocean, 0.5-300 m.

TYLASTER Danielssen & Koren, 1881

Danielssen & Koren 1881: 186; 1884b: 64.

Type species: *Tylaster willei* Danielssen & Koren, 1881.

meridionalis Mortensen, 1933a: 249. A synonym of *Spoladaster brachyactis* (H.L. Clark, 1923) according to A.M. Clark (1984).

willei Danielssen & Koren, 1881

Danielssen & Koren 1881: 186; 1884b: 64; Ludwig 1900: 459; Grieg 1907: 32; Djakonov 1933: 44; 1950a: 60 [1968: 51].

Range: Norwegian Sea, W to W Greenland and E to Kara Sea, 520-2920 m.

Family ARCHASTERIDAE Viguier

Viguier 1878: 235; Sladen 1889: 2; Fisher 1911c: 18; Spencer & Wright 1966: U56; A. M. Clark & Courtman-Stock 1976: 59.

Most of the taxa once included have been referred to the Astropectinidae and Benthopectinidae leaving only the type genus.

Genus-group names: **Archaster**, Astropus.

ARCHASTER Müller & Troschel, 1840 [with probable synonym *Astropus* Gray, 1840]

Müller & Troschel, 1840a: 104; 1840b: 323; 1842: 65; Perrier 1875: 343 [1876: 263]; Sladen 1889: 120 (restricted); H.L. Clark 1946: 76; Sukarno & Jangoux 1977: 819.

Type species: *Archaster typicus* Müller & Troschel, 1840.

agassizi Verrill, 1880, see *Plutonaster* (Astropectinidae).

americanus Verrill, 1880, see *Astropecten* (Astropectinidae).

andromeda: Möbius & Bütschli, 1875, see *Psilaster* (Astropectinidae)

angulatus Müller & Troschel, 1842 (with synonym *Archaster laevis* H. L. Clark, 1938)

Müller & Troschel, 1842: 66; Simpson & Brown 1910: 48; Fisher 1919a: 181; Domantay 1972: 57; Jangoux 1973a: 14; A.M. Clark & Courtman-Stock 1976: 59; Sukarno & Jangoux 1977: 830; Jangoux & de Ridder, 1987: 88 (type not traced).

Range: 'Java; Isle de France'* but probably Western Australia according to Sukarno & Jangoux (1977), Philippine Is, South China Sea, East Africa S to Mozambique, 0-90 m.

angulatus: Michelin 1845: 24; Möbius 1859: 50; de Loriol 1885: 78; Tortonese 1956: 189, non *Archaster angulatus* Müller & Troschel, 1842, = *Archaster lorioli* Sukarno & Jangoux, 1977.

arcticus: Perrier 1878, see *Leptychaster* (Astropectinidae)

bairdi Verrill, 1882, see *Mediaster* (Goniasteridae)

bifrons Wyville Thomson, 1873, see *Plutonaster* (Astropectinidae)

christi (Düben & Koren, 1846) Perrier: 1875: 347. A synonym of *Psilaster andromeda* (Müller & Troschel, 1842) according to Sladen (1889) (Astropectinidae).

christi Studer 1884: 48, non *Astropecten christi* Düben & Koren, = *Psilaster acuminatus* Sladen, 1889 according to Sladen (1889).

coronatus Perrier, 1884: 262. A synonym of *Cheiraster (Christopheraster) mirabilis* Perrier, 1881 according to A.M. Clark (1981) (Benthopectinidae).

dawsoni Verrill, 1880, see *Cheiraster (Luidiaster)* (Benthopectinidae)

echinulatus Perrier, 1875, see *Cheiraster (Barbadosaster)* (Benthopectinidae).

efflorescens Perrier, 1884, see *Plutonaster* (Astropectinidae)

excavatus Wyville Thomson, 1876: 71. A synonym of *Leptychaster kerguelenensis* E. A. Smith, 1876 according to Smith (1879) (Astropectinidae).

florae Verrill, 1878, see *Psilaster* (Astropectinidae)

formosus Verrill, 1884: 383. A synonym of *Paragonaster subtilis* (Perrier, 1881) (Goniasteridae) according to Downey (1973).

grandis Verrill, 1884, see *Dytaster* (Astropectinidae)

hesperus Müller & Troschel, 1840, see *Craspidaster* (Astropectinidae)

insignis Perrier, 1884, see *Dytaster* (Astropectinidae)

laevis H.L. Clark, 1938: 75; 1946: 80. A synonym of *Archaster angulatus* (Müller & Troschel, 1842 according to Sukarno & Jangoux (1977).

longobrachialis Danielssen & Koren, 1877 (as variety of *A. parelii),* see *Pseudarchaster* (Goniasteridae).

*lorioli Sukarno & Jangoux, 1977
    Sukarno & Jangoux 1977: 834,
    Range: Mauritius*, Maldive Is.
    [Probably conspecific with *Archaster mauritianus* Martens, 1866; but see also *Astropus longipes.]*

lucifer Valenciennes in Perrier, 1875: 228 [1876: 44] (cited only as MS name in synonymy of *Iconaster longimanus* (Möbius, 1859), so a NOMEN NUDUM.

magnificus Bell, 1881, see *Tethyaster* (Astropectinidae)

*mauritianus Martens, 1866: 84. A synonym of *Archaster angulatus* (Müller & Troschel, 1842 according to Perrier (1875). [But from locality must rather be conspecific with *A. lorioli* Sukarno & Jangoux, 1977, which it antedates, so suppression by the ICZN is desirable. See also *Astropus longipes*].

mirabilis Perrier, 1881, see *Cheiraster (Luidiaster)* (Benthopectinidae)

nicobaricus Behn in Möbius, 1859: 13. A synonym of *Archaster typicus* (Müller & Troschel, 1840) according to Perrier (1875).

parelii: M. Sars, 1861, see *Pseudarchaster* (Goniasteridae)

pulcher Perrier, 1881, see *Persephonaster* (Astropectinidae)

robustus Verrill, 1885: 383. A synonym of *Bathybiaster vexillifer* (Wyville

Thomson, 1873) (Astropectinidae) according to Koehler (1909).

sepitus Verrill, 1885, see *Cheiraster (Cheiraster)* (Benthopectinidae)

simplex Perrier, 1881, see *Benthopecten* (Benthopectinidae)

subinermis: Perrier, 1878, see *Tethyaster* (Astropectinidae)

*tenuis Bell, 1894

Bell 1894: 402.

Range: Macclesfield Bank, S China Sea.

[The holotype appears to be an *Astropecten*, and is probably conspecific either with *A. eucnemis* Fisher, 1919 or *A. luzonicus* Fisher, 1913. However, the name *tenuis* hardly deserves to be revived and should be suppressed.]

tenuispinus: M. Sars, 1861, see *Pontaster* (Benthopectinidae)

typicus Müller & Troschel, 1840 (with synonyms *Astropecten stellaris* Gray, 1840 and *Archaster nicobaricus* Behn in Möbius, 1859)

Müller & Troschel 1840a: 104; 1840b: 323; 1842: 65; Sluiter 1895: 52; Hayashi 1939: 419; A.H. Clark 1949a: 74; Domantay 1972: 57; Sukarno & Jangoux 1977: 822; Janssen et al. 1984: 51.

Range: Celebes*, W to Andaman Is, N to Ryu Kyu Is, E to W Polynesia, S to northern Australia.

typicus: A.M. Clark & Davies, 1965: 598, non *Archaster typicus* Müller & Troschel, 1840 = *A. lorioli* Sukarno & Jangoux, 1977 according to Sukarno & Jangoux.

[typicus: Rho, 1971: 69 (lapsus), non *Archaster typicus* Müller & Troschel, 1840 = an asteriid aff. *Marthasterias.*]

vexillifer Wyville Thomson, 1873, see *Bathybiaster* (Astropectinidae)

ASTROPUS Gray, 1840 (subgenus of *Astropecten)*

Gray 1840: 182.

Type species: *Astropecten (Astropus) longipes* Gray, 1840.

[Probably a synonym of *Archaster* Müller & Troschel, 1840 if the supposition given below is correct.]

*longipes (Gray, 1840)

Gray 1840: 182.

Type locality 'Isle of France' (Mauritius).

[Though assumed by Sladen (1889) followed by Döderlein (1917), to be conspecific with *Asterias* [i.e. *Chaetaster*] *longipes* Retzius, 1805 as noted under Chaetasteridae, three factors (detailed under Chaetasteridae) indicate that it was almost certainly a distinct new species of Gray and probably conspecific with *Archaster lorioli* Sukarno & Jangoux, 1977. Since it antedates both *A. lorioli* and *A. mauritianus* Martens, 1866 formal suppression of the name *Astropus longipes* by the ICZN is desirable.]

Family GONIASTERIDAE Forbes

Goniasteriae Forbes 1841: 77.

Goniasteridae: Perrier 1875: 185 [1876: 1]; Verrill 1899: 145-149; Fisher
1911c: 174; Mortensen 1927: 78-80; A.M. Clark 1962a: 22-23; Bernasconi 1964b: 252-253; Tortonese 1965: 155-156; Spencer & Wright 1966:
U56; Halpern 1970b: 193-278 (review); Downey 1973: 46; McKnight
1975: 171-195; Walenkamp 1976: 16; Carrera-Rodriguez & Tommasi
1977: 91; Blake 1987: 519.

Pentagonasteridae Perrier 1884: 231; Sladen 1889: 260-264.

Antheneidae Perrier 1884: 231.

Pentopliidae H.E.S. Clark, 1971: 545.

Genus-group names: Antheniaster, **Anthenoides**, Aphroditaster, **Apollonaster, Astroceramus**, Astrogonium, **Astrothauma, Atelorias, Calliaster, Calliderma, Ceramaster, Chitonaster, Circeaster, Cladaster,
Cryptopeltaster, Diplasiaster**, Dorigona, **Eugoniaster**, Euhippasteria,
Evoplosoma, Fisheraster, **Floriaster, Gephyriaster, Gigantaster, Gilbertaster, Glyphodiscus, Goniaster, [Goniodiscaster], Hippasteria,
[Hoplaster], Iconaster**, Isaster, **Johannaster**, Leptogonaster, **Lithosoma, Litonotaster, Lydiaster, Mabahissaster, Mariaster, Mediaster,
Milteliphaster, [Nectria], [Nectriaster], Nehippasteria**, Nereidaster,
Notioceramus, Nymphaster, Ogmaster, Paragonaster, Peltaster, **Pentagonaster, Pentoplia, Pergamaster, Perissogonaster**, Phaneraster, Philonaster, **Pillsburiaster, Plinthaster, Pontioceramus, Progoniaster,
Pseudarchaster, Pseudoceramaster, Pseudogoniodiscaster**, Pyrenaster, **Rosaster, Sibogaster, Siraster, Sphaeriodiscus, Stellaster, Stellasteropsis**, Stephanaster, **Styphlaster, Tessellaster, Toraster, Tosia,**
Tosiaster.

ANTHENIASTER Verrill, 1899
    Verrill 1899: 173.
    Type species: *Anthenoides sarissa* Alcock, 1893.
    A synonym of *Anthenoides* Perrier, 1881 according to Fisher (1919a).
epixanthus Fisher, 1906, see *Anthenoides*
sarissa: Verrill, 1899, see *Anthenoides*
ANTHENOIDES Perrier, 1881 (with synonyms *Leptogonaster* Sladen,
    1889 and *Antheniaster* Verrill, 1899)
    Perrier 1881: 23; 1884: 246; Macan 1938: 401; Bernasconi 1963b: 20;
    Halpern 1970b: 272; Downey 1973: 48.
    Type species: *Anthenoides piercei* Perrier, 1881.
brasiliensis Bernasconi, 1956: 33; also 1958a: 131; 1960: 20. A synonym of
    Anthenoides piercei Perrier, 1881 according to Walenkamp (1976).
cristatus (Sladen, 1889)
    Sladen 1889: 327 (as *Leptogonaster*).
    Fisher 1919a: 329; Macan 1938: 403; McKnight 1973a: 192 (as *Anthenoi-
    des*).

241

Range: Philippine Is*, New Zealand, Gulf of Aden, 164-510 m.

dubius H.L. Clark, 1938

H.L. Clark 1938: 91; 1946: 95.

Range: NW Australia.

epixanthus (Fisher, 1906)

Fisher 1906a: 1067 (as *Antheniaster*).

Hayashi 1952: 152 (as *Anthenoides*).

Range: Hawaiian Is*, W Japan, 100-488 m.

granulosus Fisher, 1913

Fisher 1913a: 647; 1919a: 333; Döderlein 1924: 65; Fell 1958: 12; Baker & Clark 1970: 4.

Range: Moluccas*, N New Zealand, 485-526 m.

laevigatus Liao & A.M. Clark, 1989

Liao & Clark 1989: 37.

Range: S of Hainan I, S China Sea, 107-170 m.

lithosorus Fisher, 1913

Fisher 1913a: 647; 1919a: 336.

Range: S China Sea near Hong Kong, 380 m.

marleyi Mortensen, 1925

Mortensen 1925b: 149; 1933a: 245; Macan 1938: 402.

Range: Natal*, Zanzibar Channel, 183-274 m.

piercei Perrier, 1881 (with synonym *Anthenoides brasiliensis* Bernasconi, 1956)

Perrier 1881: 23; 1884: 247; Verrill 1915: 113; H.L. Clark 1941: 49; Halpern 1970b: 272; Downey 1973: 48.

Range: Barbados, Windward Is*, N to North Carolina and S to southern Brazil (c.30°S), 20-844 m.

regulosus Fisher, 1913

Fisher 1913a: 648; 1919a: 338; Jangoux 1981a: 464.

Range: Philippine Is, 194-217 m.

sarissa Alcock, 1893

Alcock 1893a: 99; Macan 1938: 401.

Verrill 1899: 174 (as *Antheniaster*).

Range: Andaman Is, 230-456 m.

tenuis Liao & A.M. Clark, 1989

Liao & Clark 1989: 39.

Range: E of Hainan I, S China Sea, 205 m.

APHRODITASTER Sladen, 1889 *(Aphroditaster* Sladen in Thomson & Murray, 1885: 612 a nomen nudum, no species named)

Sladen 1889: 116; Verrill 1899: 195.

Type species: *Aphroditaster gracilis* Sladen, 1889.

A synonym of *Pseudarchaster* according to Mortensen (1927).

gracilis Sladen, 1889, see *Pseudarchaster*

microceramus Fisher, 1913, see *Pseudarchaster*

APOLLONASTER Halpern, 1970

Halpern 1970a: 9.

Type species: *Apollonaster yucatanensis* Halpern, 1970.

yucatanensis Halpern, 1970

Halpern 1970a: 9.

Range: Yucatan, SE Mexico, 40-165 m.

ASTROCERAMUS Fisher, 1906

Fisher 1906a: 1056; 1919a: 309; Döderlein 1924: 60; H.L. Clark 1941: 45.

Type species: *Astroceramus callimorphus* Fisher, 1906.

brachyactis H.L. Clark, 1941

H.L. Clark 1941: 45.

Range: Cuba*, atlantic Colombia and Venezuela, 431-1066 m.

cadessus Macan, 1938

Macan 1938: 388.

Range: Maldive Is, 229 m.

callimorphus Fisher, 1906

Fisher 1906a: 1056.

Range: Hawaiian Is, 233 m.

fisheri Koehler, 1909

Koehler 1909a: 79; Döderlein 1924: 60.

Range: S India*, East Indies, 409-519 m.

lionotus Fisher, 1913

Fisher 1913a: 643; 1919a: 310; Jangoux 1981a: 465.

Range: Philippine Is, 379-407 m.

sphaeriostictus Fisher, 1913

Fisher 1913a: 644; 1919a: 313.

Range: Philippine Is, 295 m.

ASTROGONIUM Müller & Troschel, 1842

Müller & Troschel 1842: 52.

Type species: none designated.

A composite of *Goniaster* L. Agassiz, 1836, *Hippasteria* Gray, 1840, *Pentagonaster* Gray, 1840 and *Tosia* Gray, 1840 according to Verrill (1899) and Fisher (1906). Verrill (1899: 149) noted 'If it were to be used in the modern system, it should be restricted to the group containing *A*. [i.e. *Asterias*] *granularis.*' He went on to reject it altogether as artificial and proposed a new name, *Ceramaster* for the *granularis* group, now widely used. However, Verrill also noted that *Astrogonium* is 'not very different from' *Goniaster* Agassiz. [Final disposal of *Astrogonium* as a synonym of *Goniaster* can be achieved by designation of *Goniaster cuspidatus* Gray, cited as *Astrogonium cuspidatum* by Müller & Troschel, 1842: 56, as type species of *Astrogonium* since this is a synonym of *G. tessellatus* (Lamarck,

1816), the type species of *Goniaster* L. Agassiz, 1836.]

abnormale: Sladen, 1889: 748, also Farquhar 1897: 310. A synonym of *Pentagonaster pulchellus* Gray, 1840 according to Mortensen (1925a).

aculeatum Barrett, 1857: 47. A synonym of *Hippasteria plana* Gray, 1840 [i.e. *H. phrygiana* (Parelius, 1768)] according to Sladen (1889).

aequabile Koehler, 1907b: 37. A synonym of *Pseudarchaster gracilis* (Sladen, 1889) according to Downey (1973).

annectens Perrier, 1894: 343. A synonym of *Astrogonium fallax* Perrier, 1884 [i.e. *Pseudarchaster parelii* (Düben & Koren, 1846)] according to Koehler (1909).

aphrodite Perrier, 1894: 342, 354. A synonym of *Pseudarchaster gracilis* (Sladen, 1889) according to Downey (1973).

articulatum Valenciennes in Perrier 1875: 276 [1876: 91] cited only in synonymy of *Anthenea pentagonula* (Lamarck, 1816) (Oreasteridae).

astrologorum Müller & Troschel, 1842: 54. A variety or forma of *Tosia australis* Gray, 1840 according to Ludwig (1912).

australe: Müller & Troschel, 1842, non *Tosia australis* Gray, 1840 = *Tosia magnifica* Müller & Troschel, 1842 according to A.M. Clark (1953).

boreale Barrett, 1857: 47. A synonym of *Pentagonaster* [i.e. *Ceramaster*] *granularis* (Retzius, 1783) according to Sladen (1889).

bourgeti: Sladen, 1889, see *Sphaeriodiscus*

crassimanum Möbius, 1859: 8. A synonym of *Pentagonaster duebeni* Gray, 1847 according to A.M. Clark (1953).

cuspidatum (Gray, 1840) Müller & Troschel, 1842: 56. A synonym of *Goniaster tessellatus* (Lamarck, 1816) according to Halpern (1970b).

discus: Perrier, 1894, see *Pseudarchaster*

duebeni: Sladen, 1889, see *Pentagonaster*

dubium Perrier, 1869: 89 [277]. A synonym of *Goniaster cuspidatus* Gray, 1840 [i.e. *G. tessellatus* (Lamarck, 1816)] according to Sladen (1889).

emili Perrier, 1869: 84. A synonym of *Tosia magnifica* (Müller & Troschel, 1842) according to A.M. Clark (1953).

eminens Koehler 1907b: 34. A synonym of *Pseudarchaster gracilis* (Sladen, 1889) according to Halpern (1972).

fallax Perrier, 1885c: 37; also 1894: 347. A synonym of *Pseudarchaster parelii* (Düben & Koren, 1846) according to Halpern (1972).

*fonki Philippi, 1858
    Philippi 1858: 267.
    Conspecific with *Porania magellanica* Studer, 1876 (Poraniidae) (currently treated as a subspecies of *P. antarctica* E. A. Smith, 1876) following Madsen (1956), which it antedates. [But Philippi's types lost so suppression by the ICZN desirable.]

geometricum Müller & Troschel, 1842: 54. A synonym of *Tosia australis* Gray, 1840 according to Ludwig (1912).

gracile: Perrier, 1894, see *Pseudarchaster*

granulare: Müller & Troschel, 1842, see *Ceramaster*

greeni Bell, 1889b: 433. A synonym of *Ceramaster grenadensis* (Perrier, 1881) according to Halpern in Downey (1973).

gunni: Sladen, 1889. A forma of *Pentagonaster duebeni* Gray, 1847 according to A. M. Clark (1953).

huttoni Farquhar 1897: 194, a NOMEN NUDUM, undescribed and provisional, = *Astrogonium abnormale* Gray, 1866 [i.e. *Pentagonaster pulchellus* Gray, 1840 according to Farquhar (1898).

hystrix Perrier, 1894: 345. A synonym of *Pseudarchaster parelii* (Düben & Koren, 1846) according to Halpern (1972).

inaequale Gray, 1847, see *Sphaeriodiscus*

intermcdium (Sladen, 1889) Perrier, 1894: 342. A synonym of *Pseudarchaster parelii* (Düben & Koren, 1846) according to Fisher (1911c).

jordani: Koehler, 1909, see *Pseudarchaster*

lamarcki Müller & Troschel, 1842, see under *Pentagonaster*

longobrachiale: Koehler, 1907. A variety of *Astrogonium* [i.e. *Pseudarchaster*] *parelii* (Düben & Koren, 1846).

magnificum Müller & Troschel, 1842, see *Tosia*

marginatum Koehler, 1909a: 71. A synonym of *Pseudarchaster gracilis* (Sladen, 1889) according to Halpern (1972).

meridionale E. A. Smith, 1876, see *Odontaster* (Odontasteridae)

miliare Gray, 1847, see *Diplodontias* (Odontasteridae)

mozaicum: Koehler, 1909, see *Pseudarchaster*

necator Perrier, 1894: 350. A synonym of *Pseudarchaster gracilis* (Sladen, 1889) according to Downey (1973).

ornatum Müller & Troschel, 1842: 55. A synonym of *Tosia magnifica* (Müller & Troschel, 1842) according to A.M. Clark (1953).

parelii: Koehler, 1907, see *Pseudarchaster*

patagonicum Perrier, 1891a: K125. A synonym of *Pseudarchaster discus* Sladen, 1889 according to Fisher (1940).

paxillosum Gray, 1847, see *Diplodontias* (Odontasteridae)

phrygianum: Müller & Troschel, 1842, see *Hippasteria*

pretiosus Döderlein, 1902, referred to *Pseudarchaster* by Goto (1914) followed by Hayashi (1938a) but to *Dipsacaster* (Astropectinidae) by Döderlein himself (1921) followed by Hayashi (1973).

pulchellum: Müller & Troschel, 1842, see *Pentagonaster*

roseum: Koehler, 1909, see *Pseudarchaster*

rugosum Hutton, 1872: 7. A synonym of *Asterodon* [i.e. *Diplodontias*] *miliaris* (Gray, 1847) according to Mortensen (1925).

semilunatus (Martens, 1866: 86) Perrier, 1885: 37. A synonym of *Goniaster tessellatus* (Lamarck, 1816) according to Halpern (1970b).

souleyeti Dujardin & Hupé, 1862: 397. A synonym of *Iconaster longimanus*

(Möbius, 1859) according to Lütken (1871).
tessellatum: Perrier, 1894, see *Pseudarchaster*
tuberculatum Gray, 1847, see *Toraster*
ASTROTHAUMA Fisher, 1913
  Fisher 1913a: 645; 1919a: 320.
  Type species: *Astrothauma euphylacteum* Fisher, 1913.
euphylacteum Fisher, 1913
  Fisher 1913a: 645; 1919a: 320; Chang & Liao 1964: 59.
  Range: Philippine Is*, S China Sea, 200-290 m.
ATELORIAS Fisher, 1911d
  Fisher 1911d: 424; 1919a: 342.
  Type species: *Atelorias anacanthus* Fisher, 1911.
anacanthus Fisher, 1911
  Fisher 1911d: 424; 1919a: 343.
  Range: Celebes*, Philippine Is, 935-1110 m.
CALLIASTER Gray, 1840
  Gray, 1840: 280; Döderlein 1922: 47; A.M. Clark & Courtman-Stock 1976: 60.
  Type species: *Calliaster childreni* Gray, 1840.
acanthodes H.L. Clark 1923
  H.L. Clark 1923a: 264; Mortensen 1933a: 244; A.M. Clark & Courtman-Stock 1976: 60.
  Range: S coast of South Africa* to Natal, 130-411 m.
baccatus Sladen, 1889
  Sladen 1889: 280; Bell 1905: 247; H.L. Clark 1923a: 264; Mortensen 1933a: 244; A. M. Clark & Courtman-Stock 1976: 61.
  Range: False Bay, South Africa* eastwards along south coast, 23-146 m.
childreni Gray, 1840
  Gray 1840: 280; Goto 1914: 429; A.M.C. in Clark & Rowe 1971: 40 (as *Calliaster*).
  Perrier 1875: 216 [1876: 31] (as *Pentagonaster*).
  Range: 'Japan' (Gray 1866), S China.
corynetes Fisher, 1913
  Fisher 1913a: 644; 1919a: 316.
  Range: Philippine Is, 353 m.
elegans Döderlein, 1922
  Döderlein 1922: 49; 1924: 62; A.H. Clark 1952: 284.
  Range: Flores, Indonesia*, Marshall Is, 113-137 m.
erucaradiatus Livingstone, 1936
  Livingstone 1936: 383.
  Range: N.S.W., Australia, 91 m.
mamillifer Alcock, 1893
  Alcock 1893b: 172.

Range: Andaman Sea, Bay of Bengal, 448-494 m.
pedicellaris Fisher, 1906
  Fisher 1906a: 1061.
  Range: Hawaiian Is, 238-276 m.
quadrispinus Liao, 1989
  Liao 1989: 23.
  Range: S from Macao, S China Sea, 88 m.
regenerator Döderlein, 1922
  Döderlein 1922: 50; 1924: 62.
  Range: Molo Strait, Indonesia, 98-165 m.
spinosus H. L. Clark, 1916
  H. L. Clark 1916: 44; 1946: 88; Jangoux 1981a: 465.
  Range: SE Australia*, Philippine Is, 146-407 m.
CALLIDERMA Gray, 1847
  Gray 1847: 76.
  Type species: *Calliderma emma* Gray, 1847.
emma Gray, 1847
  Gray 1847a: 197; 1847b: 343; Goto 1914: 431.
  Perrier 1975: 226 [1876: 41] (as *Pentagonaster*).
  Range: Type locality unknown. Unpublished records from Gulf of Car-
  pentaria, N Australia [Rowe (pers. comm.)] and Hainan I., S China Sea
  [Liao (pers. comm.)].
grayi Bell, 1881a: 95. A forma of *Odontaster penicillatus* (Philippi, 1870)
  according to Fisher (1940).
spectabilis Fisher, 1906
  Fisher 1906a: 1058.
  Range: Hawaiian Is, 143-390 m.
CERAMASTER Verrill, 1899 (with synonym *Philonaster* Koehler, 1909
  (as subgenus of *Pentagonaster), possibly also Tosiaster* Verrill, 1914)
  Verrill 1899: 161; Fisher 1911c: 204; Verrill 1914a: 289; Halpern 1970b:
  212; Downey 1973: 49; McKnight 1973a: 178; A. M. Clark & Courtman-
  Stock 1976: 61.
  Fisher 1906a: 1054 (as subgenus of *Tosia* Gray, 1840).
  Type species: *Asterias granularis* Retzius, 1783, by subsequent designa-
  tion by Verrill (1899).
affinis (Perrier, 1884: 243) Fisher 1911c: 165. A synonym of *Ceramaster
  grenadensis* (Perrier, 1881) according to Halpern (1970b).
ammophilus: Halpern, 1970b, see *Sphaeriodiscus*
arcticus (Verrill, 1909)
  Verrill 1909: 63 (as *Tosia*).
  Fisher 1911c: 219; 1919a: 260; Djakonov 1950a: 50 [1968: 41]; Baranova
  1957: 162 (as *Ceramaster*).
  Verrill 1914a: 292 (as *Tosiaster*).

Range: Aleutian Is, 0-186 m.

Type species of *Tosiaster* Verrill, 1914 but still referred to *Ceramaster* by Fisher (1919) and Djakonov (1950) followed by Baranova (1957) without comment on *Tosiaster.*

balteatus (Sladen, 1891: 688) Mortensen 1927: 82. A synonym of *Ceramaster grenadensis* (Perrier, 1881) according to Halpern in Downey (1973).

bowersi (Fisher, 1906)

Fisher 1906a: 1049 (as *Nereidaster).*

Fisher 1919a: 240, 241 (as *Ceramaster).*

Range: Hawaiian Is, 411-593 m.

chondriscus H.L. Clark 1923a: 258. A synonym of *Ceramaster patagonicus euryplax* H. L. Clark, 1923 according to A.M. Clark (1974).

clarki Fisher, 1910

Fisher 1910a: 552; 1911c: 217; Baranova 1957: 162.

Range: S Bering Sea to S California, 630-1100 m.

cuenoti (Koehler, 1909) (virtually **new comb.**)

Koehler 1909b: 68 (as *Pentagonaster (Tosia)).*

Fisher 1919a: 260 (as *Ceramaster,* by inference).

Range: off Sri Lanka, 1840 m.

euryplax H.L. Clark, 1923a. A subspecies of *Ceramaster patagonicus* (Sladen, 1889).

fisheri Bernasconi, 1963. A subspecies of *Ceramaster patagonicus* (Sladen, 1889).

granularis (Retzius, 1783) (with synonyms *Astrogonium boreale* Barrett, 1857, *Pentagonaster eximius* Verrill, 1894 and *P. simplex* Verrill, 1895)

Retzius 1783: 238 (as *Asterias).*

Müller & Troschel, 1842: 57; Düben & Koren 1846: 246; Verrill 1885: 542 (as *Astrogonium).*

Lütken 1865: 146 (as *Goniaster).*

Perrier 1875: 250 [1876: 65]; Danielssen & Koren 1881: 268; 1884b: 58; Ludwig 1900: 456; Grieg 1905: 1 (as *Pentagonaster).*

Verrill 1899: 162 (as *Tosia (Ceramaster)).*

Verrill 1914a: 290; Koehler 1924: 175; Mortensen 1927: 81; Djakonov 1950a: 47 [1968: 39]; Tortonese 1955: 675; Tortonese & A.M. Clark 1956: 348; Harvey et al. 1988: 161 (as *Ceramaster).*

Range: 'St Croix'* (?New Brunswick, E Canada), circumpolar N Boreal, S to Long Id, U.S.A., Rockall Trough and Norway, also N Pacific, 40-2185 m.

grenadensis (Perrier, 1881) (with synonyms *Pentagonaster affinis* Perrier, 1884, *P. gosselini* Perrier, 1884, *P. haesitans* Perrier, 1885, *P. deplasi* Perrier, 1885, *P. crassus* Perrier, 1885, *Astrogonium greeni* Bell, 1889, *P.*

248

balteatus Sladen, 1891, *P. hystricis* Marenzeller, 1893 and *P. kergroheni* Koehler, 1895).

Perrier 1881: 19; 1884: 232 (as *Pentagonaster*).

Verrill 1899: 162 (as *Tosia (Ceramaster)*).

Verrill 1915: 222; H.L. Clark 1941: 38; Halpern 1970b: 213; Downey 1973: 49 (as *Ceramaster*).

Range: Grenada, Windward Is*, Florida to northern Brazil, the Azores, W of Ireland S to the Gulf of Guinea and Mediterranean, 200-2500 m.

hystricis (Marenzeller, 1893: 4) Koehler, 1924: 176. A synonym of *Ceramaster grenadensis* (Perrier, 1881) according to Halpern in Downey (1973).

japonicus (Sladen, 1889)

Sladen 1889: 272; Goto 1914: 313 (as *Pentagonaster*).

Verrill 1899: 179 (as *Mediaster*).

Fisher 1911c: 206; Djakonov 1950a: 46 [1968: 39]; Baranova 1957: 161 (as *Ceramaster*).

Range: N Japan Sea to Bering Sea and SE to Oregon, U.S.A., 194-1410 m.

lennoxkingi McKnight, 1973

McKnight 1973a: 178.

Range: southern New Zealand, 252-1080 m.

leptoceramus (Fisher 1905)

Fisher 1905: 306 (as *Tosia*).

Fisher 1911c: 210; Blake 1973: 51 (as *Ceramaster*).

Range: southern California, 395-1170 m.

misakiensis (Goto, 1914)

Goto 1914: 332 (as *Astrogonium*).

Hayashi 1973b: 13 (as *Ceramaster*).

Range: Sagami Bay area, SE Japan, 560 m.

mortenseni (Koehler, 1909)

Koehler 1909b: 74 (as *Pentagonaster (Philonaster)*).

Fisher 1919a: 260; Downey 1973: 49 (as *Ceramaster*).

Range: Bay of Bengal, off Burma, 1755 m.

patagonicus (Sladen, 1889) (with synonyms *Pentagonaster austrogranularis* Perrier, 1891 and *Ceramaster chondriscus* H.L. Clark, 1923, also subspecies *euryplax* H. L. Clark, 1923, *productus* Djakonov, 1950 and *fisheri* Bernasconi, 1963)).

patagonicus patagonicus (Sladen, 1889)

Sladen 1889: 269 (as *Pentagonaster*).

Verrill 1899: 184 (as *Mediaster*).

Fisher 1911c: 214; 1940: 118; [?]Tommasi 1970: 12 (as *Ceramaster*).

Range: Magellan Strait*, Falkland Is, N to S Brazil [?], 100-450 m.

patagonicus euryplax H.L. Clark, 1923

H.L. Clark 1923a: 262; 1926a: 9; Mortensen 1933a: 243; A.M. Clark 1952a: 204 (as variety).

A.M. Clark 1974: 435; A.M.C. in Clark & Courtman-Stock 1976: 61 (as subspecies).

Range: W and S coasts of South Africa, 156-462 m.

patagonicus fisheri Bernasconi, 1963

Bernasconi 1963: 288.

Range: Gulf of California [? to Bering Sea].

patagonicus productus Djakonov, 1950 *(Ceramaster patagonicus productus* Djakonov, 1949: 19 a nomen nudum, insufficiently characterized)

Djakonov 1950b: 50; 1950a: 46 [1968: 40]; Baranova 1957: 161.

Range: Okhotsk Sea.

placenta: Fisher, 1911, see *Peltaster*

?scotiocryptus: Halpern, 1970b, see *Sphaeriodiscus*

smithi Fisher, 1913

Fisher 1913a: 640; 1919a: 257.

Range: Philippine Is, 686-1010 m.

stellatus Djakonov, 1950 *(Ceramaster stellatus* Djakonov, 1949: 20 a nomen nudum, insufficiently characterized)

Djakonov 1950b: 48; 1950a: 49 [1968: 41]; Baranova 1957: 161.

Range: Okhotsk and Bering Seas, 110-500 m.

trispinosus H.L. Clark, 1923

H.L. Clark 1923a: 260; A.M. Clark & Courtman-Stock 1976: 62.

Range: SW of Cape Town, South Africa, 420 m.

CHITONASTER Sladen, 1889

Sladen 1889: 282.

Type species: *Chitonaster cataphractus* Sladen, 1889.

cataphractus Sladen, 1889

Sladen 1889: 283; A.M. Clark 1962a: 22.

Range: off Queen Mary Land, Antarctica, 3610 m.

johannae Koehler, 1907

Koehler 1907a: 144; 1908a: 542.

Range: off South Orkney Is, Weddell Quadrant, Southern Ocean, 3246 m.

CIRCEASTER Koehler, 1909 (with synonym *Lydiaster* Koehler, 1909)

Koehler 1909b: 83; Halpern 1970b: 265.

Type species: *Circeaster marcelli* Koehler, 1909, by subsequent designation by Halpern (1970).

americanus (A.H. Clark, 1916) (with synonym *Circeaster occidentalis* H. L. Clark, 1941)

A.H. Clark 1916c: 141 (as *Lydiaster).*

Halpern 1970b: 265; Downey 1973: 55 (as *Circeaster).*

Range: Gulf of Mexico*, ESE to Guyana, 500-1450 m.

250

johannae (Koehler, 1909)
    Koehler 1909b: 92 (as *Lydiaster*).
    Halpern 1970b: 265 (as *Circeaster*).
    Range: off Sri Lanka, 730 m.
magdalenae Koehler, 1909
    Koehler 1909b: 88.
    Range: SW of Bombay, India, 1670-1700 m.
marcelli Koehler, 1909
    Koehler 1909b: 84.
    Range: S of India, 1925 m.
occidentalis H.L. Clark 1941: 46. A synonym of *Circeaster americanus* A.
    H. Clark, 1916 according to Halpern (1970b).
CLADASTER Verrill, 1899
    Verrill 1899: 175; Fisher 1911c: 221; Bernasconi 1963b: 13.
    Type species: *Cladaster rudis* Verrill, 1899.
analogus Fisher, 1940
    Fisher 1940: 123; Bernasconi 1963b: 14; 1973: 287.
    Range: Falkland Is*, northern Argentina, 147-151 m.
imperialis: Fisher 1940: 125, see *Hippasteria*
macrobrachius H.L. Clark, 1923
    H.L. Clark 1923a: 268; 1926a: 12; A.M. Clark & Courtman-Stock 1976:
    62.
    Range: SW of Cape Point, South Africa, 380-570 m.
rudis Verrill, 1899
    Verrill 1899: 176; Fisher 1911c: 223; 1940: 125; Bernasconi 1963: 14.
    Range: Blake Plateau, NE Florida* (also unpublished records from Yuca-
    tan Channel, SW of Cuba, identified by Halpern), 150-900 m.
validus Fisher, 1910
    Fisher 1910a: 552; 1911c: 222; Baranova 1957: 162.
    Range: Aleutian Is, 517 m.
CRIPTOPELTASTER Codoceo & Andrade, 1981, lapsus for
CRYPTOPELTASTER Fisher 1905
    Fisher 1905: 311; 1911c: 237.
    Type species: *Cryptopeltaster lepidonotus* Fisher, 1905.
lepidonotus Fisher 1905
    Fisher 1905: 311; 1911c: 238.
    Range: California, 487 m.
philippii Codoceo & Andrade, 1981
    Codoceo & Andrade 1981: 381.
    Range: Chile.
DIPLASIASTER Halpern, 1970
    Halpern 1970a: 7.
    Type species: *Plinthaster productus* A.H. Clark, 1917.

productus (A. H. Clark, 1917)

A. H. Clark 1917b: 67 (as *Plinthaster*).

Halpern: 1970a: 7 (as *Diplasiaster*).

Range: Cuba*, Florida Strait and Venezuela, 78-567 m.

DONGONA Gray, 1866, figure caption, lapsus for

DORIGONA Gray, 1866: 7.

Type species: *Dorigona reevesi* Gray, 1866, a synonym of *Ogmaster capella* (Müller & Troschel, 1842) by subsequent virtual designation by Sladen (1889) (see Fisher, 1919a: 262, 305).

A synonym of *Ogmaster* Martens, 1865 according to Fisher (1917d).

longimana Gray, 1866, see *Iconaster*

reevesi Gray, 1866: 7. A synonym of *Ogmaster capella* (Müller & Troschel, 1842) according to Sladen (1889).

DORIGONA: Perrier, 1885c: 39; 1894: 365, non *Dorigona* Gray, 1866, mostly = *Nymphaster* Sladen, 1889 according to Verrill (1899).

arenata: Perrier, 1885c and 1894, see *Nymphaster*

belli Koehler 1909b: 58. A synonym of *Nymphaster moebii* (Studer, 1884) according to Macan (1938).

confinis Koehler, 1910, see *Rosaster*

jacqueti Perrier, 1894: 383. A synonym of *Nymphaster arenatus* (Perrier, 1881) according to Macan (1938).

ludwigi Koehler, 1909b: 61. A synonym of *Nymphaster moebii* (Studer, 1884) according to Macan (1938).

pentaphylla Alcock, 1893, see *Lithosoma*

prehensilis Perrier, 1885: 39; 1894: 32. A synonym of *Nymphaster arenatus* (Perrier, 1881) according to Macan (1938).

subspinosa: Perrier, 1894, see *Nymphaster*

ternalis: Perrier, 1885c: 39; also 1894: 371. A synonym of *Nymphaster arenatus* (Perrier, 1881) according to Halpern (1970b).

ternalis: Koehler, 1909b: 54, non *Dorigona ternalis* (Perrier, 1881), = *Nymphaster moebii* (Studer, 1884) according to Macan (1938).

EUGONIASTER Verrill, 1899

Verrill 1899: 172.

Type species: *Pentagonaster investigatoris* Alcock, 1893.

A synonym of *Plinthaster* Verrill, 1899 according to Halpern (1970b).

doederleini: Macan 1938, see *Plinthaster*

ephemeralis Macan, 1938, see *Plinthaster*

investigatoris: Verrill 1899, see *Plinthaster*

EUHIPPASTERIA Dons, 1938 (as subgenus of *Hippasteria* Gray, 1840)

Dons 1938: 17.

Type species: *Asterias phrygiana* Parelius, 1768, so coordinate with and a synonym of *Hippasteria* Gray, 1840.

EVOPLOSOMA Fisher, 1906
  Fisher 1906a: 1065; Koehler 1909b: 96.
  Type species: *Evoplosoma forcipifera* Fisher, 1906.
augusti Koehler, 1909
  Koehler 1909b: 96.
  Range: Sri Lanka, 733 m.
forcipifera Fisher, 1906
  Fisher 1906a: 1065.
  Range: Hawaiian Is, 930-1250 m.
scorpio Downey, 1981
  Downey 1981: 561; Gage et al. 1983: 280.
  Range: NE Bay of Biscay*, Rockall Trough, 1600-1900 m.
virgo Downey, 1982
  Downey 1982: 772.
  Range: NW Gulf of Mexico, 2060 m.
FISHERASTER Halpern, 1970
  Halpern 1970a: 2.
  Type species: *Aphroditaster microceramus* Fisher, 1913.
  A synonym of *Pseudarchaster* Sladen, 1889 according to Aziz & Jangoux
  (1984c).
microceramus: Halpern, 1970a, see *Pseudarchaster*
FLORIASTER Downey, 1980
  Downey 1980: 105.
  Type species: *Floriaster maya* Downey, 1980.
maya Downey, 1980
  Downey 1980: 105.
  Range: Yucatan Channel, SW of Cuba, 933-1025 m.
GEPHYREASTER Fisher, 1910, referred from Radiasteridae (Paxillosida)
  by Blake (1987). See pt. 1, pp. 294-295.
GIGANTASTER Döderlein, 1924
  Döderlein 1924: 65.
  Type species: *Gigantaster weberi* Döderlein, 1924.
weberi Döderlein, 1924
  Döderlein 1924: 66.
  Range: E Java Sea, Indonesia, 538-700 m.
GILBERTASTER Fisher, 1906
  Fisher 1906: 1062.
  Type species: *Gilbertaster anacanthus* Fisher, 1906.
anacanthus Fisher, 1906
  Fisher 1906: 1063.
  Range: Hawaiian Is, 462-700 m.
brodei McKnight, 1973
  McKnight 1973a: 192.

Range: NE New Zealand, 750-913 m.
GLYPHODISCUS Fisher, 1917
Fisher 1917e: 173; 1919a: 306 (as subgenus of *Iconaster* Sladen, 1889).
Rowe 1989a: 273 (as genus).
Type species: *Iconaster perierctus* Fisher, 1913.
mcknighti Rowe, 1989
Rowe 1989a: 273.
Range: Off Norfolk I, NE of New Zealand, 450-475 m.
perierctus (Fisher, 1913)
Fisher 1913a: 642 (as *Iconaster*).
Fisher 1917e: 173; 1919a: 306 (as *Iconaster (Glyphodiscus)*).
Rowe 1989a: 273 (as *Glyphodiscus*).
Range: Philippine Is, 178 m.
GONIASTER L. Agassiz, 1836 (with synonyms *Astrogonium* Müller & Troschel, 1842 [after new type designation] and *Phaneraster* Perrier, 1894).
L. Agassiz 1836a: 143; 1836b: 191; Gray 1840: 280; Lütken 1865: 143; Verrill 1899: 150; Koehler 1909a: 87; Fisher 1911c: 167; Verrill 1915: 102; Tortonese 1937: 55; Halpern 1970a: 256.
Type species: *Asterias tessellatus* Lamarck, 1816.
abbensis Forbes, 1843: 280. A synonym of *Hippasteria plana* Gray, 1840 [i.e. *H. phrygiana* (Parelius. 1768)] according to Sladen (1889).
acutus (Heller, 1863) Lütken, 1865: 145. A synonym of *Pentagonaster* [i.e. *Peltaster*] *placenta* Müller & Troschel, 1842 according to Ludwig (1897).
africanus Verrill, 1871b: 131; 1899: 156. A synonym of *Goniaster cuspidatus* Gray, 1840 [i.e. *G. tessellatus* (Lamarck, 1816)] according to Madsen (1950).
africanus: Koehler, 1914: 169, non *Goniaster africanus* Verrill, 1871, = *Plinthaster dentatus* (Perrier, 1884) according to Halpern (1970b).
americanus Verrill, 1871b: 130; 1899: 151; 1915: 104. A synonym of *Goniaster cuspidatus* Gray, 1840 [i.e. *G. tessellatus* (Lamarck, 1816)] according to Madsen (1950).
articulatus Lütken, 1865: 147. A synonym of *Goniodiscaster scaber* (Möbius, 1859) according to Döderlein (1935) and Madsen (1959b).
articulatus (L. Agassiz, MS) Gray, 1866: 9. A synonym of *Anthenea pentagonula* (Lamarck, 1816) according to Perrier (1875) but others, notably Goto, 1914: 442, disagree.
belcheri Lütken, 1865. A forma of *Stellaster equestris* (Retzius, 1805) according to Döderlein (1935).
capella: Martens, 1865 (as *Goniaster (Ogmaster)*), see *Ogmaster*
cuspidatus Gray, 1840: 280; also Tortonese 1937: 31; Madsen 1950: 209. A

synonym of *Goniaster tessellatus* (Lamarck, 1816) according to Halpern (1970b).

duebeni: Lütken, 1865, see *Pentagonaster*

equestris (Linnaeus, 1758) L. Agassiz, 1836: 191. A synonym of *Astrogonium* [i.e. *Hippasteria*] *phrygiana* (Parelius, 1768) according to Müller & Troschel (1842).

equestris: Martens, 1865 (as *Goniaster (Stellaster)*), see *Stellaster*

granularis: Lütken, 1865, see *Ceramaster*

hispidus M. Sars, 1872, see *Poraniomorpha* (Poraniidae)

incei: Lütken, 1865. A forma of *Stellaster equestris* (Retzius, 1805) according to Döderlein (1935).

lamarcki Verrill, 1899: 157, see under *Pentagonaster*

longimanus: Lütken, 1865, see *Iconaster*

muelleri Martens, 1865: 359 (as *Goniaster (Stellaster)*). A synonym of *Stellaster equestris* forma *childreni* Gray, 1840 according to Döderlein (1935).

muelleri: Lütken, 1871: 248, non *Goniaster muelleri* Martens, 1865, = *Ogmaster capella* (Müller & Troschel, 1842) according to Döderlein (1935).

multiporum Hoffman in Rowe, 1974: 214 (as *Goniaster (Goniodiscus)*). A synonym of *Culcita novaeguineae* Müller & Troschel, 1842 (Oreasteridae) according to Rowe.

nidarosiensis Storm, 1881, see *Peltaster*

obtusangulus: Perrier, 1875, see *Pseudoreaster* (Oreasteridae)

pentagonulus: Martens, 1867: 111, non *Anthenea pentagonula* (Lamarck, 1816), = *Anthenea regalis* Koehler, 1910 according to Döderlein (1915).

phrygianus: Norman, 1865, see *Hippasteria*

placenta: Marenzeller, 1875, see *Peltaster*

placentaeformis (Heller, 1863) Lütken, 1865: 145. A synonym of *Pentagonaster* [i.e. *Peltaster*] *placenta* (Müller & Troschel, 1842) according to Ludwig (1897).

pleyadella: Martens, 1865, see *Goniodiscaster*

*regularis Gray, 1840: 280. From *Pentagonaster regularis* of Linck, 1733 pl. 13, fig. 22 and Seba 3 pl. 8, fig. 4. [No subsequent references traced.]

*sebae Gray, 1840: 280 (from Seba's pl.8, fig.2). A synonym of *Culcita novaeguineae* according to Fisher (1919). [But locality St Thomas, presumably West Indies, not Indo-West Pacific.]

sebae (Müller & Troschel, 1842, as *Goniodiscus,* from Seba's pl.8, figs 7,8) Martens, 1866: 86, non *G. sebae* Gray, 1840, = *Culcita novaeguineae* Müller & Troschel, 1842 according to Goto (1914).

semilunatus Martens, 1866: 86. From *Pentagonaster semilunatus* Linck, 1733. A synonym of *Goniaster tessellatus* Lamarck, 1816 according to Halpern (1970b).

templetoni Forbes, 1839: 118. A synonym of *Porania pulvillus* O. F. Müller, 1776 (Poraniidae) according to Perrier (1875).

tessellatus (Lamarck, 1816) (with synonyms *Goniaster cuspidatus* Gray, 1840, *G. semilunatus* Martens, 1866, *Astrogonium dubium* Perrier, 1869, *G. africanus* Verrill, 1871 and *G. americanus* Verrill, 1871).

Lamarck 1816: 552 (as *Asterias,* varieties C and D only).

Agassiz 1836a: 143; 1836b: 191; Fisher 1911c: 167; Halpern 1970b: 256 (validated by Agassiz's restriction of Lamarck, 1816); Downey 1973: 46 (as *Goniaster*).

Range: North Carolina to Brazil, Morocco to Gabon. (Also from various localities in the Indo-West Pacific by Halpern (1970b) but second-hand and almost certainly mistakes, H. L. Clark's of 1909 from Australia being repudiated by him in 1946.)

tuberculatus: Martens, 1869, see *Toraster*

tuberculosus Martens, 1865, see under *Stellaster*

GONIASTER: Perrier, 1875: 268, non *Goniaster* L. Agassiz, 1836, = *Pseudoreaster* Verrill, 1899, nom. nov.

GONIASTER (OGMASTER) Martens, 1865, see *Ogmaster*

[*GONIODISCASTER* H. L. Clark, 1909, see Oreasteridae].

HIPPASTERIA Gray, 1840 (with synonyms *Euhippasteria* Dons, 1938 and probably also *Nehippasteria* Dons, 1938, as *subgenera)*

Gray 1840: 279; Perrier 1875: 371 [1876: 86]; Fisher 1911c: 223; Verrill 1914a: 300; Koehler 1924: 178; Mortensen 1927: 88; Dons 1938: 17; Fisher 1940: 125; Djakonov 1950a: 51 [1968: 42]; Bernasconi 1964b: 253; A. M. Clark & Courtman-Stock 1976: 63.

Type species: *Hippasteria europaea* Gray, 1840, a synonym of *Asterias phrygiana* Parelius, 1768, by virtual monotypy, all four nominal species included by Gray being synonymous with *H. phrygiana.*

aculeata Djakonov, 1950: 56. A forma of *Hippasteria leiopelta* Fisher, 1910. [Would be a junior secondary homonym of *aculeata* Barrett, 1857 (as *Astrogonium)* if that had not been synonymized with *Hippasteria phrygiana* (Parelius, 1768) by Sladen (1889).]

argentinensis Bernasconi, 1961
Bernasconi 1961b: 1.
Range: northern Argentina, c.38°S, 108-162 m.

armata Fisher, 1911c: 230. A forma of *Hippasteria leiopelta* according to Fisher but of *H. spinosa* Verrill, 1909 according to Djakonov (1950).

californica Fisher, 1905
Fisher 1905: 310; 1911c: 233.
Range: Southern California* to Washington.

caribaea Verrill, 1899
Verrill 1899: 174.
Range: Not Caribbean but NE Florida, c.31°N, 490 m.

256

colossa Djakonov, 1950 *(Hippasteria colossa* Djakonov, 1949: 21 a nomen nudum, insufficiently characterized)
Djakonov 1950a: 55 [1968: 47]; 1952a: 411; Baranova 1957: 162.
Range: Bering Sea, 238 m.

cornuta Gray, 1840: 279. A synonym of *Hippasteria plana* (Gray, 1840) [i.e. *H. phrygiana* (Parelius, 1768)] according to Sladen (1889).

derjugini Djakonov, 1950 *(Hippasteria derjugini* Djakonov, 1949: 22 a nomen nudum, insufficiently characterized)
Djakonov 1950a: 55 [1968: 46]; 1952a: 412.
Range: Okhotsk Sea, 192 m.

europaea Gray, 1840: 279. A synonym of *Astrogonium* [i.e. *Hippasteria*] *phrygiana* (Parelius, 1768) according to Müller & Troschel, 1842.

falklandica Fisher, 1940
Fisher 1940: 125; Bernasconi 1973a: 287.
Range: Falkland Is*, northern Argentina, 225-251 m.

heathi Fisher, 1905
Fisher 1905: 319; 1911c: 231.
Range: Alaska, 377-454 m.

hyadesi Perrier, 1891a: K128; A.M. Clark 1962a: 22. A synonym of *Hippasteria plana* (Gray, 1840) [i.e. *H. phrygiana* (Parelius, 1768)] according to Koehler (1926, *Echinoidea* of the Australasian Antarctic Expedition, p. 107; noted by A.M.C., 1962).

imperialis Goto, 1914
Goto 1914: 338; Hayashi 1952: 338; 1973b: 6, 15.
Fisher 1940: 125 (as *?Cladaster*).
Range: Sagami Bay area, southern Japan, 640 m.
Possibly a *Cladaster* according to Fisher (1940) but retained as a valid species of *Hippasteria* by Hayashi (1952 and 1973).

*insignis Dons, 1938: 16 (as *Hippasteria (Nehippasteria)*). [Possibly a synonym of *Hippasteria phrygiana* (Parelius, 1768) but Madsen (pers. comm., 1987) thinks it to be distinct.]

johnstoni (Gray, 1836) Gray, 1840: 279. A synonym of *Astrogonium* [i.e. *Hippasteria*] *phrygiana* (Parelius, 1768) according to Müller & Troschel (1842).

kurilensis Fisher, 1911
Fisher 1911c: 226 (as *H. spinosa kurilensis*).
Djakonov 1950a: 56 [1968: 48] (as species).
Range: Kurile Is, Okhotsk Sea, 165-500 m.

leiopelta Fisher, 1910 (with formae *aculeata* and *longimana* Djakonov, 1950)
Fisher 1910a: 553; 1911c: 227; Djakonov 1950a: 56 [1968: 47]; 1952a: 413 (with formae *aculeata* and *longimana*).
Range: S Bering Sea, Okhotsk Sea, Tartar Strait.

257

leiopelta armata Fisher 1911, see *Hippasteria spinosa*

longimana Djakonov, 1950: 56. A forma of *Hippasteria leiopelta*.

magellanica Perrier, 1888: 764. A NOMEN NUDUM, undescribed.

mammifera Djakonov, 1950 *(Hippasteria mammifera* Djakonov, 1949: 21 a nomen nudum, insufficiently characterized)

Djakonov 1950a: 54 [1968: 45]; 1952a: 409.

Range: Okhotsk Sea, 97 m.

nozawai Goto, 1914

Goto 1914: 344.

Range: Hokkaido, northern Japan.

pacifica Ludwig, 1905: 138. Probably a synonym of *Cryptopeltaster lepido-notus* Fisher, 1905 according to Fisher (1911).

pedicellaris Djakonov, 1950 *(Hippasteria pedicellaris* Djakonov, 1949: 21 a nomen nudum, insufficiently characterized)

Djakonov 1950a: 54 [1968: 45]; 1952a: 410.

Range: Okhotsk Sea.

philippinensis Domantay & Roxas, 1938: 209. A synonym of *Culcita novae-guineae* Müller & Troschel, 1842 (Oreasteridae) according to A.H. Clark (1949).

***phrygiana** (Parelius, 1768) (with synonyms *Asterias johnstoni* Gray, 1836, *Hippasteria cornuta* Gray, 1840, *H. europaea* Gray, 1840, *H. plana* Gray, 1840, *Goniaster abbensis* Forbes, 1843, *Astrogonium aculeatum* Barrett, 1857, *H. hyadesi* Perrier, 1891 and [possibly] *H. (Nehippasteria) insignis* Dons, 1938, also subspecies *capensis* Mortensen, 1933 and [?] variety *aculeatum* Barrett, 1843).

[See under *Stellaster equestris* for nomenclatural note.]

phrygiana phrygiana (Parelius, 1768)

Parelius 1768: 425 (as *Asterias).*

Müller & Troschel 1842: 52 (as *Astrogonium).*

Norman 1865: 123; 1869: 313 (as *Goniaster,* with variety *aculeatum* Barrett, 1857).

Verrill 1874: 413; 1885b: 542; Bell 1891a: 2; Ludwig 1900: 457; Mortensen 1927: 88; A.H. Clark 1949b: 373; Djakonov 1950a: 53 [1968: 45]; Buchanan 1966: 25.

Range: boreal N. Atlantic, S to Cape Cod and northern British Isles, 20-860 m.

phrygiana capensis Mortensen, 1933

Mortensen 1933a: 245; A.M. Clark 1952: 207 (as variety).

A.M.C. in Clark & Courtman-Stock 1976: 63 (as subspecies).

Range: SW of Cape Town*, W of Cape Province, 310-980 m.

plana Gray, 1840: 279 (from Linck, 1733); also Sladen 1889: 341. A synonym of *Hippasteria phrygiana* (Parelius, 1768) according to Mortensen (1927).

258

spinosa Verrill, 1909 (with forma *armata* (Fisher, 1911, of *H. leiopelta)*
    Verrill 1909: 63; Fisher 1911c: 224; Goto 1914: 349; Verrill 1914a: 301;
    Djakonov 1950a: 53 [1968: 45] (subspecies *kurilensis* Fisher, 1911, raised
    to species but forma *armata* Fisher transferred from *H. leiopelta); Bara-
    nova 1957: 162.
    Range: British Columbia, Canada*, S California to Bering Sea, 50-512
    m.
strongylactis H.L. Clark, 1926
    H.L. Clark 1926a: 13; A.M. Clark & Courtman-Stock 1976: 63.
    Range: west coast of South Africa, 320-980 m.
trojana Fell, 1958
    Fell 1958: 11; McKnight 1967a: 300.
    Range: Chatham Is, E of New Zealand*, Campbell Plateau, 400-690 m.
*HIPPASTERIA (NEHIPPASTERIA) Dons, 1938
    Dons 1938: 16.
    Type species: *Hippasteria (Nehippasteria) insignis* Dons, 1938. But see
    under *Hippasteria* and *H. insignis.*
[*HOPLASTER* Perrier in Milne Edwards, 1882: 48; also Verrill, 1899: 197,
    see Odontasteridae]
ICONASTER Sladen, 1889 Sladen 1889: 261; Fisher 1917d: 168; 1919a:
    303; Aziz & Jangoux 1984b: 189.
    Type species: *Astrogonium longimanum* Möbius, 1859.
corindonensis Aziz & Jangoux 1984
    Aziz & Jangoux 1984b: 187.
    Range: Macassar Strait, 490 m.
elegans Jangoux, 1981
    Jangoux 1981a: 467.
    Range: Philippine Is, 186-208 m.
gardineri Bell, 1909, see *Nymphaster*
longimanus (Möbius, 1859) (with synonym *Astrogonium souleyeti* Dujardin
    & Hupé, 1862)
    Möbius 1859: 7 (as *Astrogonium).*
    Gray 1866: 7; H.L. Clark 1916: 36 (as *Dorigona).*
    Lütken 1865: 144 (as *Goniaster).*
    Perrier 1875: 228 [1876: 44] (as *Pentagonaster).*
    Sladen 1889: 261; Fisher 1919a: 303; Döderlein 1924: 56; Engel 1938a: 3;
    H.L. Clark 1946: 95; A.M. Clark & Rowe 1971: 32; Aziz & Jangoux
    1984b: 189 (as *Iconaster).*
    Range: SE Arabia, Philippine Is, Indonesia to northern Australia; 30-85
    m.
pentaphylla Koehler, 1909, see *Lithosoma*
ICONASTER (GLYPHODISCUS) Fisher 1917. Raised to generic rank by
    Rowe (1989a).

periectus: Aziz & Jangoux 1984b: 189, lapsus for *perierctus*
perierctus Fisher, 1913, see *Glyphodiscus*
ISASTER Verrill, 1894: 257. A junior homonym of *Isaster* Desor, 1858, =
  Mediaster Stimpson, 1857 according to Verrill (1899).
bairdi: Verrill, 1894, see *Mediaster*
JOHANNASTER Koehler, 1909
  Koehler 1909b: 7.
  Type species: *Johannaster superbus* Koehler, 1909.
giganteus Goto, 1914, see *Mariaster*
superbus Koehler, 1909
  Koehler 1909b: 8.
  Range: Arabian Sea SW of Bombay, 1670-1700 m.
LEPTOGONASTER Sladen, 1889 *(Leptogonaster* Sladen in Thomson &
  Murray, 1885: 616 a nomen nudum, diagnosed but no species named).
  Sladen 1889: 326.
  Type species: *Leptogonaster cristatus* Sladen, 1889.
  A synonym of *Anthenoides* Perrier, 1881 according to Fisher (1919).
cristatus Sladen, 1889, see *Anthenoides*
LITHOSOMA Fisher, 1911
  Fisher 1911d: 422; 1919a: 298; Macan 1938: 387; McKnight 1973a: 189.
  Type species: *Lithosoma actinometra* Fisher, 1911.
actinometra Fisher, 1911
  Fisher 1911d: 422; 1919a: 298; Döderlein 1924: 58 *(actinometrae).*
  Range: Philippine Is*, Borneo, 208-540 m.
  A possible synonym of *Lithosoma pentaphylla* (Alcock, 1893) according
  to Fisher (1919a).
actinometrae Döderlein, 1924: 58, lapsus for *actinometra*
brevipes Döderlein, 1924. A variety of *Lithosoma pentaphylla* (Alcock,
  1893).
japonica Hayashi, 1952
  Hayashi 1952: 149.
  Range: Seto, W Japan.
novaezealandiae McKnight, 1973
  McKnight 1973a: 189.
  Range: S of New Zealand, c.49°S, 169°E*, to northern New Zealand,
  413-820 m.
ochlerotatus Macan, 1938
  Macan 1938: 386.
  Range: Zanzibar area, 280 m.
penichra Fisher, 1917
  Fisher 1917c: 90; 1919a: 301.
  Range: Philippine Is, 517-800 m.
pentaphylla (Alcock, 1893) (with possible synonym *Lithosoma actinometra*

Fisher, 1911 and variety *brevipes* Döderlein, 1924)

Alcock 1893a: 93 (as *Dorigona*).

Koehler 1909b: 64 (as *Iconaster*).

Fisher 1919a: 301; Döderlein 1924: 59 (variety *brevipes*) (as *Lithosoma*).

Range: Andaman Sea, 495 m; variety *brevipes* from Timor, Indonesia, 520 m.

LITONOTASTER Verrill, 1899

Verrill, 1899: 171; Halpern 1969a: 129; 1970b: 252.

Type species: *Pentagonaster intermedius* Perrier, 1884.

africanus Halpern, 1969

Halpern 1969a: 134.

Range: off São Thomé, Gulf of Guinea, 2525 m.

intermedius (Perrier, 1884)

Perrier 1884: 243 (as *Pentagonaster*).

Verrill 1899: 172; H.L. Clark 1941: 43; Madsen 1951: 88; Halpern 1969a: 130; Downey 1973: 56 (as *Litonotaster*).

Range: Gulf of Mexico*, Cuba to Guyana, 1960-3530 m.

rotundigranulum Halpern, 1969

Halpern 1969a: 139; 1970b: 252.

Range: Florida Strait*, Guyana, 1050-1630 m.

tumidus H.L. Clark, 1920

H.L. Clark 1920: 83.

Range: East Pacific, c.06°S, 83°W* and off Peru, 4060-5200 m.

LYDIASTER Koehler, 1909

Koehler 1909b: 91.

Type species: *Lydiaster johannae* Koehler, 1909.

A synonym of *Circeaster* Koehler, 1909 according to Halpern (1970b).

americanus A.H. Clark, 1916, see *Circeaster*

johannae Koehler, 1909, see *Circeaster*

MABAHISSASTER Macan, 1938

Macan 1938: 391.

Type species: *Mabahissaster zengi* Macan, 1938.

zengi Macan, 1938

Macan 1938: 391.

Range: Zanzibar area, 183-194 m.

MARIASTER A.H. Clark, 1916

A.H. Clark 1916c: 116.

Type species: *Johannaster giganteus* Goto, 1914.

giganteus (Goto, 1914)

Goto 1914: 361 (as *Johannaster*).

A.H. Clark 1916c: 116 (as *Mariaster*).

Range: Sagami Bay, SE Japan, 160-1120 m.

261

MEDIASTER Stimpson, 1857 (including *Isaster* Verrill, 1894, an invalid homonym)

Stimpson 1857: 530; Perrier 1894: 377; Verrill 1899: 178; Fisher 1911c: 196; Verrill 1914a: 295; 1915: 108; Fisher 1919a: 255; Macan 1938: 369; H. L. Clark 1946: 83; Bernasconi 1963b: 11; Halpern 1970b: 202.

Type species: *Mediaster aequalis* Stimpson, 1857.

abyssi Ludwig, 1905. A variety of *Mediaster elegans* Ludwig, 1905.

aequalis Stimpson, 1857

Stimpson 1857: 530; Fisher 1911c: 198; Ziesenhenne 1937: 214; Blake 1973: 49.

Dujardin & Hupé 1862: 365 (as *Ophidiaster*).

Range: Alaska to Lower California, 16-293 m.

agassizi Verrill, 1899: 181. A synonym of *Mediaster pedicellaris* (Perrier, 1881) according to Halpern (1970b).

arcuatus (Sladen, 1889)

Sladen 1889: 277 (as *Pentagonaster*).

Verrill 1899: 159 (as *Mediaster*).

Range: Japan, 630-1023 m.

australiensis H. L. Clark, 1916

H. L. Clark 1916: 39; 1946: 83.

Range: Bass Strait, N of Tasmania.

bairdi (Verrill, 1882) (with synonym *Mediaster stellatus* Perrier, 1891)

Verrill 1882: 139 (as *Archaster*).

Verrill 1894: 258 (as *Isaster*).

Verrill 1899: 181; 1914a: 298; Gray, Downey & Cerame-Vivas 1968: 150; A. M.C. in Harvey et al. 1988: 162 (as *Mediaster*).

Range: off New Jersey, U.S.A.*, N to Newfoundland and from the Rockall Trough, NE Atlantic, 640-1590 m.

boardmani (Livingstone, 1934)

Livingstone 1934: 177 (as *Pseudarchaster*).

H. L. Clark 1946: 83 (as *Mediaster*).

Range: N. S. W., Australia.

brachiatus Goto, 1914

Goto 1914: 354; Hayashi 1952: 145.

Range: Japan, 493-1000 m.

capensis H. L. Clark, 1923 (with subspecies *durbanensis* Mortensen, 1933)

capensis capensis H. L. Clark 1923

H. L. Clark 1923a: 256; Mortensen 1933a: 241; A. M. Clark & Courtman-Stock 1976: 64.

Range: False Bay*, Cape Town area, 38-170 m.

capensis durbanensis Mortensen, 1933

Mortensen 1933a: 242 (as variety).

A. M. Clark & Courtman-Stock 1976: 64 (as subspecies).

Range: Off Durban, 410 m.

dawsoni McKnight, 1973

McKnight 1973a: 175.

Range: S of New Zealand*, to northern New Zealand, 601-1280 m.

durbanensis Mortensen, 1933a. A subspecies of *Mediaster capensis* H. L. Clark, 1923.

elegans Ludwig, 1905 (with variety *abyssi* Ludwig, 1905)

Ludwig 1905: 125.

Range: W from Panama, 1790 m.

florifer: Fisher, 1919a *(?Mediaster),* see *Rosaster*

japonicus: Verrill, 1899, see *Ceramaster*

monacanthus H.L. Clark, 1916, see *Nectriaster* [Oreasteridae]

murrayi Macan, 1938

Macan 1938: 470.

Range: Zanzibar area, 256-640 m.

ornatus Fisher, 1906

Fisher 1906a: 1046; 1919a: 256; Döderlein 1924: 52; Macan 1938: 470.

Range: Hawaiian Is*, Philippine Is, East Indies, Maldive area and Arabian Sea, 523-1630 m.

patagonicus Verrill, 1899, see *Ceramaster*

pedicellaris (Perrier, 1881) (with synonyms *Mediaster agassizi* Verrill, 1899 and *M. trindadensis* Bernasconi, 1957)

Perrier 1881: 23; 1884: 173, 245 (as *Goniodiscus).*

Perrier 1884: 168 (as *Pentagonaster).*

Verrill 1899: 182; 1915: 109; H.L. Clark 1941: 34; Halpern 1970b: 202 (as *Mediaster).*

Range: Barbados*, NW to Florida Straits, SE to S Trinidad Id, Brazil, 197-576 m.

praestans Livingstone, 1933

Livingstone 1933: 21; H.L. Clark 1946: 83.

Range: Queensland, Australia.

roseus Alcock, 1893, see *Pseudarchaster*

sladeni Benham, 1909

Benham 1909b: 94 [12]; Fell 1958: 10; McKnight 1967a: 300; Baker & Clark 1970: 4.

Range: Oamaru, New Zealand*, N and E New Zealand, 45-585 m.

stellatus Perrier, 1891b: 268; also Grieg 1932: 21. A synonym of *Mediaster bairdi* (Verrill, 1882) according to Verrill (1899).

tenellus Fisher, 1905

Fisher 1905: 307; 1911c: 202.

Range: Southern California.

Probably intergrades with *Mediaster transfuga* Ludwig, 1905 according to Fisher (1911).

transfuga Ludwig, 1905

Ludwig 1905: 120.

Range: off Panama, 900 m.

trinadensis Halpern, 1970b: 207, lapsus for

trindadensis Bernasconi, 1957: 33; 1958a: 133. A synonym of *Mediaster pedicellaris* (Perrier, 1881) according to Halpern (1970b).

MILTELIPHASTER Alcock, 1893

Alcock 1893a: 91.

Type species: *Milteliphaster woodmasoni* Alcock, 1893.

wanganellensis H.E.S. Clark, 1982

H.E.S. Clark 1982: 35.

Range: N New Zealand.

woodmasoni Alcock, 1893

Alcock 1893a: 91; Macan 1938: 391.

Range: Andaman Sea*, Maldive area, 229-530 m.

[NECTRIA Gray, 1840 and *NECTRIASTER* H.L. Clark, 1946, see Oreasteridae]

NEHIPPASTERIA Dons, 1938. A subgenus [probably a synonym] of *Hippasteria* Gray, 1840.

insignis Dons, 1938: 16, see *Hippasteria (Nehippasteria)*

NEREIDASTER Verrill, 1899 (either a distinct genus or a subgenus of *Nymphaster* Sladen, 1889)

Verrill 1899: 186.

Type species: *Nymphaster symbolicus* Sladen, 1889.

A synonym of *Rosaster* Perrier, 1894 according to Fisher (1917d).

bipunctus: Verrill, 1899, see *Rosaster*

bowersi Fisher, 1906, see *Ceramaster*

symbolicus: Verrill, 1899, see *Rosaster*

NOTIOCERAMUS Fisher, 1940

Fisher 1940: 119.

Type species: *Notioceramus anomalus* Fisher, 1940.

anomalus Fisher, 1940

Fisher 1940: 119.

Range: Clarence Id, Weddell Quadrant, Antarctica, 342 m.

NYMPHASTER Sladen, 1889 *(Nymphaster* Sladen in Thomson & Murray, 1885: 612 a nomen nudum, diagnosed but no species named).

Sladen 1889: 294; Fisher 1913a: 633; H.L. Clark 1916: 36; Fisher 1917d: 167; 1919a: 261; Halpern 1970b: 222.

Type species: *Nymphaster protentus* Sladen, 1889, by subsequent designation by Fisher (1917d), a synonym of *Pentagonaster arenatus* Perrier, 1881.

albidus Sladen, 1889: 306. A synonym of *Nymphaster arenatus* (Perrier, 1881) according to Halpern (1970b).

alcocki Ludwig, 1900

Ludwig in Chun 1900: 493.

Range: Gulf of Aden, 1470 m.

arenatus (Perrier, 1881)(with synonyms *Pentagonaster ternalis* Perrier, 1881, *Dorigona prehensilis* Perrier, 1885, *Nymphaster albidus* Sladen, 1889, *N. basilicus* Sladen, 1889, *N. protentus* Sladen, 1889, *Pentagonaster (Dorigona) jacqueti* Perrier, 1894 and *Podosphaeraster crassus* Cherbonnier, 1974).

Perrier 1881: 21; 1884: 236 (as *Pentagonaster*).

Perrier 1885c: 39; 1894: 379; Koehler 1895a: 451; Grieg 1932: 19 (as *Dorigona*).

Sladen 1889: 752; Macan 1938: 374; H.L. Clark 1941: 39; Halpern 1970b: 223 (lectotype); Downey 1973: 58; Rowe 1985: 313 (as *Nymphaster*).

Range: off Barbados*, Georgia, U.S.A. to northern Brazil and W of Ireland S to Morocco, the Canary and Cape Verde Islands, 225-3000 m.

arthrocnemis Fisher, 1913

Fisher 1913a: 638; 1919a: 277; Jangoux 1981a: 469.

Range: Sulawesi (Celebes)*, Philippine Is, 610-1020 m.

atopus Fisher, 1913

Fisher 1913a: 640; 1919a: 285.

Range: southern Philippine Is, 2010 m.

basilicus Sladen, 1889: 308. A synonym of *Nymphaster arenatus* (Perrier, 1881) according to Halpern (1970b).

basilicus: Alcock, 1893a: 95, non *Nymphaster basilicus* Sladen, 1889, = *N. moebii* (Studer, 1884) according to Macan (1938).

belli (Koehler, 1909a) Fisher 1919a: 266. A synonym of *Nymphaster moebii* (Studer, 1884) according to Macan (1938).

bipunctus Sladen, 1889, see *Rosaster*

diomediae Ludwig, 1905

Ludwig 1905: 128; Macan 1938: 374.

Range: Galapagos and Cocos Is, E Pacific, 700-1810 m.

dyscritus Fisher, 1913

Fisher 1913a: 635; 1919a: 266.

Range: Philippine Is, 400-510 m.

euryplax Fisher, 1913

Fisher 1913a: 634; 1919a: 264.

Range: Philippine Is, 320 m.

florifer Alcock, 1893, see *Rosaster*

gardineri (Bell, 1909)

Bell 1909: 22 (as *Iconaster*).

Fisher 1919a: 269 (as *Nymphaster*).

Range: Saya da Malha Bank, W Indian Ocean, 220 m.

habrotatus Fisher, 1913

   Fisher 1913a: 639; 1919a: 282.

   Range: Philippine Is, 620-1350 m.

jacqueti (Perrier, 1894) Koehler, 1924: 182. A synonym of *Nymphaster arenatus* (Perrier, 1881) according to Macan (1938).

leptodomus Fisher 1913

   Fisher 1913a: 637; 1919a: 272.

   Range: Philippine Is, 393-413 m.

ludwigi (Koehler, 1909) Fisher 1919a: 276. A synonym of *Nymphaster moebii* (Studer, 1884) according to Macan (1938).

meseres Fisher 1913

   Fisher 1913a: 639; 1919a: 280.

   Range: Philippine Is, 620 m.

moebii (Studer, 1884) (with synonyms *Dorigona belli* Koehler, 1909 and *D. ludwigi* Koehler, 1909)

   Studer, 1884: 35 (as *Pentagonaster (Dorigona)*).

   Sladen 1889: 869; Döderlein 1924: 55; Macan 1938: 375; H.L. Clark 1946: 87 (as *Nymphaster*).

   Range: Zanzibar area to northern Australia, 360-2390 m.

moluccanus Fisher, 1913

   Fisher 1913a: 637; 1919a: 274.

   Range: Moluccas, 420-500 m.

mucronatus Fisher, 1913

   Fisher 1913a: 636; 1919a: 269.

   Range: Philippine Is, 366 m.

nora Alcock, 1893

   Alcock 1893a: 96; Fisher 1919a: 280.

   Range: Andaman Sea, 880 m.

pentagonus H.L. Clark, 1916

   H.L. Clark 1916: 36.

   Range: Great Australian Bight, S Australia c.130°E, 458-823 m.

prehensilis (Perrier, 1885) Sladen, 1889: 752. A synonym of *Nymphaster arenatus* (Perrier, 1881) according to Macan (1938).

protentus Sladen, 1889: 303; also Bell 1889b: 434. A synonym of *Nymphaster arenatus* (Perrier, 1881) according to Macan (1938).

protentus: Alcock, 1893a: 95, non *Nymphaster protentus* Sladen, 1889, = *N. moebii* (Studer, 1884) according to Macan (1938).

subspinosus (Perrier, 1881)

   Perrier 1881: 21; 1884: 234 (as *Pentagonaster*).

   Perrier 1885c: 39; 1894: 375; Koehler 1895b: 451 (as *Dorigona*).

   Sladen 1889: 752; Verrill 1899: 185; H.L. Clark 1941: 40; Halpern 1970b: 228 (lectotype) (as *Nymphaster*).

   Range: Barbados*, West Indies, Bay of Biscay, 298-400 m.

Treated as a synonym of *Nymphaster arenatus* (Perrier, 1881) by Macan (1938) and Madsen (1951) but a valid species according to H.L. Clark (1941) and Halpern (1970b).

subspinosus: Bell, 1892: 75, non *Nymphaster subspinosus* (Perrier, 1881), = *N. arenatus* (Perrier, 1881) according to Halpern (1970b).

symbolicus Sladen, see *Rosaster*

ternalis (Perrier, 1881) Sladen, 1889: 732. A synonym of *Nymphaster arenatus* (Perrier, 1881) according to Halpern (1970a).

NYMPHASTER (NEREIDASTER) Verrill, 1899: 186. A synonym of *Rosaster* Perrier, 1894 according to Fisher (1917d).

OGMASTER Martens, 1865 (with synonym *Dorigona* Gray, 1866)
Martens 1865: 359 (as subgenus of *Goniaster*).
Sladen 1889: 261; Fisher 1919a: 262, 305 (as genus).
Type species: *Goniodiscus capella* Müller & Troschel, 1842.

capella (Müller & Troschel, 1842) (with synonym *Dorigona reevesi* Gray, 1866)
Müller & Troschel 1842: 61 (as *Goniodiscus*).
Martens 1865: 359 (as *Goniaster (Ogmaster)*).
Sladen 1889: 261; Leipoldt 1895: 649; Koehler 1910a: 79; H.L. Clark 1916: 47; Fisher 1919a: 262, 305; Döderlein 1935: 101; Jangoux in Guille & Jangoux 1978: 53.
Range: S China Sea* to N Australia, Bay of Bengal and Red Sea, shallow - 143 m.

PARAGONASTER Sladen, 1889 *(Paragonaster* Sladen in Thomson & Murray, 1885: 617 a nomen nudum, no species named)
Sladen 1889: 310; Perrier 1894: 355; Fisher 1911c: 163; Halpern 1972: 373; Downey 1973: 56.
Type species: *Paragonaster ctenipes* Sladen, 1889, by subsequent designation by Fisher (1919a).

breviradiatus Macan, 1938. A subspecies of *Paragonaster ctenipes* Sladen, 1889.

chinensis Liao, 1983
Liao 1983: 367.
Range: Yellow Sea, China, 50-64 m.

ctenipes Sladen, 1889 (with subspecies *hypacanthus* Fisher, 1913 and *breviradiatus* Macan, 1938)

ctenipes ctenipes Sladen, 1889
Sladen 1889: 311; Fisher 1919a: 229; Döderlein 1924: 51; Jangoux 1981a: 469.
Range: Arafura Sea*, Indonesia, Philippine Is, 217-256 m.

ctenipes breviradiatus Macan, 1938
Macan 1938: 362.
Range: Zanzibar to Maldives, 183-293 m.

ctenipes hypacanthus Fisher, 1913
   Fisher 1913a: 627; 1919a: 228; Blake 1973: 50.
   Range: Philippine Is, 208-314 m.
cylindratus Sladen, 1889: 314. A synonym of *Paragonaster subtilis* (Perrier, 1881) according to Grieg (1932).
elongatus (Perrier, 1885) Perrier, 1894: 362. A synonym of *Paragonaster subtilis* (Perrier, 1881) according to Koehler (1909a).
formosus (Verrill, 1884) Verrill, 1894: 257. A synonym of *Paragonaster subtilis* (Perrier, 1881) according to Grieg (1932).
grandis H.L. Clark, 1941
   H.L. Clark 1941: 32; Halpern 1972: 378.
   Range: Cuba, 311-466 m.
hypacanthus Fisher, 1913. A subspecies of *Paragonaster ctenipes* Sladen, 1889.
ridgwayi McKnight, 1973
   McKnight 1973a: 172.
   Range: Tasman Sea, N of New Zealand (c.36.5°S, 170.5°E), 2110-2160 m.
stenostichus Fisher 1913
   Fisher 1913a: 627; 1919a: 232; Döderlein 1924: 50.
   Range: Philippine Is*, Indonesia, 315-363 m.
strictus Perrier, 1894: 363. A synonym of *Paragonaster subtilis* (Perrier, 1881) according to Macan (1938).
subtilis (Perrier, 1881) (with synonyms *Archaster formosus* Verrill, 1884, *Pentagonaster elongatus* Perrier, 1885, *Paragonaster cylindratus* Sladen, 1889 and *P. strictus* Perrier, 1894).
   Perrier 1881: 26; 1884: 242; 1885c: 41 (as *Goniopecten).*
   Perrier 1894: 358; Koehler 1909a: 86; Grieg 1932: 20; Halpern 1972: 374; Downey 1973: 57; Sibuet 1975: 292.; Gage et al. 1983: 280.
   Range: Gulf of Mexico*, N American Basin, Azores area, Rockall Trough S to Gulf of Guinea, 1845-4700 m.
tenuiradiis Alcock, 1893
   Alcock 1893a: 97.
   Range: Bay of Bengal, 3200 m.
PELTASTER Verrill, 1899 (with possible synonym *Sphaeriodiscus* Fisher, 1910).
   Verrill 1899: 168; Tortonese & A.M. Clark 1956: 348; Halpern 1970b: 234; Downey 1973: 50.
   Type species: *Peltaster hebes* Verrill, 1899, a synonym of *Goniaster nidarosiensis* Storm, 1881.
cycloplax Fisher, 1913
   Fisher 1913a: 641; 1919a: 290.
   Range: Philippine Is, 219 m.

268

hebes Verrill, 1899: 169. A synonym of *Peltaster nidarosiensis* (Storm, 1881) according to Grieg (1905) and Halpern (1970b), though meantime thought to be distinct by H.L. Clark (1941).

micropeltus (Fisher, 1906)
Fisher 1906a: 1054 (as *Tosia (Ceramaster)*).
Fisher 1911c: 205; 1940: 119 (as *Peltaster*).
McKnight 1973a: 180 (possibly referable to *Pillsburiaster*).
Range: Hawaiian Is, 570-1460 m.

nidarosiensis (Storm, 1881) (with synonyms *Pentagonaster vincenti* Perrier, 1885 and *Peltaster hebes* Verrill, 1899).
Storm 1881: 90 (as *Goniaster*).
Storm 1888: 61 (as *Pentagonaster*).
Grieg 1905: 1; Mortensen 1927: 79; Tortonese & Clark 1956: 348; Halpern 1970b: 235; Downey 1973: 50 (as *Peltaster*).
Range: Norway*, southern Iceland, Bay of Biscay, Canary Is, also Georgia to Cuba and Lesser Antilles. 95-1110 m.

placenta (Müller & Troschel, 1842) (with synonyms *Goniodiscus acutus* Heller, 1863, *G. placentaeformis* Heller, 1863, *Pentagonaster mirabilis* Perrier, 1875, *P. minor* Koehler, 1895 and *P. planus* Verrill, 1895).
Müller & Troschel 1842: 59 (as *Goniodiscus*).
Perrier 1878: 21, 84; Ludwig 1897: 157; Koehler 1895b: 454 (as *Pentagonaster*).
Verrill 1899: 161 (as *Tosia (Ceramaster)*).
Fisher 1911c: 205; Koehler 1921: 42 (as *Ceramaster*).
Fisher 1919a: 260 (as *?Sphaeriodiscus*).
Tortonese & A.M. Clark 1956: 347; Madsen 1958: 90 (as *Sphaeriodiscus*).
Halpern 1970b: 238; Downey 1973: 51; Tortonese 1983: 27; 1984: 99 (as *Peltaster*).
Range: Mediterranean, Bay of Biscay, W Africa, Cape Cod area S to Caribbean, 10-605 m.

planus (Verrill, 1895) Verrill, 1899: 170. A synonym of *Sphaeriodiscus* [i.e. *Peltaster*] *placenta* (Müller & Troschel, 1842) according to Tortonese & A.M. Clark (1956).

planus: Boone, 1933 pl.43, non *Pentagonaster planus* Verrill, 1895, = *Ceramaster granularis* (Retzius, 1783) according to Halpern (1970b).

PENTAGONASTER Gray, 1840 (with synonym *Stephanaster* Ayres, 1851).
Gray 1840: 280; Perrier 1875: 190 [1876: 6] (pt); Sladen 1889: 264 (pt); Verrill 1899: 157; Fisher 1911c: 166; Ludwig 1912: 8; H.L. Clark 1946: 88; A.M. Clark 1953: 398.
Type species: *Pentagonaster pulchellus* Gray, 1840.

abnormalis Gray, 1866: 11; also Ludwig 1912: 13. A synonym of *Pentago-*

naster pulchellus Gray, 1840 according to Mortensen (1925a).

*abyssalis Ludwig, 1900

   Ludwig in Chun 1900: 493.

   Range: Equator Channel, S of Maldive Is, 2250 m.

   [Almost certainly not congeneric with *Pentagonaster pulchellus*.]

acutus (Heller, 1863) Perrier, 1878: 21, 84. A synonym of *Peltaster placenta* (Müller & Troschel, 1842) according to Ludwig (1897).

affinis Perrier, 1884: 243. A synonym of *Ceramaster grenadensis* (Perrier, 1881) according to Halpern (1970b).

alexandri Perrier, 1881, see *Rosaster*

ammophilus Fisher, 1906a, see *Sphaeriodiscus* [though *Ceramaster* according to Halpern (1970b)]

*annandalei Koehler, 1909, see under *Pillsburiaster*

arcuatus Sladen, 1889, see *Mediaster*

arenatus Perrier, 1881, see *Nymphaster*

astrologorum Perrier, 1875: 196 [1976: 12]. A forma of *Tosia australis* Gray, 1840 according to A. M. Clark (1953).

attenuatus Marenzeller, 1895a: 190. Lapsus for *Ophidiaster* (i.e. *Hacelia*) *attenuatus* [Ophidiasteridae].

auratus (Gray, 1847) Perrier, 1875: 204 [1876: 20]. A synonym of *Tosia magnifica* (Müller & Troschel, 1842) according to A. M. Clark (1953).

australis: Perrier, 1875, see *Tosia*

austrogranularis Perrier, 1891a: U127 *(Pentagonaster austro-granularis* Perrier, 1888: 764 a nomen nudum, undescribed). A synonym of *Ceramaster patagonicus* (Sladen, 1889) according to Madsen (1956).

balteatus Sladen, 1891: 688; also Ludwig 1897: 186. A synonym of *Ceramaster granularis* (Retzius, 1783) according to Verrill (1899) and most following authors but of *C. grenadensis* (Perrier, 1881) according to Halpern (1970, thesis) and in Downey (1973).

belcheri: Perrier 1878: 85 (as *Pentagonaster (Stellaster)).* A forma of *Stellaster equestris* (Retzius, 1805) according to Döderlein (1935).

belli Studer, 1884: 31. A synonym of *Odontaster penicillatus* (Philippi, 1870) (Odontasteridae) according to Madsen (1956).

borealis (Barrett, 1857) Perrier, 1878: 84. A synonym of *Pentagonaster* [i.e. *Ceramaster*] *granularis* (Retzius, 1783) according to Sladen (1889).

childreni: Perrier, 1875, see *Calliaster*

concinnus Sladen, 1891: 690; also Ludwig 1897: 187. A synonym of *Plinthaster dentatus* (Perrier, 1884) according to Halpern (1970b).

coppingeri Bell, 1884: 128. A synonym of *Goniodiscaster rugosus* (Perrier, 1875) according to A. M. C. in Clark & Rowe (1971). (Oreasteridae).

crassimanus (Möbius, 1859) Ludwig, 1912: 15; also H. L. Clark 1946: 89. A synonym of *Pentagonaster duebeni* Gray, 1847 according to A. M. Clark (1953).

270

crassus Perrier, 1885b: 884; 1885c: 34. A synonym of *Ceramaster grenadensis* (Perrier, 1881) according to Halpern (1970, thesis) and in Downey (1973).

cuenoti Koehler, 1909b, see *Ceramaster*

dentatus Perrier, 1884, see *Plinthaster*

deplasi Perrier 1885b: 886; 1885c: 34. A synonym of *Ceramaster grenadensis* (Perrier, 1881) according to Halpern (1970, thesis) and in Downey (1973).

dilatatus Perrier, 1875, see *Diplodontias* (Odontasteridae)

doederleini Koehler, 1909b, see *Plinthaster*

duebeni Gray, 1847 (with synonyms *Astrogonium crassimanum* Möbius, 1859, *Pentagonaster gunni* Perrier, 1875 [forma at most] and *P. stibarus* H.L. Clark, 1914).

Gray 1847: 79; Ludwig 1912: 18; H.L. Clark 1928: 380; 1946: 88; A.M. Clark 1953: 400; 1966: 308; Shepherd 1968: 308.

Lütken, 1865: 144 (as *Goniaster*).

Sladen 1889: 748 (as *Astrogonium*).

Perrier 1894: 403 (as *Stephanaster*).

Range: Fremantle area, Western Australia* to SE Australia, ?S Queensland, 0-200 m.

elongatus Perrier, 1885c: 38. A synonym of *Paragonaster subtilis* (Perrier, 1881) according to Halpern (1970, thesis) and in Downey (1973).

emma: Perrier, 1875, see *Calliderma*

equestris: Perrier, 1875 (as *Pentagonaster (Stellaster),* see *Stellaster*

ernesti Ludwig, 1905, see *Pillsburiaster*

*excellens Ludwig, 1900

Ludwig in Chun 1900: 494.

Range: off Somalia, East Africa, 630 m.

[Almost certainly not congeneric with *Pentagonaster pulchellus* Gray.]

eximius Verrill, 1894: 264. A synonym of *Ceramaster granularis* (Retzius, 1783) according to Halpern (1970, thesis).

fallax: Sladen, 1889, see *Pseudarchaster*

*fonki: Perrier, 1894, see under *Astrogonium*

*gibbosus Perrier, 1875

Perrier 1875: 219 [1876: 35].

Range: Unknown.

[No subsequent records. Holotype untraced in BM(NH) collections.]

gosselini Perrier, 1885b: 886; 1885c: 35. A synonym of *Ceramaster grenadensis* (Perrier, 1881) according to Halpern (1970, thesis) and in Downey (1973).

grandis (Gray, 1847) Sladen 1889: 744. A synonym of *Tosia magnifica* (Müller & Troschel, 1842) according to A.M. Clark (1953).

grandis Perrier, 1885b: 886; 1885c: 35. A secondary (though transitory) homonym of *Pentagonaster grandis* (Gray, 1847, as *Tosia*) Sladen, 1889 according to Sladen (1889) and renamed *Pentagonaster perrieri*, a synonym of *Plinthaster dentatus* (Perrier, 1884) according to Halpern (1970b).

granularis: Danielssen & Koren, 1884, see *Ceramaster*

granulatus Bell, 1905: 246. A synonym of *Tosia* [i.e. *Toraster*] *tuberculata* (Gray, 1847) according to Mortensen (1933a).

granulosus Perrier, 1878 (as *Pentagonaster (Stellaster)*, a NOMEN NUDUM, undescribed.

greeni (Bell, 1889b) Bell 1892: 74. A synonym of *Ceramaster grenadensis* (Perrier, 1881) according to Halpern (1970, thesis) and in Downey (1973).

grenadensis Perrier, 1881, see *Ceramaster*

gunni Perrier, 1875: 203. A forma of *Pentagonaster duebeni* Gray, 1847 according to A. M. Clark (1953).

haesitans Perrier, 1885c: 36. A synonym of *Ceramaster grenadensis* according to Halpern (1970, thesis) and in Downey (1973).

hispidus: Perrier, 1878, see *Poraniomorpha* (Poraniidae)

hystricis Marenzeller, 1893: 67 *(Pentagonaster hystricis* Marenzeller, 1891: 445 a nomen nudum, undescribed). A synonym of *Ceramaster grenadensis* (Perrier, 1881) according to Halpern (1970, thesis) and in Downey (1973).

inaequalis: Sladen, 1889, see *Sphaeriodiscus*

incei: Perrier, 1875: 237 [1876: 53] (as *Pentagonaster (Stellaster)*. A forma of *Stellaster equestris* (Retzius, 1805) according to Döderlein (1935).

intermedius Perrier, 1884, see *Litonotaster*

intermedius Alcock, 1893a: 90. A junior primary homonym of *Pentagonaster intermedius* Perrier, 1884 with which it is not conspecific according to Downey (1973). [Identity uncertain.]

investigatoris Alcock, 1893, see *Plinthaster*

jacqueti Perrier in Milne-Edwards, 1882: 47 (as *Pentagonaster (Dorigona))*. A synonym of *Nymphaster arenatus* (Perrier, 1881) according to Macan (1938).

japonicus Sladen, 1889, see *Ceramaster*

kergroheni Koehler, 1895a: 453. A synonym of *Ceramaster grenadensis* (Perrier, 1881) according to Halpern (1970, thesis) and in Downey (1973).

*lamarcki (Müller & Troschel, 1842)

Müller & Troschel 1842: 56; Dujardin & Hupé 1862: 393 (as *Astrogonium)*.

Gray 1866: 11 (as *Tosia*).

Perrier 1875: 213 [1876: 18]; Sladen 1889: 746 (as *Pentagonaster*).

Perrier 1894: 388 (as *Phaneraster*).

Verrill 1899: 157 (as *Goniaster*).

Range: Probably 'les mers des Indes' according to Dujardin & Hupé (1862).

[Identity uncertain; type in Paris Museum.]

lepidus Sladen, 1889: 275. Probably a synonym of *Hoplaster spinosus* Perrier, 1882 (Odontasteridae) according to Verrill (1899), confirmed by A.M.C. in Gage et al. (1983).

longimanus: Perrier, 1875, see *Iconaster*

magnificus: Perrier, 1894, see *Tosia*

mammillatus: Perrier, 1875, see under *Tosia*

meridionalis: E. A. Smith, 1879, see *Odontaster* (Odontasteridae)

miliaris: Perrier, 1875, see *Diplodontias* (Odontasteridae)

minimus Perrier, 1875: 207 [1876: 22]. A synonym of *Tosia tubercularis* Gray, 1847 according to Livingstone (1932) itself a synonym of *Tosia nobilis* Müller & Troschel, 1843) according to A.M. Clark (1953).

minor Koehler, 1895a: 451. A synonym of *Pentagonaster* [i.e. *Peltaster*] *placenta* (Müller & Troschel, 1842) according to Ludwig (1897).

mirabilis Perrier, 1875: 224 [1876: 39]. A synonym of *Pentagonaster* [i.e. *Peltaster*] *placenta* (Müller & Troschel, 1842) according to Ludwig (1897).

misakiensis Goto, 1914, see *Ceramaster*

moebii Studer, 1884 (as *Pentagonaster (Dorigona))*, see *Nymphaster*

mortenseni Koehler, 1909b: 74 (as *Pentagonaster (Philonaster)*, see *Ceramaster.*

muelleri: Perrier, 1875: 230 [1876: 44] (as *Pentagonaster (Dorigona))*, non *Goniaster (Stellaster) muelleri* Martens, 1865, = *Ogmaster capella* (Müller & Troschel, 1842) according to Fisher (1919a).

nidarosiensis: Storm, 1888, see *Peltaster*

nobilis: Perrier, 1894, see *Tosia*

ornatus (Müller & Troschel, 1842) Sladen, 1889: 746. A synonym of *Tosia magnifica* (Müller & Troschel, 1842) according to A.M. Clark (1953)

parvus Perrier, 1881, see *Tosia*

patagonicus Sladen, 1889, see *Ceramaster*

paxillosus: Perrier. 1875, see *Diplodontias* (Odontasteridae)

paxillosus: Bell, 1881: 95, non *Astrogonium paxillosum* Gray, 1847, = *Odontaster penicillatus* (Philippi, 1870) (Odontasteridae) according to Fisher (1940).

pedicellaris: Perrier, 1884 (: 168 non 173), see *Mediaster*

perrieri Sladen, 1889: 746 (nom. nov. for *Pentagonaster grandis* Perrier, 1885, non *Tosia grandis* Gray, 1847). A synonym of *Plinthaster dentatus* (Perrier, 1884) according to Halpern (1970b).

placenta: Perrier, 1878, see *Peltaster*

planus Verrill, 1895: 135. A synonym of *Peltaster placenta* (Müller &

Troschel, 1842) according to Halpern (1970b).

pulchellus Gray, 1840 (with synonyms *Stephanaster elegans* Ayres, 1851 and *Pentagonaster abnormalis* Gray, 1866)

Gray 1840: 280; Benham 1909b: 93 [11]; Ludwig 1912: 9; Mortensen 1925a: 281; Farquhar 1927: 237; A.M. Clark 1953: 398; Fell 1958: 12; McKnight 1967a: 300.

Müller & Troschel 1842: 55; Farquhar 1894: 200 (as *Astrogonium*).

Perrier 1894: 403 (as *Stephanaster*).

Range: New Zealand, Chatham Is, 0-73 (?220) m.

*pulvinus Alcock, 1893

Alcock 1893a: 90; Goto 1914: 337.

Verrill 1899: 162 (as *Tosia*).

Fisher 1919a: 260 (as '*Pentagonaster*').

Range: Indian waters, c.2200 m.

Not congeneric with *Pentagonaster pulchellus* Gray, 1840 according to Fisher (1919) (by implication).

semilunatus (Martens, 1866) Perrier, 1875: 208 [1876: 24] (from Linck, 1733). A synonym of *Goniaster tessellatus* (Lamarck, 1816) according to Halpern (1970b).

simplex Verrill, 1895: 135. A synonym of *Ceramaster granularis* (Retzius, 1783) according to Halpern (1970, thesis).

singularis: Perrier, 1875: 222 [1876: 38], see *Diplodontias* (Odontasteridae)

spinulosus (Gray, 1847) Perrier, 1875: 218 [1876: 34]. A synonym of *Culcita novaeguineae* Müller & Troschel, 1842 (Oreasteridae) according to Goto (1914).

squamulosus Studer, 1884, see *Stellaster*

stibarus H.L. Clark, 1914: 136; 1923b: 238. A synonym of *Pentagonaster duebeni* Gray, 1847 according to A.M. Clark (1953).

subspinosus Perrier, 1881, see *Nymphaster*

ternalis Perrier, 1881: 20. A synonym of *Nymphaster arenatus* (Perrier, 1881) according to Halpern (1970b).

tubercularis (Gray, 1847) Perrier, 1894: 390. A synonym of *Tosia nobilis* (Müller & Troschel, 1843) according to A.M. Clark (1953).

tuberculatus (Gray, 1847) Perrier, 1875, see *Toraster*

tuberculosus: Perrier, 1878, see *Stellaster*

validus Bell, 1884: 129. A synonym of *Goniodiscaster pleyadella* (Lamarck, 1816) according to H.L. Clark (1921).

verrucosus: Perrier, 1878, see *Cycethra* (Ganeriidae)

vincenti Perrier, 1885b: 886; 1885c: 34; 1894: 396. A synonym of *Peltaster nidarosiensis* (Storm, 1881) according to Halpern (1970, thesis) and in Downey (1973).

PENTAGONASTER (PHILONASTER) Koehler, 1909

Koehler 1909b: 78.

Type species: *Pentagonaster (Philonaster) mortenseni* Koehler, 1909.
A synonym of *Ceramaster* Verrill, 1899 according to Fisher (1919a).
mortenseni Koehler, 1909, see *Ceramaster*
PENTOPLIA H.E.S. Clark, 1971
  H.E.S. Clark 1971: 545 (Pentopliidae); Blake 1987: 519 (Goniasteridae).
  Type species: *Pentoplia felli* H.E.S. Clark, 1971.
felli H.E.S. Clark, 1971
  H.E.S. Clark 1971: 546.
  Range: N of S Orkney Is, Southern Ocean, 2580-4020 m.
PERGAMASTER Koehler, 1920
  Koehler 1920: 237; Fisher 1940: 121.
  Type species: *Pergamaster tessellatus* Koehler, 1920, a synonym of
  Pentagonaster incertus Bell, 1908.
incertus (Bell, 1908) (with synonym *Pergamaster tessellatus* Koehler,
  1920)
  Bell, 1908: 9 (as *Pentagonaster*).
  Fisher 1940: 121; A.M. Clark 1962a: 23; McKnight 1976: 25 (as *Pergamaster*).
  Range: Ross Sea*. probably circumpolar, 175-3100 m.
synaptorus Fisher, 1940
  Fisher 1940: 122.
  Range: Clarence Id, Weddell Quadrant, Antarctica, 342 m.
tessellatus Koehler, 1920: 238. Probably a synonym of *Pergamaster incertus*
  (Bell, 1908) according to Fisher (1940), confirmed by Madsen (1951).
triseriatus H.E.S. Clark, 1963
  H.E.S. Clark 1963a: 46; 1963b: 43.
  Range: Ross Sea, 370-385 m.
PERISSOGONASTER Fisher, 1913
  Fisher 1913a: 628; 1919a: 235.
  Type species: *Perissogonaster insignis* Fisher, 1913.
insignis Fisher, 1913
  Fisher 1913a: 628; 1919a: 236; Jangoux 1981a: 469.
  Range: Philippine Is, 265-464 m.
PHANERASTER Perrier, 1894
  Perrier 1894: 387.
  Type species: '*Pentagonaster semilunatus* Linck, 1733' [i.e. *Asterias tessellata* Lamarck, 1816].
  A synonym of *Goniaster* L. Agassiz, 1836 according to Verrill (1899).
lamarckii: Perrier, 1894: 388, see under *Pentagonaster*
semilunatus (Martens, 1866) Perrier, 1894: 388. A synonym of *Goniaster tessellatus* (Lamarck, 1816) according to Halpern (1970b).
PHILONASTER Koehler, 1909b: 78 (as *Pentagonaster (Philonaster)*. A
  synonym of *Ceramaster* Verrill, 1899 according to Fisher (1919a) (by
  implication).

275

mortenseni Koehler, 1909, see *Ceramaster*

PILLSBURIASTER Halpern, 1970
   Halpern 1970a: 2; McKnight 1973a: 180.
   Type species: *Pillsburiaster geographicus* Halpern, 1970.
annandalei (Koehler, 1909) *(?Pillsburiaster)*
   Koehler, 1909b: 65 (as *Pentagonaster*).
   McKnight, 1973a: 183 (as *?Pillsburiaster*).
   Range: Sri Lanka area, 1570-1610 m.
aoteanus McKnight, 1973
   McKnight 1973a: 180.
   Range: New Zealand, 820-910 m.
ernesti (Ludwig, 1905)
   Ludwig 1905: 136 (as *Pentagonaster*).
   Halpern 1970a: 3 (as *Pillsburiaster*).
   Range: W of Panama, 2150 m.
geographicus Halpern, 1970
   Halpern 1970a: 3.
   Range: Gulf of Guinea to Liberia, 1465-1990 m.
maini McKnight, 1973
   McKnight 1973a: 183.
   Range: Tasman Sea, NW of New Zealand, c.2000 m.

PLINTHASTER Verrill, 1899 (with synonyms *Pyrenaster* Verrill, 1899 and
   Eugoniaster Verrill, 1899)
   Verrill 1899: 161; Fisher 1906a: 1052 (as *Tosia (Plinthaster)*).
   Fisher 1910c: 172; 1911c: 165; Halpern 1970b: 244 (as genus).
   Type species: *Pentagonaster perrieri* Sladen, 1889, a synonym of *Pentagonaster dentatus* Perrier, 1884, by subsequent designation by Fisher (1910).
ceramoidea (Fisher, 1906)
   Fisher 1906a: 1052 (as *Tosia (Plinthaster)*).
   Range: Hawaiian Is, 420-510 m.
comptus (Verrill, 1899) Fisher, 1911: 165. A synonym of *Plinthaster dentatus* (Perrier, 1884) according to Halpern (1970b).
dentatus (Perrier, 1884) (with synonyms *Pentagonaster grandis* Perrier, 1885, *P. perrieri* Sladen, 1889, *P. concinnus* Sladen, 1891, *Tosia (Plinthaster) compta* Verrill, 1899 and *Tosia (Plinthaster) nitida* Verrill, 1899)
   Perrier 1884: 242; Sladen 1889: 265; Farran 1913: 10; Grieg 1932: 21 (as *Pentagonaster*).
   Verrill 1899: 167 (as *Pyrenaster*).
   Fisher 1910c: 172; 1911c: 165; Verrill 1915: 107; H.L. Clark 1941: 42; Halpern 1970b: 244 (lectotype); Downey 1973: 52; Gage et al. 1983: 280 (as *Plinthaster*).
   Range: off Grenada, Lesser Antilles*, North Carolina to Gulf of Mexico

and northern Brazil, also Azores, N to the Rockall Trough and S to the Cape Verde Is, 230-2910 m.

dentatus: Gray, Downey & Cerame-Vivas, 1968: 151, non *Plinthaster dentatus* (Perrier, 1884), = *Tosia parva* (Perrier, 1881) according to Halpern (1969b).

doederleini (Koehler, 1909)

Koehler 1909b: 71 (as *Pentagonaster*).

Macan 1938: 382 (as *Eugoniaster*).

Range: Laccadive area, S of India, 2160 m.

To *Plinthaster* by synonymy of the type species of *Eugoniaster*.

ephemeralis Macan, 1938

Macan 1938: 383.

Range: off Zanzibar, E Africa, 640-660 m.

To *Plinthaster* by synonymy of the type species of *Eugoniaster*.

investigatoris (Alcock, 1893)

Alcock 1893a: 88 (as *Pentagonaster*).

Verrill 1899: 173 (as *Eugoniaster*).

Halpern 1970b: 244 (as *Plinthaster*).

Range: Bay of Bengal, 1380 m.

nitidus (Verrill, 1899) Fisher, 1911c: 165. A synonym of *Plinthaster dentatus* (Perrier, 1884) according to Halpern (1970b).

perrieri (Sladen, 1889) Fisher, 1911c: 165. A synonym of *Plinthaster dentatus* (Perrier, 1884) according to Farran (1913).

productus A.H. Clark, 1917, see *Diplasiaster*

singletoni McKnight, 1973

McKnight 1973a: 185.

Range: New Zealand, 710-815 m.

PONTIOCERAMUS Fisher, 1911

Fisher 1911d: 420; 1919a: 294.

Type species: *Pontioceramus grandis* Fisher, 1911.

grandis Fisher, 1911

Fisher 1911d: 421; 1919a: 294; Jangoux 1981a: 471.

Range: Philippine Is, c.220-300 m.

PROGONIASTER Döderlein, 1924

Döderlein 1924: 61.

Type species: *Progoniaster atavus* Döderlein, 1924.

atavus Döderlein, 1924

Döderlein 1924: 61.

Range: West Irian (New Guinea), Indonesia, 32 m.

PSEUDARCHASTER Sladen, 1889 *(Pseudarchaster* Sladen in Thomson & Murray, 1885: 617 a nomen nudum, no species named) (with synonyms *Aphroditaster* Sladen, 1889 and *Fisheraster* Halpern, 1970).

Sladen 1889: 109; Verrill 1899: 189; Fisher 1911c: 179; Koehler 1924:

180; Mortensen 1927: 86; Macan 1938: 355; Fell 1958: 8; Bernasconi 1963b: 4; Halpern 1972: 360; Downey 1973: 59; A.M. Clark & Courtman-Stock 1976: 64.

Type species: *Pseudarchaster discus* Sladen, 1889, by subsequent designation by Verrill, 1899.

abernethyi Fell, 1958 (with possible synonym *Pseudarchaster garricki* Fell, 1958)

Fell 1958: 10; 1963b: 29; Baker & H.E.S. Clark 1970: 5.

Range: Cook Strait to northern New Zealand, 56-115 m.

abnormale '(Bell, date ?)' in Benham, 1909, ?lapsus for *Pentagonaster abnormale* Gray, 1866, = *P. pulchellus* Gray, 1840 according to Mortensen (1925a) but listed as *Pseudarchaster* by Macan (1938).

aequabile (Koehler, 1907) Macan, 1938: 355. A synonym of *Pseudarchaster gracilis* (Sladen, 1889) according to Halpern (1972).

alascensis Fisher, 1905

Fisher 1905: 303.

Fisher 1911c: 180 (as subspecies of *Pseudarchaster parelii* (Düben & Koren, 1846)

Range: Alaska, Bering Sea.

A valid species according to Halpern (1972).

annectens (Perrier, 1894) Verrill, 1899: 195. A synonym of *Pseudarchaster parelii* (Düben & Koren, 1846) according to Halpern (1972).

aphrodite (Perrier, 1894) Verrill, 1899: 195. A synonym of *Pseudarchaster gracilis* (Sladen, 1889) according to Halpern (1972).

boardmani Livingstone, 1934, see *Mediaster*

brachyactis H.L. Clark, 1923a: 254; also Madsen 1950: 211; A.M. Clark & Courtman-Stock 1976: 65. A synonym of *Pseudarchaster tessellatus* Sladen, 1889 according to Halpern (1972).

concinnus Verrill 1894: 250; 1899: 193; also H.L. Clark 1941: 31. A synonym of *Pseudarchaster gracilis* (Sladen, 1889) according to Halpern (1972).

discus Sladen, 1889 (with synonym *Astrogonium patagonicum* Perrier, 1891)

Sladen 1889: 110; Verrill 1899: 195; Fisher 1940: 117; Bernasconi 1963b: 5.

Perrier, 1891a: K125 (as *Astrogonium*).

Range: southern Chile*, Falkland-Magellan area and Argentina, 140-283 m.

dissonus Fisher, 1910

Fisher 1910a: 551; 1911c: 192; Baranova 1957: 160.

Range: Bering Sea to Oregon, U.S.A., 1437-2200 m.

diversigranulatus Macan, 1938

Macan 1938: 359.

278

Range: Gulf of Aden, off Arabia, 1950-2310 m.

eminens (Koehler, 1907) Macan, 1938: 355. A synonym of *Pseudarchaster gracilis* (Sladen, 1889) according to Halpern (1972).

fallax (Perrier, 1885) Verrill, 1899: 190. A synonym of *Pseudarchaster parelii* (Düben & Koren, 1846) according to Halpern (1972).

garricki Fell, 1958

Fell 1958: 8; 1962: 29; McKnight 1967a: 300; Aziz & Jangoux 1984c: 258.

Range: Cook Strait, southern New Zealand, 165-1000 m.

Possibly a synonym of *Pseudarchaster abernethyi* Fell, 1958 according to Baker & H.E.S. Clark (1970).

gracilis (Sladen, 1889) (with synonyms *Pseudarchaster concinnus* Verrill, 1894, *Astrogonium aphrodite* Perrier, 1894, *A. necator* Perrier, 1894, *P. ordinatus* Verrill, 1899, *A. eminens* Koehler, 1907, *A. aequabile* Koehler, 1907 and *A. marginatum* Koehler, 1909).

Sladen 1889: 117; Verrill 1899: 195; Fisher 1919a: 227 (as *Aphroditaster*).

Mortensen 1927: 86; Macan 1938: 355; Halpern 1972: 360; Downey 1973: 59; Gage et al. 1983: 279; Aziz & Jangoux 1984c: 257; Harvey et al. 1988: 162 (as *Pseudarchaster*).

Range: Azores*, equatorial Atlantic N to Cape Cod and to the Rockall Trough on the two sides of the Atlantic, 320-2740 m.

granuliferus Verrill, 1899: 192. A synonym of *Pseudarchaster parelii* (Düben & Koren, 1846) according to Halpern (1972).

*hispidus Verrill, 1899

Verrill 1899: 191.

Range: 'West Indies', 1100 m.

'Not a typical *Pseudarchaster*' according to Verrill (1899). Type material immature.

huttoni (Farquhar, 1897) (as *Astrogonium huttoni*) Macan 1938: 356. A NOMEN NUDUM, provisional name, unmerited according to Farquhar (1898).

hystrix (Perrier, 1894) Verrill 1899: 195. A synonym of *Pseudarchaster parelii* (Düben & Koren, 1846) according to Halpern (1972).

insignis Verrill, 1895, as variety of *Pseudarchaster intermedius,* see below

intermedius Sladen, 1889: 115. A synonym of *Pseudarchaster parelii* (Düben & Koren, 1846) according to Fisher (1911c) (presumably also variety *insignis*).

jordani Fisher, 1906

Fisher 1906a: 1038; 1919a: 220; Jangoux 1981a: 471.

Koehler 1909b: 49 (as *Astrogonium*).

Range: Hawaiian Is* to the Moluccas, Borneo and S of India, 590-1980 m.

longobrachialis Danielssen & Koren, 1877. A variety of *Pseudarchaster parelii* (Düben & Koren, 1846).

macdougalli McKnight, 1973
McKnight 1973a: 172.
Range: Tasman Sea, NE of New Zealand, 2150 m.

marginatus (Koehler, 1909a) Macan, 1938: 356. A synonym of *Pseudarchaster gracilis* according to Halpern (1972).

microceramus (Fisher, 1913)
Fisher 1913a: 626; 1919a: 225 (as *Aphroditaster*).
Halpern 1970a: 3 (as *Fisheraster*).
Aziz & Jangoux 1984c: 257 (as *Pseudarchaster*).
Range: Sulawesi (Celebes), Indonesia, 1020 m.

mozaicus Wood-Mason & Alcock, 1891
Wood-Mason & Alcock 1891: 432; Alcock 1893b: 171; Macan 1938: 357.
Koehler 1909b: 50 (as *Astrogonium*).
Range: off SE India* and Bay of Bengal, to Zanzibar, 342-1800 m.

myobrachius Fisher, 1906
Fisher 1906a: 1037.
Range: Hawaiian Is, 780-1240 m.

necator (Perrier, 1894) Verrill 1899: 195. A synonym of *Pseudarchaster gracilis* (Sladen, 1889) according to Halpern (1972).

obtusus Hayashi, 1973
Hayashi 1973a: 4; 1973b: 54.
Range: SE Japan, 190-230 m.

oligoporus Fisher, 1913
Fisher 1913a: 625; 1919a: 222.
Range: Sulawesi (Celebes), Indonesia, c.2000 m.

ordinatus Verrill, 1899: 194. A synonym of *Pseudarchaster gracilis* (Sladen, 1889) according to Halpern (1972).

ornatus Djakonov, 1950 *(Pseudarchaster ornatus* Djakonov, 1949 a nomen nudum, insufficiently characterized).
Djakonov 1950b: 53; 1950a: 45 [1968: 38].
Range: Okhotsk Sea, 1640 m.

parelii (Düben & Koren, 1846) (with synonyms *Astrogonium fallax* Perrier, 1885, *Pseudarchaster intermedius* Sladen, 1889 with variety *insignis* Verrill 1895, *A. annectens* Perrier, 1894, *A. hystrix* Perrier, 1894, *P. granuliferus* Verrill, 1899 and variety *arcticus* of *P. tessellatus,* possibly also *P. tessellatus* Sladen, 1889 itself; also variety *longobrachialis* (Danielssen & Koren, 1877)).
Düben & Koren 1846: 247 (as *Astropecten*).
M. Sars 1861: 35; Perrier 1875: 347 [1876: 267] Verrill 1885: 543 (as *Archaster*).

Danielssen & Koren 1877: 17; 1884b: 88 *(Archaster parelii* var. *longobrachialis).*
Sladen 1889: 102 (as *Plutonaster (Tethyaster)).*
Koehler 1907b: 31; Grieg 1932: 44 (as *Astrogonium).*
Koehler 1909a: 75 *(Astrogonium parelii* var. *longobrachiale).*
Süssbach & Breckner 1911: 202 (as *Tethyaster).*
Fisher 1911c: 180; Farran 1913: 13; Koehler 1924: 180; Mortensen 1927: 27; Hayashi 1938a: 115; Djakonov 1950a: 44 [1968: 37]; Baranova 1957: 159; Halpern 1972: 366 (subspecies *alascensis* Fisher, 1905 restored to specific rank); Gage et al. 1983: 279; A.M.C. in Harvey et al. 1988: 162 (as *Pseudarchaster).*
Range: SW Norway*, N Norway, W and S to the Rockall Trough and Bay of Biscay, also S Greenland to E Florida, 75-2965 m.
patagonicus (Perrier, 1891) Macan 1938: 356. A synonym of *Pseudarchaster discus* Sladen, 1889 according to Fisher (1940).
pectinifer Ludwig, 1905
Ludwig 1905: 106; H.L. Clark 1920: 82; Döderlein 1924: 49; Macan 1938: 356.
Range: off Peru, E Pacific*, Indonesia, 725-1865 m.
pretiosus: Goto, 1914, see *Dipsacaster* (Astropectinidae)
pulcher Ludwig, 1905
Ludwig 1905: 110; Macan 1938: 356.
Range: Galapagos Is*, W of Mexico, 700-1620 m.
pusillus Fisher, 1905
Fisher 1905: 304.
Range: Southern and Lower California, 98-700 m.
roseus (Alcock, 1893)
Alcock 1893a: 98 (as *Mediaster).*
Verrill 1899: 196; Fisher 1919a: 225; Aziz & Jangoux 1984c: 257 (as *Pseudarchaster).*
Koehler 1909b: 49 (as *Astrogonium).*
Range: Laccadive Sea.
spatuliger Mortensen, 1934: 6. A synonym of *Craspidaster hesperus* (Müller & Troschel, 1840) (Astropectinidae) according to A.M. Clark (1982b).
tessellatus Sladen, 1889 (with synonym *Pseudarchaster brachyactis* H. L. Clark, 1923)
Sladen 1889: 112; H.L. Clark 1923a: 253; 1926a: 9; Mortensen 1933a: 240; Halpern 1972: 370.
Range: South Africa to Angola, 155-720 m.
Possibly conspecific with *Pseudarchaster parelii* (Düben & Koren, 1846) according to Halpern (1972).
verrilli Ludwig, 1905
Ludwig 1905: 116.

Range: Gulf of Panama, 1000 m.
PSEUDOCERAMASTER Jangoux, 1981
    Jangoux 1981a: 471.
    Type species: *Pseudoceramaster regularis* Jangoux, 1981.
regularis Jangoux, 1981
    Jangoux 1981a: 472.
    Range: Philippine Is, 174-204 m.
PSEUDOGONIODISCASTER Livingstone, 1930
    Livingstone 1930: 15; H.L. Clark 1946: 93.
    Type species: *Pseudogoniodiscaster wardi* Livingstone, 1930.
wardi Livingstone, 1930
    Livingstone 1930: 16.
    Range: Queensland.
PYRENASTER Verrill, 1899
    Verrill 1899: 166.
    Type species: *Pentagonaster dentatus* Perrier, 1884.
    A synonym of *Plinthaster Verrill,* 1899 according to Fisher (1911c).
affinis (Perrier, 1884) Verrill, 1899: 168. A synonym of *Ceramaster grena-*
    densis (Perrier, 1881) according to Halpern (1970b).
dentatus: Verrill, 1899, see *Plinthaster*
ROSASTER Perrier, 1894 (with synonym *Nereidaster* Verrill, 1899)
    Perrier 1894: 386; Fisher 1911c: 164; Verrill 1915: 110; Fisher 1919a:
    240; Halpern 1970b: 208.
    Type species: *Pentagonaster alexandri* Perrier, 1881.
alexandri (Perrier, 1881)
    Perrier 1881: 22; 1884: 238 (as *Pentagonaster).*
    Perrier 1894: 387; Verrill 1915: 111; Halpern 1970b: 208; Downey 1973:
    54 (as *Rosaster).*
    Range: Barbados*, Gulf of Mexico to northern Brazil, 68-443 (?-3470)
    m.
attenuatus Liao, 1984
    Liao 1984: 478, 480.
    Range: East China Sea, 250 m.
bipunctus (Sladen, 1889)
    Sladen 1889: 301 (as *Nymphaster).*
    Verrill 1899: 187 (as *Nereidaster* or *Nymphaster (Nereidaster)).*
    Fisher 1919a: 253 (as *Rosaster).*
    Range: N of New Guinea, 275 m.
breviradiata Sladen 1889. A variety of *Rosaster symbolicus* (Sladen, 1889).
cassidatus Macan, 1938
    Macan 1938: 365.
    Range: Maldive area, 230-290 m.
confinis (Koehler, 1910) (with subspecies *timorensis* Döderlein, 1924)

confinis confinis (Koehler, 1910)
Koehler 1910a: 57 (as *Dorigona*).
Fisher 1919a: 243 (as *Rosaster*).
Range: Andaman Sea, 120 m.
confinis timorensis Döderlein, 1924
Döderlein 1924: 54.
Range: Timor, S Indonesia, 216 m.
endilius McKnight, 1975
McKnight 1975b: 54.
Range: Tasman Sea, NW of New Zealand, 840 m.
Near *R. mimicus* Fisher, 1913, according to H.E.S. Clark (1982).
florifer (Alcock, 1893)
Alcock 1893a: 94 (as *Nymphaster*).
Fisher 1919a: 242 (as *?Mediaster*).
Macan 1938: 365 (as *Rosaster*).
Range: Andaman Sea, 236-456 m.
mamillatus Fisher, 1913
Fisher 1913a: 632; 1919a: 247.
Range: Philippine Is, 112 m.
mimicus Fisher, 1913
Fisher 1913a: 632; 1919a: 250; H.E.S. Clark 1982: 38.
Range: Philippine Is*, New Zealand, Chatham Rise, c.200-920 m.
nannus Fisher 1913
Fisher 1913a: 631; 1919a: 244; Döderlein 1924: 53.
Range: Philippine Is*, Indonesia, 54-93 m.
symbolicus (Sladen, 1889) (with variety *breviradiata* Sladen, 1889)
Sladen 1889: 297, 300 (as *Nymphaster* with variety *breviradiata*).
Verrill 1899: 187 (as *Nereidaster* or *Nymphaster (Nereidaster)*).
Fisher 1913a: 630; 1919a: 253; Chang & Liao 1964: 59; Jangoux 1981a:
472 (as *Rosaster*).
Range: Philippine Is to Torres Strait, 50-256 m.
timorensis Döderlein, 1924. A subspecies of *Rosaster confinis* (Koehler,
1910).
SIBOGASTER Döderlein, 1924
Döderlein 1924: 63.
Type species: *Sibogaster digitatus* Döderlein, 1924.
digitatus Döderlein, 1924
Döderlein 1924: 63
Range: S of Flores, Indonesia, 960 m.
SIRASTER H.L. Clark, 1915
H.L. Clark 1915: 86.
Type species: *Siraster tuberculatus* H.L. Clark, 1915.
squamulosus: H.L. Clark, 1915, non *Stellaster squamulosus* (Studer, 1884),

= *Siraster tuberculatus* H.L. Clark, 1915 according to A.M.C. in Clark & Rowe (1971).

tuberculatus H.L. Clark, 1915

H.L. Clark 1915: 86; A.M. Clark & Rowe 1971: 32, 40, 49; Jangoux & Aziz 1984: 864.

Range: Sri Lanka (Ceylon)*, Maldives and Seychelles, 44 m.

*SPHAERIODISCUS Fisher, 1910

Fisher 1910c: 171; 1911c: 167; Tortonese & A.M. Clark 1956: 343.

Type species: *Stephanaster (Pentagonaster) bourgeti* Perrier, 1885.

A possible synonym of *Peltaster* Verrill, 1899 according to Halpern (1970b). [However, he took into consideration only *Sphaeriodiscus placenta* (Müller & Troschel, 1842), which he referred to *Peltaster,* rather than the type species.

ammophilus Fisher, 1906

Fisher 1906a: 1051 (as *Pentagonaster*).

Fisher 1919a: 290 (as *Sphaeriodiscus*).

Range: Hawaiian Is, 403-470 m.

Referred to *Ceramaster* by Halpern (1970b) when doubting the validity of *Sphaeriodiscus.*

bourgeti (Perrier, 1885)

Perrier 1885b: 885; 1885c: 31; 1894: 403 (as *Stephanaster*).

Sladen 1889: 748 (as *Astrogonium*).

Fisher 1910c: 171; 1919a: 290; H.L. Clark 1926a: 10; Tortonese & A.M. Clark, 1956: 349; Madsen 1958: 92; A.M. Clark & Courtman-Stock 1976: 65.

Range: N of Cape Verde Is*, Natal, South Africa, 405-760 m.

inaequalis (Gray, 1847)

Gray 1847: 79 (as *Astrogonium*).

Sladen 1889: 746 (as *Pentagonaster*).

A.M. Clark 1953: 397; Madsen 1958: 93; A.M. Clark 1976: 250.

Range: New Guinea or Amboina*, S Molucca Is.

maui McKnight, 1973

McKnight 1973a: 187.

Range: Tasman Sea, NW of New Zealand, 1010-1030 m.

mirabilis A.M. Clark, 1976

A.M. Clark 1976: 248.

Range: Amsterdam or St Paul Islands, S Indian Ocean, 200-300 m.

placenta: Fisher, 1919a, see *Peltaster*

scotocryptus Fisher, 1913

Fisher 1913a: 641; 1919a: 287; Jangoux 1981a: 472.

Range: Philippine Is, c.45-450 m.

Referred to *Ceramaster* by Halpern (1970b) when doubting the validity of *Sphaeriodiscus* but not followed by Jangoux (1981).

STELLASTER Gray, 1840

Gray 1840: 277; Döderlein 1935: 86; H.L. Clark 1946: 96 (Döderlein's treatment unsatisfactory); A.M. Clark & Rowe 1971: 49.

Type species: *Stellaster childreni* Gray, 1840, currently cited as a synonym of *Asterias equestris* Retzius, 1805 but deserving restoration.

andamana and

bandana Döderlein, 1935. Formae of *Stellaster equestris* (Retzius, 1805).

belcheri Gray, 1847: 76. A synonym or forma of *Stellaster equestris* (Retzius, 1805) according to Döderlein (1935).

*childreni Gray, 1840

Gray 1840: 278.

Range: Indo-West Pacific.

Treated as a synonym or forma of *Stellaster equestris* (Retzius, 1805) by Döderlein (1935). [But unjustifiably and invalidly (see under *equestris* below) and should be revived.]

convexus Jangoux, 1981

Jangoux 1981a: 473.

Range: Philippine Is, 143-178 m.

crassa and

elongata Döderlein, 1935. Formae of *Stellaster equestris* (Retzius, 1805).

*equestris (Retzius, 1805) (a 'super-species' according to Döderlein (1935) with synonyms or forms: *andamana, bandana, belcheri, childreni, crassa, elongata, gracilis, incei, indica, latior, semoni* and *tenuispina,* needing re-evaluation; also synonym *Goniaster (Stellaster) muelleri* Martens, 1865).

Retzius 1805: 12 (as *Asterias*).

Müller & Troschel 1842: 62, 128; Goto 1914: 411; Döderlein 1935: 91; Jangoux in Guille & Jangoux 1978: 53.

[Investigation indicates that Döderlein's use of the name '*Asterias equestris* Retzius, 1820' (in fact 1805) is invalid. If it had been a new species of Retzius, the specific name would be a junior primary homonym of *Asterias equestris* Linnaeus, 1758. However, Retzius cited Linnaeus' 12th edition, p.1100 as source. *A. equestris* Linnaeus, though described as having 5 tubercles, was supposedly from the Mediterranean and of the seven figures of Linck cited, though one of the four recognisable ones is probably a *Ceramaster* and another *Goniaster,* the two best are of *Hippasteria phrygiana* (Parelius), which agrees with the diagnosis of Retzius in having laterally spinose marginals. It is therefore impossible to use *equestris* Linnaeus for the Indo-Pacific species, so *Stellaster childreni* Gray, 1840, widely used up to 1935 should be revived in its place. The name *Asterias equestris* Linnaeus, 1758 needs to be formally suppressed by the ICZN in order to validate the long-accepted *Hippasteria phrygiana* (Parelius, 1768).]

gibbosus Döderlein, 1916: 412; also 1935: 89. A synonym of *Stellaster inspinosus* H. L. Clark, 1916 according to H. L. Clark (1946).

gracilis Möbius, 1859. A synonym or forma of *Stellaster equestris* (Retzius, 1805) according to Döderlein (1935).

*granulosus (Perrier, 1878)(as *Pentagonaster (Stellaster)*) Sladen 1889: 750. A NOMEN NUDUM, undescribed.

incei Gray, 1847: 76; also Döderlein 1896: 307; H.L. Clark 1946: 97. A synonym or forma of *Stellaster equestris* (Retzius, 1805) according to Döderlein (1935) but retained as a valid species by H.L. Clark (1946).

indica Döderlein, 1935. A forma of *Stellaster equestris* (Retzius, 1805).

inspinosus H.L. Clark, 1916 (with synonym *Stellaster gibbosus* Döderlein, 1916)

H.L. Clark 1916: 48; Döderlein 1935: 91; H.L. Clark 1946: 98.

Range: Western Australia.

latior Döderlein, 1935. A forma of *Stellaster equestris* (Retzius, 1805).

megaloprepes H.L. Clark 1914: 141. A synonym of *Stellaster princeps* Sladen, 1889. according to H.L. Clark (1938).

princeps Sladen, 1889 (with synonym *Stellaster megaloprepes* H.L. Clark, 1914)

Sladen 1889: 323; Livingstone 1932a: 246; H.L. Clark 1938: 98; 1946: 97.

Range: Western Australia and Torres Strait.

semoni Döderlein, 1935. A forma of *Stellaster equestris* (Retzius, 1805).

squamulosus (Studer, 1884)

Studer 1884: 33 (as *Pentagonaster (Stellaster)*).

Döderlein 1935: 88 (as *Stellaster*).

Range: NW Australia, 110 m.

squamulosus: Koehler, 1910a: 81, non *Stellaster squamulosus* (Studer, 1884), = *Stellaster equestris* forma *childreni* Gray according to Döderlein (1935) but more likely = *Siraster tuberculatus* H.L. Clark, 1915 according to A.M.C. in Clark & Rowe (1971).

sulcatus Möbius, 1859: 2. A synonym of *Archaster* [i.e. *Craspidaster*] *hesperus* (Müller & Troschel, 1840) (Astropectinidae) according to Perrier (1875).

tenuispina Döderlein, 1935. A forma of *Stellaster equestris* (Retzius, 1805).

tuberculosus (Martens, 1865)

Martens 1865: 358 (as *Goniaster (Stellaster)*).

Perrier 1878: 85 (as *Pentagonaster (Stellaster)*).

Döderlein 1935: 88 (as *Stellaster*).

Range: China.

STELLASTEROPSIS Dollfus, 1936

Dollfus 1936: 151; Macan 1938: 395.

Type species: *Stellasteropsis fouadi* Dollfus, 1936.

colubrinus Macan, 1938
   Macan 1938: 395.
   Range: Gulf of Oman, Gulf of Aden, NE Africa, 13.5-200 m.
fouadi Dollfus, 1936
   Dollfus 1936: 151.
   Range: Gulf of Suez.
tuberculiferus Macan 1938
   Macan 1938: 398.
   Range: Arabian coast, 38 m.
STEPHANASTER Ayres, 1851
   Ayres 1851: 118; Perrier 1894: 402.
   Type species: *Stephanaster elegans* Ayres, 1851.
   A synonym of *Pentagonaster* Gray, 1840, by synonymy of the type
   species, according to Dujardin & Hupé (1862), Perrier's usage invalid.
astrologorum: Perrier, 1894. A forma of *Tosia australis* Gray, 1840 according to H. L. Clark (1946).
australis: Perrier, 1894, see *Tosia*
bourgeti: Perrier, 1894, see *Sphaeriodiscus*
duebeni: Perrier, 1894, see *Pentagonaster*
elegans Ayres, 1851: 118. A synonym of *Pentagonaster pulchellus* Gray,
   1840 according to Dujardin & Hupé (1862).
gunni: Perrier, 1894. A forma of *Pentagonaster duebeni* Gray, 1847 according to A. M. Clark (1953).
procyon (Cuvier, 1836) Perrier, 1894: 403. Cited as a synonym of *Tosia
   australis* Gray, 1840 by Ludwig (1912) but a NOMEN NUDUM according to A. M. Clark (1953).
pulchellus: Perrier, 1894, see *Pentagonaster*
STYPHLASTER H.L. Clark, 1938
   H.L. Clark 1938: 88.
   Type species: *Styphlaster notabilis* H.L. Clark, 1938.
notabilis H.L. Clark, 1938
   H.L. Clark 1938: 89.
   Range: NW Australia, 13-16 m.
TESSELLASTER H.L. Clark, 1941
   H.L. Clark 1941: 36; Halpern 1970b: 218.
   Type species: *Tessellaster notabilis* H.L. Clark, 1941.
notabilis H.L. Clark, 1941
   H.L. Clark 1941: 37; Halpern 1970b: 218; Downey 1973: 53.
   Range: Cuba*, SW Florida to Leeward Is. 329-575 m.
TORASTER A.M. Clark, 1952
   A.M. Clark 1952a: 205.
   Type species: *Astrogonium tuberculatum* Gray, 1847.
tuberculatus Gray, 1847 (with synonym *Pentagonaster granulatus* Bell,
   1905)

Gray 1847: 79 (as *Astrogonium*).

Perrier 1875: 222 [1876: 38]; Bell 1905: 246 (as *Pentagonaster*).

H. L. Clark 1923a: 266; 1926a: 11; Mortensen 1933a: 243 (as *Tosia*).

A. M. Clark 1952a: 205; A. M. Clark & Courtman-Stock 1976: 66 (as *Toraster*).

Range: Natal*, to western Cape Province, South Africa, 75-366 m.

TOSIA Gray, 1840

Gray 1840: 281; 1847: 80; Verrill 1899: 148, 158, 160 (emended); Fisher 1906a: 1052; Ludwig 1912: 22; H. L. Clark 1946: 93; A. M. Clark 1953: 396.

Type species: *Tosia australis* Gray, 1840.

affinis (Perrier, 1884: 168) Perrier, 1884: 183[!]. A synonym of *Ceramaster grenadensis* (Perrier, 1881) according to Halpern (1970b).

arctica Verrill, 1909, see *Ceramaster*

astrologorum: Verrill, 1899. A variety of *Tosia australis* Gray, 1840 according to H. L. Clark (1938).

aurata Gray, 1847: 80; also Ludwig 1912: 34. A synonym of *Tosia magnifica* (Müller & Troschel, 1842) according to A. M. Clark (1953).

australis Gray, 1840 (with synonym *Astrogonium geometricum* Müller & Troschel, 1842 and variety or forma *astrologorum* (Müller & Troschel, 1842)).

Gray 1840: 281; Ludwig 1912: 23; H. L. Clark 1928: 381; 1938: 381; 1946: 94; A. M. Clark 1953: 404; Shepherd 1968: 741.

Perrier 1875: 200 [1876: 15] (as *Pentagonaster*).

Perrier 1894: 403 (as *Stephanaster*).

Range: southern Australia, 1-10 m.

clugreta Walenkamp, 1976

Walenkamp 1976: 61; Jangoux & de Ridder 1987: 89.

Range: Suriname (Dutch Guiana), 130 m.

ceramoidea Fisher, 1906a, see *Plinthaster*

compta Verrill, 1899: 163. A synonym of *Plinthaster dentatus* (Perrier, 1884) according to Halpern (1970b).

eximia (Verrill, 1894) Verrill, 1899: 161. A synonym of *Ceramaster granularis* (Retzius, 1783) according to Halpern (1970, thesis).

gosselini (Perrier, 1884) Verrill 1899: 162 (as *Tosia (Ceramaster)*). A synonym of *Ceramaster grenadensis* (Perrier, 1881) according to Halpern (1970b).

grandis Gray, 1847: 80. A synonym of *Tosia magnifica* (Müller & Troschel, 1842) according to Ludwig (1912).

granularis: Verrill, 1899, see *Ceramaster*

greenei (Bell, 1889) Verrill 1899: 161, lapsus for *greeni*. A synonym of *Ceramaster grenadensis* (Perrier, 1881) according to Halpern (1970, thesis) and in Downey (1973).

288

grenadensis: Verrill, 1899, see *Ceramaster*

haesitans (Perrier, 1885) Verrill, 1899: 162. A synonym of *Ceramaster grenadensis* (Perrier, 1881) according to Halpern in Downey (1973).

lamarcki: Gray, 1866, see *Pentagonaster*

leptocerama Fisher, 1905, see *Ceramaster*

magnifica (Müller & Troschel, 1842) (with synonyms *Astrogonium ornatum* Müller & Troschel, 1842, *Tosia grandis* Gray, 1847, *T. aurata* Gray, 1847 and *A. emili* Perrier, 1869).

> Müller & Troschel 1842: 53 (as *Astrogonium*).
> Perrier 1894: 390 (as *Pentagonaster*).
> Ludwig 1912: 36; A.M. Clark 1953: 408; Shepherd 1968: 742; Jangoux & de Ridder 1987: 90 (as *Tosia*).
> Range: South Australia, Tasmania and Victoria, 5-200 m.

*mammillata (Müller & Troschel, 1842)

> Müller & Troschel, 1842: 61 (as *Goniodiscus*).
> Philippi 1857: 132 (as *Linckia*).
> Perrier 1875: 223 [1876: 39]; Sladen 1889: 746 (as *Pentagonaster*).
> Verrill 1899: 162 (as *Tosia (Ceramaster)*).
> Range: Unknown.
> [Validity doubtful.]

micropelta Fisher, 1906, see *Peltaster*

mirabilis (Perrier, 1875) Verrill, 1899: 161 (as *Tosia (Ceramaster)*). A synonym of *Pentagonaster* [i.e. *Peltaster*] *placenta* (Müller & Troschel, 1842) according to Ludwig (1897).

nitida Verrill, 1899: 165 (as *Tosia (Plinthaster)*). A synonym of *Plinthaster dentatus* (Perrier, 1884) according to Halpern (1970b).

nobilis (Müller & Troschel, 1843) (with synonyms *Tosia tubercularis* Gray, 1847, *Pentagonaster minimus* Perrier, 1875 and probably *T. rubra* Gray, 1847).

> Müller & Troschel 1843: 116 (as *Astrogonium*).
> Ludwig 1912: 30; A.M. Clark 1953: 406 (as *Tosia*).
> Range: Western Australia.

parva (Perrier, 1881)

> Perrier 1881: 19; 1884: 231 (as *Pentagonaster (Tosia)*).
> H.L. Clark 1898: 5 (as *Pentagonaster*).
> Halpern 1969b: 503; Downey 1973: 54; Walenkamp, 1976: 61; Carrera-Rodriguez & Tommasi 1977: 96; Walenkamp 1979: 32 (as *Tosia*).
> Range: Carolina to southern Brazil, 30-600 m.

perrieri (Sladen, 1889) Verrill 1899: 161. A synonym of *Plinthaster dentatus* (Perrier, 1884) according to Halpern (1970b).

placenta: Verrill, 1899, see *Peltaster*

pulvinus: Verrill, 1899, see *Pentagonaster*

queenslandensis Livingstone, 1932

Livingstone 1932a: 243; 1932b: 381; H.L. Clark 1946: 94; A.M. Clark 1953: 411; Jangoux 1984: 282; 1986a: 122.

Range: Great Barrier Reef*, New Caledonia, 4-6 m.

rubra Gray, 1847: 81. Probably a synonym of *Tosia nobilis* (Müller & Troschel, 1843) according to A.M. Clark (1953).

simplex (Verrill, 1895) Verrill, 1899: 161 (as *Tosia (Ceramaster)*). A synonym of *Ceramaster granularis* (Retzius, 1783) according to Halpern (1970, thesis).

tubercularis Gray, 1847: 80; also Livingstone 1932b: 378. A synonym of *Tosia nobilis* (Müller & Troschel, 1843) according to Ludwig (1912).

tuberculata: H.L. Clark, 1923, see *Toraster*

TOSIASTER Verrill, 1914

Verrill 1914a: 292.

Type species: *Tosia arctica* Verrill, 1909.

Type species retained in *Ceramaster* by Fisher (1919a), Djakonov (1950b) and Baranova (1957) without mention of *Tosiaster.*

arcticus: Verrill, 1914, see *Ceramaster*

Family SPHAERASTERIDAE Schöndorf

Schöndorf 1906: 256; A.M. Clark & Wright 1962: 243; Spencer & Wright 1966: U55.

The only recent genus is **Podosphaeraster.**

PODOSPHAERASTER A.M. Clark, 1962

A.M. Clark & Wright 1962: 243; Spencer & Wright 1966: U55; Cherbonnier 1970: 206; Rowe & Nichols 1980: 289; Rowe, Nichols & Jangoux 1982: 83; Rowe 1985: 305.

Type species: *Podosphaeraster polyplax* A.M. Clark & Wright, 1962.

crassus Cherbonnier, 1974: 1731. A fragment of *Nymphaster arenatus* (Perrier, 1881) (Goniasteridae) according to Rowe (1985).

gustavei Rowe, 1985

Rowe 1985: 310.

Range: NW and N of Spain, ?Azores, 500-540 m.

polyplax A.M. Clark & Wright, 1962

A.M.C. in Clark & Wright 1962: 243; Rowe & Nichols 1980: 290; Rowe, Nichols & Jangoux 1982: 88.

Range: Macclesfield Bank, S China Sea*, Indonesia, Arafura Sea N of Australia, 72-125 m.

pulvinatus Rowe & Nichols, 1980

Rowe & Nichols 1980: 290.

Range: Guam* and Loyalty Is, 244-324 m.

thalassae Cherbonnier, 1970

290

Cherbonnier 1970: 203; 1974: 1731; Rowe 1985: 313.
Range: NW of Spain, 500-520 m.

Family ASTERODISCIDIDAE Rowe

Asterodiscididae Rowe 1977: 190; 1985: 532.
   Genus-group names: **Amphiaster**, **Asterodiscides**, Asterodiscus, **Paulia**,
   Pauliella.

AMPHIASTER Verrill
   Verrill, 1868: 372; Fisher 1911c: 168; Verrill 1914a: 294; Rowe 1977:
   211.
   Type species: *Amphiaster insignis* Verrill, 1868.
insignis Verrill, 1868
   Verrill 1868: 372; 1914a: 294; Ziesenhenne 1937: 214; Rowe 1977: 212.
   Range: Gulf of California, 55-82 m.
ASTERODISCIDES A.M. Clark, 1974 (nom. nov. for *Asterodiscus* Gray,
   1847, preoccupied)
   A.M. Clark 1974: 435; Rowe 1977: 192; 1985: 532.
   Type species: *Asterodiscus elegans* Gray, 1847.
belli Rowe, 1977
   Rowe 1977: 199 (as subspecies of *Asterodiscus elegans* Gray, 1847).
   Rowe 1985: 534 (as species).
   Range: Amirante Is, W Indian Ocean*, Madagascar, W Thailand, 30-200
   m.
cherbonnieri Rowe, 1985
   Rowe 1985: 542.
   Range: Madagascar, 50-110 m.
crosnieri Rowe, 1985
   Rowe 1985: 543.
   Range: Madagascar, 150 m.
culcitulus Rowe, 1977
   Rowe 1977: 194; 1985: 550.
   Range: Abrolhos Is, W Australia.
elegans (Gray, 1847)
   Gray 1847: 176; Fisher 1919a: 355; A.M.C. in Clark & Rowe 1971: 40 (as
   Asterodiscus).
   A.M. Clark 1974: 435; Rowe 1977: 197 (as *Asterodiscides*).
   Range: Philippine Is, W to Seychelles and Natal, 0-49 m.
fourmanoiri Rowe, 1985
   Rowe 1985: 545.
   Range: Madagascar, 55-c.110 m.
grayi Rowe, 1977

Rowe 1977: 208; 1985: 550.

Range: Queensland* and northern N.S.W., Australia, S Japan, 100-108 (?20) m.

helenotus Rowe, 1977: 202, lapsus for *helonotus* (Fisher, 1913).

helenotus: Jangoux, 1984: 280 and presumably 1986a: 122, non *Asterodiscides helonotus* (Fisher, 1913), = *A. soelae* Rowe, 1985, according to Rowe (1985).

helonotus (Fisher, 1913) (with synonym *Asterodiscus hiroi* Hayashi, 1938)

Fisher 1913c: 210; 1919a: 357 (as *Asterodiscus*).

Rowe 1977: 202 (helenotus)(as *Asterodiscides*).

Range: Philippine Is* to southern Japan, 18-81 m.

lacrimulus Rowe, 1977

Rowe 1977: 206.

Range: NW Arabian Sea, 75-175 m.

macroplax Rowe, 1985

Rowe 1985: 536.

Range: Abrolhos Is* to NW Australia, 20-80 m.

multispinus Rowe, 1985

Rowe 1985: 540.

Range: Queensland, 23-28 m.

pinguiculus Rowe 1977

Rowe 1977: 204.

Range: NW Australia, 7-27 m.

soelae Rowe, 1985

Rowe 1985: 547.

Range: Abrolhos Is*, NW Australia, New Caledonia, 20-80 m.

tessellatus Rowe, 1977

Rowe 1977: 203.

Range: near Seychelles, W Indian Ocean, 99 m.

truncatus (Coleman, 1911)

Coleman 1911: 699; Fisher 1919a: 355; Powell 1937: 78; Fell 1958: 13; Dartnall 1968: 23 (as *Asterodiscus*).

Rowe 1977: 200 (as *Asterodiscides*).

Range: South Australia E to Kermadec Is, 14-792 m.

tuberculosus (Fisher, 1906)

Fisher 1906a: 1075; 1919a: 355 (as *Asterodiscus*).

Rowe 1977: 207 (as *Asterodiscides*).

Range: Hawaiian Is, 59-396 m.

ASTERODISCUS Gray, 1847

Gray, 1847: 176; Sladen 1889: 353.

Type species: *Asterodiscus elegans* Gray, 1847.

A junior homonym of *Asterodiscus* Ehrenberg, 1839 (Protozoa), renamed *Asterodiscides* by A.M. Clark, 1974.

elegans Gray, 1847, see *Asterodiscides*

helonotus Fisher, 1913c, see *Asterodiscides*

hiroi Hayashi 1938c: 277. A synonym of *Asterodiscides helonotus* Fisher, 1913 according to Rowe (1977).

truncatus Coleman, 1911, see *Asterodiscides*

tuberculosus Fisher, 1906a, see *Asterodiscides*

PAULIA Gray, 1840 (probably including *Pauliella* Ludwig, 1905)

Gray 1840: 278; Rowe 1977: 209.

Type species: *Paulia horrida* Gray, 1840.

galapagensis Ludwig, 1905 a variety of

horrida Gray, 1840 (with probable synonym *Pauliella aenigma* Ludwig, 1905 and variety *galapagensis* Ludwig, 1905).

Gray 1840: 278; Ludwig 1905: 143 (variety *galapagensis); H.L. Clark 1910: 333; Rowe 1977: 210.

Müller & Troschel 1842: 61 (as *?Goniodiscus).

Perrier 1875: 255 [1876: 70] (as *Nidorellia).

Range: Equador*, Galapagos Is to Lower California, 22-32 m, also ?Cocos and Clarion Is, W from Central America.

*PAULIELLA Ludwig, 1905

Ludwig 1905: 151.

Type species: *Pauliella aenigma* Ludwig, 1905.

A junior homonym of *Pauliella* Munier-Chalmas, 1895. Also presumably = *Paulia* Gray, 1840, since Döderlein (1936) lists *P. aenigma* in the synonymy of *Paulia horrida*, though Rowe (1977) has doubts.

aenigma Ludwig, 1905: 151; Ziesenhenne 1937: 215. A synonym of *Paulia horrida* Gray, 1840 according to Döderlein (1936), though Rowe (1977) has doubts.

Family OREASTERIDAE Fisher

Pentacerotidae Gray 1840: 275; Perrier 1884: 165; Sladen 1889: 343; Perrier 1894: 405.

Oreasteridae Fisher 1911c: 18; Döderlein 1935: 71-110; H.L. Clark 1946: 105; Spencer & Wright 1966: U63; A.M. Clark & Courtman-Stock 1976: 66.

Goniasteridae (Oreasterinae): Döderlein 1916: 413; 1936: 295.

The earlier family-group name Pentacerotidae is based on the invalid pre-linnaean name *Pentaceros* Linck, 1733. The family has been mono-graphed by Döderlein (1935, Oreasteridae except Oreasterinae, and 1936, Oreasterinae). However, H.L. Clark (1938 and 1946), followed by others including myself (in Clark & Rowe, 1971) continued to place most of the genera dealt with by Döderlein in 1935 in the Goniasteridae as defined by Fisher (1911 and 1919a). A balance was reached by Spencer & Wright

(1966), whose scope for the Oreasteridae is now followed. They listed a large number of genera as being of uncertain position in the subfamilies of Goniasteridae and it is possible that some of these may prove better placed in the Oreasteridae when a detailed study of the limits between the two families is undertaken.

Genus-group names: **Acheronaster, Anthaster, Anthenea**, Bothriaster, **Choriaster, Culcita**, Culcitaster, **Goniodiscaster**, Goniodiscides, Goniodiscus, **Gymnanthenea, Halityle**, Hosea, Hosia, **Monachaster, Nectria, Nectriaster, Nidorellia**, Oreaster, **Pentaceraster**, Pentaceropsis, Pentaceros, **Pentaster, Poraster, Protoreaster, Pseudanthenea, Pseudoreaster**, Randasia.

ACHERONASTER H.E.S. Clark, 1982
H.E.S. Clark 1982: 39.
Type species: *Acheronaster tumidus* H.E.S. Clark, 1982.
tumidus H.E.S. Clark, 1982
H.E.S. Clark 1982: 39.
Range: Kermadec Is, N of New Zealand, 110-146 m.
ANTHASTER Döderlein, 1915
Döderlein 1915: 29.
Type species: *Oreaster valvulatus* Müller & Troschel, 1843.
valvulatus (Müller & Troschel, 1843)
Müller & Troschel, 1843: 115 (as *Oreaster*).
Sladen 1889: 345 (as *Pentaceros*).
Döderlein 1915: 30; 1935: 105; H.L. Clark 1928: 386; 1938: 102; Shepherd 1968 (as *Anthaster*).
Range: SW Australia, 3-40 m.
ANTHENEA Gray, 1840 (with synonym *Hosia* Gray, 1840 *[Hosea* Gray, 1866, lapsus])
Gray 1840: 279; Döderlein 1915: 23; 1935: 105; H.L. Clark 1938: 110; 1946: 100.
Type species: *Anthenea chinensis* Gray, 1840, ? a synonym of *Asterias pentagonula* Lamarck, 1816.
acanthodes H.L. Clark, 1938
H.L. Clark 1938: 124; Endean 1953: 53; 1965: 230.
Range: Queensland.
acuta (Perrier, 1869) Perrier, 1875: 276 [1876: 91]; also H.L. Clark 1946: 102; Endean 1965: 230. A primary homonym of *Goniodiscus acutus* Heller, 1863; see *Anthenea edmondi* A.M. Clark, 1970, also *A. sydneyensis* Döderlein, 1915.
articulata: Perrier 1875: 273 [1876: 87]; also Viguier 1878: 180, non *Goniodiscus articulatus* Perrier, 1869, = *Anthenea viguieri* Döderlein, 1915, nom. nov.

aspera Döderlein, 1915

Döderlein 1915: 28, 35; Mortensen 1934: 8; H.L. Clark 1938: 118.

Range: 'Australien'*, Queensland, Hong Kong area of S China Sea.

australiae Döderlein, 1915 (nom. nov. for *Anthenea tuberculosa* Perrier, 1875, non Gray, 1847)

Döderlein 1915: 24, 52; H.L. Clark 1946: 102.

Range: W and NW Australia.

***chinensis** Gray, 1840: 279. A synonym of *Anthenea pentagonula* (Lamarck, 1816) according to Müller & Troschel (1842) [but possibly valid in my opinion].

conjugens Döderlein, 1935

Döderlein 1935: 107; H.L. Clark 1938: 119; 1946: 103.

Range: NW Australia.

crassa H.L. Clark, 1938

H.L. Clark 1938: 124; Endean 1953: 53.

Range: Queensland.

crudelis Döderlein, 1915

Döderlein 1915: 53 (as variety of *Anthenea australiae*).

H.L. Clark 1938: 114; 1946: 103 (as species).

Range: 'Probably Australia' but dealt with under non-australian Antheneas by H.L. Clark (1938).

diazi Domantay, 1969

Domantay 1969: 50; 1972: 82.

Range: southern Philippine Is.

***edmondi** A.M. Clark, 1970

A.M. Clark 1970: 157 (nom. nov. for *Goniodiscus acutus* Perrier, 1869, non Heller, 1863).

[See references under *Anthenea acuta* above.]

Range: Tasmania N to southern Queensland.

[Itself invalidated by *Anthenea sydneyensis* Döderlein, 1915 (as variety of *A. australiae* Döderlein, 1915) but synonymized with *A. acuta* (Perrier) by H.L. Clark (1946). However, as the status of the name *sydneyensis* has been either infrasubspecific or as a synonym, there is a good case for its suppression by the ICZN.]

elegans H.L. Clark, 1938

H.L. Clark 1938: 126.

Range: NW Australia.

flavescens (Gray, 1840) (with variety *nuda* Döderlein, 1915).

Gray, 1840: 278 (as *Hosia*).

Gray 1866: 9 (as *Hosea*).

Perrier 1875: 277 [1876: 92]; Döderlein 1915: 28, 41 (with variety *nuda*); [?]H.L. Clark 1928: 384; Djakonov 1930: 246; Chen 1932: 67; A.M. Clark 1982b: 490 (as *Anthenea*).

Range: Type locality unknown, other records from South China Sea, ?Australia; variety *nuda* from Moluccas.

globifera Döderlein 1915: 25, 50, also 1926: 13, lapsus for *globigera*.

globigera Döderlein, 1915, see *Gymnanthenea*

godeffroyi Döderlein, 1915

Döderlein 1915: 29, 45; 1935: 108; H.L. Clark 1938: 120.

Range: Given as Samoa (1915) but probably Australia (1935) according to H.L. Clark (1938).

granulifera Gray, 1847, see *Goniodiscaster*

grayi Perrier, 1875

Perrier 1875: 278 [1876: 93]; A.M. Clark & Rowe 1971: 51.

Range: Philippine Is.

A synonym of *Anthenea flavescens* (Gray, 1847) according to Sladen (1889) but treated as valid by A.M.C. in Clark & Rowe (1971).

mertoni Koehler, 1910

Koehler 1910b: 268; Döderlein 1915: 28, 37; H.L. Clark 1938: 121.

Range: Aru Is, Indonesia*, N and NW Australia.

Thought to be a synonym of *Anthenea tuberculosa* Gray, 1847 by H.L. Clark (1921) but a valid species according to H.L. Clark (1938).

mexicana A.H. Clark, 1916

A.H. Clark 1916a: 56.

Range: W Mexico (no details).

nuda Döderlein, 1915: 41. A variety of *Anthenea flavescens* (Gray, 1840).

obesa H.L. Clark, 1938

H.L. Clark 1938: 127.

Range: SW Australia.

obtusangula: Döderlein, 1915, see *Pseudoreaster*

***pentagonula** (Lamarck, 1816) (with presumed synonyms *Anthenea chinensis* Gray, 1840 and *Goniodiscus articulatus* Perrier, 1869).

Lamarck 1816: 554 (as *Asterias*).

Lütken 1864: 147 (as *Goniaster*).

Müller & Troschel 1842: 57 (pt)(as *Goniodiscus*).

Perrier 1875: 275 [1876: 90]; Döderlein 1915: 28, 32; H.L. Clark 1938: 115 (as *Anthenea*).

Range: Type locality 'les mers australes', coll. Peron & Lesueur but most attributed specimens from the S China Sea.

Validity doubtful but widely quoted; needs neotype in conformity with Döderlein (1915) if holotype untraceable.

pentagonula: Studer, 1884: 37, non *Anthenea pentagonula* sensu Döderlein, 1915, = *Anthenea mertoni* Koehler, 1910 according to Döderlein (1915).

polygnatha H.L. Clark, 1938

H.L. Clark 1938: 128.

Range: NW Australia.

296

regalis Koehler, 1910

Koehler 1910a: 82; Döderlein 1915: 28, 38; H.L. Clark 1938: 115; A. M. Clark & Rowe 1971: 32, 52; Aziz & Jangoux 1984a: 133.

Range: northern Bay of Bengal*, Sri Lanka, Philippine Is, W Java, 30-50 m.

rudis Koehler, 1910

Koehler 1910a: 86; Döderlein 1915: 28, 40; H.L. Clark 1938: 116.

Range: Mergui Archipelago, Burma*, Sri Lanka, Arabian Gulf.

sibogae Döderlein, 1915

Döderlein 1915: 29, 47; 1935: 106; H.L. Clark 1938: 122.

Range: Lesser Sunda Is, Indonesia*, Queensland, 16-36 m.

spinulosa (Gray, 1847) Sladen, 1889: 758 *(?Anthenea)*. A synonym of *Culcita novaeguineae* Müller & Troschel, 1842 according to Goto (1914).

*sydneyensis Döderlein, 1915

Döderlein 1915: 53 (as variety of *Anthenea australiae* Döderlein, 1915)).

Range: Sydney Harbour.

A synonym of *Anthenea acuta* (Perrier, 1869), which name is invalid and was replaced by *A. edmondi* (q.v.) A.M. Clark, 1970, unaware of the availability of *sydneyensis* as a replacement name.

tuberculata Döderlein, 1915: 29, lapsus for

tuberculosa Gray, 1847 (with variety *vanstraeleni* Engel, 1938).

Gray 1847: 77; Döderlein 1915: 43; H.L. Clark 1928: 385; Engel 1938: 4 (variety *vanstraeleni); Döderlein 1935: 106; H.L. Clark 1946: 104.

Range: northern Australia; var. *vanstraeleni* from Indonesia.

tuberculosa: Perrier 1875: 273 [1876: 88], non *Anthenea tuberculosa* Gray, 1847, = *Anthenea australiae* Döderlein, 1915, nom. nov.

tuberculosa: H.L. Clark, 1921: 29, non *Anthenea tuberculosa* Gray, 1847, = *Anthenea sibogae* Döderlein, 1915 according to H.L. Clark (1938).

vanstraeleni Engel, 1938a. A variety of *Anthenea tuberculosa* Gray, 1847.

*viguieri Döderlein, 1915 Döderlein 1915: 28, 34; 1926: 12; H.L. Clark 1946: 105.

Range: Amboina, South Molucca Is [*?], Queensland.

Said to be a nom. nov. for *Anthenea articulata* (Perrier, 1875) [non *Goniodiscus articulatus* Perrier, 1869] from the Seychelles but described from Amboina specimen, originally *Goniodiscus pentagonulus:* Müller & Troschel, 1842.

*BOTHRIASTER Döderlein, 1916

Döderlein 1916: 415.

Type species: *Bothriaster primigenius* Döderlein, 1916.

*primigenius Döderlein, 1916

Döderlein 1916: 417; also 1935: 85.

Range: East Indies, 27-45 m.

[Validity of genus and species doubtful, diagnostic characters attributable to immaturity; holotype lost according to Jangoux (pers. comm.).]

CHORIASTER Lütken, 1869

Lütken 1869: 35; 1871: 243; Goto 1914: 605; Fisher 1919a: 367.

Type species: *Choriaster granulatus* Lütken, 1869.

granulatus Lütken, 1869 (with possible synonym *Culcita niassensis* Sluiter, 1895)

Lütken 1869: 35; Goto 1914: 604; Fisher 1919a: 367; Domantay & Roxas 1938: 217; Hayashi 1938d: 424; Chang & Liao 1964: 61; A.M. Clark 1967a: 37; Liao 1980: 154; Marsh & Marshall 1983: 675; Jangoux 1986a: 124.

Range: Fiji*, S China Sea, New Caledonia, northern Australia, Red Sea, East Africa, 5-40 m.

*niassensis (Sluiter, 1895)

Sluiter 1895: 58 (as *Culcita*).

H.L. Clark 1921: 32 (as *Choriaster*).

Range: Sumatra, W Indonesia.

Possibly a synonym of *Choriaster granulatus* Lütken, 1869 according to H. L. Clark (1921).

CULCITA L. Agassiz, 1836 (with synonyms *Goniodiscus* Müller & Troschel, 1842 [q.v.] and *Goniodiscides* Fisher, 1906).

L. Agassiz 1836: 192 [25]; Hartlaub 1893: 65; Döderlein 1896: 314; Goto 1914: 515; Döderlein 1935: 108; Livingstone 1932b: 265.

Type species: *Asterias discoidea* Lamarck, 1816, a synonym of *Asterias schmideliana* Retzius, 1805.

acutispina Sladen, 1889: 352, lapsus for

acutispinosa Bell, 1883. A variety of *Culcita novaeguineae* Müller & Troschel, 1842 according to Goto (1914).

africana Döderlein, 1896: 315. A variety of *Culcita schmideliana* (Retzius, 1805).

arenosa Perrier, 1869. A variety of *Culcita novaeguineae* Müller & Troschel, 1842 according to Döderlein (1896).

borealis Süssbach & Breckner, 1911, see *Poraniomorpha (Culcitopsis)* (Poraniidae).

ceylonica Döderlein, 1896: 315. A variety of *Culcita schmideliana* (Retzius, 1805).

coriacea Müller & Troschel, 1842

Müller & Troschel, 1842: 38; Tortonese 1980: 316.

Range: Red Sea.

coriacea: Peters, 1852: 177, non *Culcita coriacea* Müller & Troschel, 1842, probably = *Culcita schmideliana* (Retzius, 1805) according to A.M.C. in Clark & Rowe (1971).

discoidea: L. Agassiz, 1836: 192 [25]; also Müller & Troschel 1840b: 323;

1842: 37. A synonym of *Culcita schmideliana* (Retzius, 1805) according to Gray (1840). [Though a synonym of *C. novaeguineae* Müller & Troschel, 1842 according to Sluiter (1889)].

*grex Müller & Troschel, 1842

Müller & Troschel, 1842: 39; Hartlaub 1893: 87; Döderlein 1896: 313; Livingstone 1932b: 270; Domantay & Roxas 1938: 215; Jangoux & de Ridder 1987: 89 (type extant).

Range: Type locality unknown, later records from W Pacific and Andaman Is.

Validity doubtful, probably a forma of *Culcita novaeguineae* Müller & Troschel, 1842 according to Döderlein (1896).

leopoldi Engel, 1938a: 10. A variety of *Culcita novaeguineae* Müller & Troschel, 1842.

niassensis Sluiter, 1895, see *Choriaster*

novaeguineae Müller & Troschel, 1842 (with synonyms *Goniodiscus sebae* Müller & Troschel, 1842, *Hosia spinulosa* Gray, 1847, *Culcita pentangularis* Gray, 1847, *C. pulverulenta* Perrier, 1869 and *Hippasteria philippinensis* Domantay & Roxas, 1938, also varieties *arenosa* Perrier, 1869, *acutispinosa* Bell, 1883, *plana* Hartlaub, 1892 and *leopoldi* Engel, 1938; see also *Goniaster (Goniodiscus) multiporum* Hoffman [but ?MS name])

Müller & Troschel, 1842: 38; Sluiter 1895: 57; Döderlein 1896: 310; H. L. Clark 1908: 201; Goto 1914: 507; Fisher 1919a: 360; H.L. Clark 1921: 32; Djakonov 1930: 247; Livingstone 1932a: 265; Engel 1938b: 10 (variety *leopoldi*); Domantay & Roxas 1938: 215; Liao 1980: 154; Jangoux & de Ridder 1987: 90.

Range: New Guinea*, West Pacific, Andaman Is.

novaeguineae: H.L. Clark 1923a: 273, non *Culcita novaeguineae* Müller & Troschel, 1842, = *C. schmideliana* Retzius, 1805 according to A.M.C. in Clark & Courtman-Stock (1976).

plana Hartlaub 1892: 74, 84. A variety of *Culcita novaeguineae* Müller & Troschel, 1842 according to Döderlein (1896). Holotype a syntype of *C. novaeguineae* according to Jangoux & de Ridder (1987).

pentangularis Gray, 1847: 74 [195]. A synonym of *Culcita novaeguineae* Müller & Troschel, 1842 according to Sladen (1889).

pulverulenta Perrier 1869: 68 [260]. A synonym of *Culcita novaeguineae* Müller & Troschel, 1842 according to Perrier (1875).

schmideliana (Retzius, 1805) (with synonym *Goniodiscus studeri* de Loriol, 1885 and varieties *africana* and *ceylonica* both of Döderlein, 1896)

Retzius 1805: 11 (as *Asterias*).

Gray 1840: 276; Perrier 1875: 266 [74]; Döderlein 1896: 315 (with varieties); Simpson & Brown 1910: 53; A.M.C. in Clark & Rowe 1971: 41; Jangoux 1973: 18; A. M. Clark & Courtman-Stock 1976: 67; Marsh & Marshall 1983: 675; Jangoux 1985b: 31 (as *Culcita*).

Range: Indian Ocean [?Andaman Is], Indonesia.

veneris Perrier, 1879, see *Spoladaster* (Poraniidae).

veneris: Bell, 1905: 248, non *Culcita veneris* Perrier, 1879, = *Spoladaster brachyactis* (H.L. Clark, 1923) (Poraniidae) according to Mortensen (1933a).

CULCITASTER H.L. Clark, 1915
H.L. Clark 1915: 144.
Type species: *Culcitaster anamesus* H.L. Clark, 1915.
A synonym of *Halityle* Fisher, 1913 according to Fisher (1919a).

anamesus H.L. Clark, 1915: 144. A synonym of *Halityle regularis* Fisher, 1913 according to Döderlein (1935) and Baker & Marsh (1976), though meantime treated as valid by H.L. Clark (1946).

GONIODISCASTER H.L. Clark, 1909
H.L. Clark 1909a: 530; Döderlein 1935: 74.
Type species: *Asterias pleyadella* Lamarck, 1816.

acanthodes H.L. Clark. 1938
H.L. Clark 1938: 84; 1946: 91.
Range: NW Australia, 12-14 m.

australiae Tortonese, 1937
Tortonese 1937b: 293; H.L. Clark 1938: 80; 1946: 91.
Range: NW Australia, 0-9 m.

bicolor H.L. Clark, 1938
H.L. Clark 1938: 87.
Range: NW Australia.

coppingeri (Bell, 1884) H.L. Clark, 1921: 28; Endean 1953: 53. A synonym of *Goniodiscaster rugosus* (Perrier, 1875) according to A.M.C. in Clark & Rowe (1971).

foraminatus (Döderlein, 1916)
Döderlein 1916: 415 (as *Goniodiscus*).
Döderlein 1935: 79 (as *Goniodiscaster*).
Range: NW Australia, 12-20 m.

forficulatus (Perrier, 1875)
Perrier 1875: 234 [1876: 50]; Koehler 1910a: 61 (as *Goniodiscus*).
Fisher 1919a: 324; Döderlein 1935: 78; A.H. Clark 1949a: 74; A.M. Clark & Rowe 1971: 50; Jangoux 1981a: 465.
Range: Philippine Is*, East Indies, Bay of Bengal, 30-120 m.
Possibly a synonym of *Goniodiscaster scaber* (Möbius, 1859) according to A.M.C. in Clark & Rowe (1971).

granuliferus (Gray, 1847) (with synonym *Goniodiscus stella* Möbius, 1859)
Gray 1847: 77 (as *Anthenea*).
Perrier 1875: 236 [1876: 51] (as *Goniodiscus*).
Döderlein 1935: 81; A.M. Clark & Rowe 1971: 40.

Range: 'China'*, East Indies.
insignis (Koehler, 1910)
Koehler 1910a: 66 (as *Goniodiscus*).
Döderlein 1935: 85 (as *Goniodiscaster*).
Range: Arabian Gulf, 88-90 m.
integer Livingstone, 1931
Livingstone 1931: 135; H.L. Clark 1946: 91; Endean 1953: 53.
Range: Queensland, Australia.
pleyadella (Lamarck, 1816) (with synonym *Pentagonaster validus* Bell, 1884)
Lamarck 1816: 553 (as *Asterias*).
Müller & Troschel 1842: 59 (as *Goniodiscus*).
H.L. Clark 1909a: 110 (as *Goniodiscaster*).
Range: East Indies, northern Australia, 0-9 m.
porosus (Koehler, 1910)
Koehler 1910a: 70 (as *Goniodiscus*).
Döderlein 1935: 82; A.M.C. in Clark & Rowe 1971: 49.
Range: Bay of Bengal*, northern Australia.
rugosus (Perrier, 1875) (with synonym *Pentagonaster coppingeri* Bell, 1884)
Perrier 1875: 233 [1876: 49] (as *Goniodiscus*).
A.M.C. in Clark & Rowe 1971: 50 (as *Goniodiscaster*).
Range: northern Australia.
scaber (Möbius, 1859) (with synonym *Goniaster articulatus* Lütken, 1865, possibly also *Goniodiscus forficulatus* Perrier, 1875).
Möbius, 1859: 10 (as *Goniodiscus*).
Döderlein 1926: 11; 1935: 82; Madsen 1959b: 165; Jangoux 1981a: 465; Aziz & Jangoux 1984a: 134 (as *Goniodiscaster*).
Range: Bay of Bengal*, East Indies, 12-50 m.
seriatus (Müller & Troschel, 1843) (with synonym *Pentaceros granulosus* Gray, 1840).
Müller & Troschel 1843: 117 (as *Goniodiscus*).
Döderlein 1935: 80 (as *Goniodiscaster*).
Range: W and SW Australia.
vallei (Koehler, 1910)
Koehler 1910a: 75 (as *Goniodiscus*).
Döderlein 1935: 85; A.M.C. in Clark & Rowe 1971 (as *Goniodiscaster*).
Range: eastern India*, Bay of Bengal, Sri Lanka, 28 m.
GONIODISCIDES Fisher, 1906
Fisher 1906a: 1070 (nom. nov. for *Goniodiscus* Müller & Troschel, 1842, 'untenable').
Type species: *Goniodiscus sebae* Müller & Troschel, 1842.
A synonym of *Culcita* L. Agassiz, 1836 according to Fisher (1919a).

sebae (Müller & Troschel, 1842) Fisher, 1906: 1070. A synonym of a species of *Culcita* according to Fisher (1919a).

GONIODISCUS Müller & Troschel, 1842

Müller & Troschel 1842: 58.

Type species: None designated.

Untenable and = *Culcita* L. Agassiz, 1836 according to Fisher (1919a). [Formal designation now of *Goniodiscus sebae* Müller & Troschel, 1842 as type species disposes of the name *Goniodiscus* by rendering it a synonym of *Culcita*, although many of the included names are referable to the Goniasteridae.]

acutus Heller, 1863: 420. A synonym of *Pentagonaster* [i.e. *Peltaster*] *placenta* (Müller & Troschel, 1842) (Goniasteridae) according to Ludwig (1897).

acutus Perrier, 1869: 88 [280]. A junior primary homonym of *Goniodiscus acutus* Heller, 1863, renamed *Anthenea edmondi* by A.M. Clark, 1970, but that name could be threatened by *Anthenea sydneyensis* Döderlein, 1915.

armatus: Lütken 1859, see *Nidorellia*

articulatus Perrier, 1869: 87 [279] (from Valenciennes, MS), non *Goniaster articulatus* Lütken, 1865 *(Asterias articulatus* Linnaeus, 1753, invalid), = *Anthenea pentagonula* (Lamarck, 1816) according to Perrier (1875).

articulatus: de Loriol, 1884: 638, non *Goniodiscus articulatus* Perrier, 1869, but instead *Goniaster articulatus* Lütken, 1865, = *Goniodiscaster scaber* (Möbius, 1859) according to Döderlein (1935).

capella Müller & Troschel, 1842, see *Ogmaster* (Goniasteridae)

conifer Möbius, 1859: 10. A synonym of *Nidorellia armata* Gray, 1840 according to Sladen (1889).

cuspidatus (Gray, 1840) Müller & Troschel, 1842: 60. A synonym of *Goniaster tessellatus* (Lamarck, 1816) (Goniasteridae) according to Halpern (1970).

foraminatus Döderlein, 1916, see *Goniodiscaster*

forficulatus Perrier, 1875, see *Goniodiscaster*

frascheli Philippi in Quijada, 1911: [p?]. A NOMEN NUDUM according to Madsen (1956).

'gracilis Gray, 1840' in Martens, 1866: 86, under *Randasia luzonicus* Gray. Non-existent, 'a phantom species' according to Bell (1893b).

granuliferus: Perrier, 1875, see *Goniodiscaster*

horridus: Müller & Troschel, 1842 *(?Goniodiscus),* see *Paulia* (Asterodiscididae)

insignis Koehler, 1910b, see *Goniodiscaster*

mammillatus Müller & Troschel, 1842: 61, see under *Tosia* (Goniasteridae).

michelini Perrier, 1869: 89. A synonym of *Nidorellia armata* Gary, 1840 according to Döderlein (1936).

ocelliferus: Müller & Troschel, 1842, see *Nectria*

pedicellaris Perrier, 1881, see *Mediaster* (Goniasteridae)

penicillatus Philippi, 1870, see *Odontaster* (Odontasteridae).

pentagonulus: Müller & Troschel, 1842, see *Anthenea pentagonula* also *A. viguieri.*

placenta Müller & Troschel, 1842, see *Peltaster* (Goniasteridae).

placentaeformis Heller, 1863: 419. A synonym of *Pentagonaster* [i.e. *Peltaster*] *placenta* Müller & Troschel, 1842 (Goniasteridae) according to Ludwig (1897).

pleyadella: Müller & Troschel, 1842, see *Goniodiscaster*

porosus Koehler, 1910a, see *Goniodiscaster*

regularis Müller & Troschel, 1842: 59 (from Linck, 1733 and Seba, 1758). [Apparently = *Peltaster placenta* (Müller & Troschel, 1842).]

rugosus Perrier, 1875, see *Goniodiscaster*

scaber Möbius, 1859, see *Goniodiscaster*

sebae Müller & Troschel, 1842: 58. A synonym of *Culcita novaeguineae* Müller & Troschel, 1842 according to Fisher (1919a). Types extant according to Jangoux & de Ridder (1987).

sebae: Perrier, 1875: 230 [1876: 46], non Müller & Troschel, 1842, = *Goniaster articulatus* Lütken, 1865 [i.e. *Goniodiscaster scaber* (Möbius, 1859)] according to Döderlein (1935) and Madsen (1959). [This is the species figured by Seba (1758, in pl.6, figs 7,8) which is not conspecific with the specimen described by Müller & Troschel, though they are quoted by Perrier.]

seriatus Müller & Troschel, 1843, see *Goniodiscaster*

singularis Müller & Troschel, 1842, see *Diplodontias* (Odontasteridae)

stella Möbius, 1859: 9. A synonym of *Goniodiscaster granuliferus* (Gray, 1847) according to Döderlein (1935).

stella Verrill, 1870: 284. A junior primary homonym of *Goniodiscus stella* Möbius, 1859 but = *Nidorellia armata* (Gray, 1840) according to Verrill (1868). [See reference list to account for date anomaly.]

studeri de Loriol, 1885: 49. A synonym of *Culcita schmideliana* (Retzius, 1805) according to A.M.C. in Clark & Rowe (1971).

vallei Koehler, 1910, see *Goniodiscaster*

verrucosus Philippi, 1857, see *Cycethra* (Ganeriidae)

GYMNANTHENEA H.L. Clark, 1938

H.L. Clark 1938: 105; 1946: 100.

Type species: *Anthenea globigera* Döderlein, 1915.

globigera (Döderlein, 1915)

Döderlein 1915: 29, 50 (as *Anthenea*).

H.L. Clark 1938: 106; 1946: 100 (as *Gymnanthenea*).

Range: NW* and N Australia.

laevis H.L. Clark, 1938

H.L. Clark 1938: 108; 1946: 100.

Range: Abrolhos Is, western Australia.

HALITYLE Fisher, 1913 (with synonym *Culcitaster* H.L. Clark, 1915)

Fisher 1913c: 211; 1919a: 362.

Type species: *Halityle regularis* Fisher, 1913.

anamesus (H.L. Clark, 1915) H.L. Clark 1946: 109. A synonym of *Halityle regularis* Fisher, 1913 according to Baker & Marsh (1976).

regularis Fisher, 1913 (with synonym *Culcitaster anamesus* H.L. Clark, 1915)

Fisher 1913c: 211; 1919a: 362; Döderlein 1935: 108; Domantay & Roxas 1938: 217; H. L. Clark 1946: 109; James 1976: 557; Baker & Marsh 1976: 108; Jangoux 1986a: 124.

Range: Philippine Is*, W to Madagascar, E to New Caledonia and S to western Australia, 3-90 m.

HOSEA Gray, 1866, lapsus for

HOSIA Gray, 1840: 279.

Type species: *Hosia flavescens* Gray, 1840.

A synonym of *Anthenea* Gray, 1840 according to Perrier (1875).

flavescens Gray, 1840, see *Anthenea*

spinulosa Gray, 1847: 78. A synonym of *Culcita novaeguineae* Müller & Troschel, 1842 according to Koehler (1910).

MONACHASTER Döderlein, 1916

Döderlein 1916: 412; 1935: 103; A.M.C. in Clark & Rowe 1971: 48; Tortonese 1976: 271.

Type species: *Goniodiscus sanderi* Meissner, 1892.

sanderi (Meissner, 1892) (with synonym *Monachaster umbonatus* Macan, 1938)

Meissner 1892: 185 (as *Goniodiscus*).

Döderlein 1916: 412; 1935: 103; A.M. Clark & Rowe 1971: 48; Tortonese 1976: 271;

[?]Applegate 1984: 97 (as *Monachaster*).

Range: Zanzibar*, East Africa N to Gulf of Suez, [?]Taiwan, 0-68 m.

umbonatus Macan, 1938: 399; also Tortonese 1979: 31. A possible synonym of *Monachaster sanderi* (Meissner, 1892) according to A.M.C. in Clark & Rowe (1971), confirmed by Tortonese (1976).

NECTRIA Gray, 1840

Gray 1840: 287; Fisher 1917d: 167; H.L. Clark 1928: 379; A.M. Clark 1966: 310; Shepherd 1967: 464; Zeidler & Rowe 1986: 117.

Type species: '*Asterias oculifera* Lam.' by monotypy. [This has been presumed to be a lapsus for *ocellifera* Lamarck, 1816 since it was changed to that spelling by Gray in 1866 but in 1840 Gray cited a sample in the British Museum collections which then probably housed only the species

subsequently named *Nectria ocellata* by Perrier, 1875 and this should be treated as type species.
Provided both *N. ocellifera* and *N. ocellata* continue to be regarded as congeneric, the question is academic. However, Zeidler & Rowe (1986) think a decision from the ICZN is needed.]

humilis Zeidler & Rowe, 1986
Zeidler & Rowe 1986: 124.
Range: N Tasmania, c.50-550 m.

macrobrachia H.L. Clark, 1923
H.L. Clark 1923b: 236; A.M. Clark 1966: 311; Shepherd 1967: 474; Zeidler & Rowe 1986: 121.
Range: Abrolhos Is, W Australia*, to Bass Strait, N of Tasmania, 0-180 [?350] m.

monacantha: Fisher, 1917d, see *Nectriaster*

multispina H.L. Clark 1928 (with possible synonym *Chaetaster munitus* Möbius, 1859 according to A.M.C. (1966) but see under *N. ocellifera)*
H.L. Clark 1928: 375; A.M. Clark 1966: 314; Shepherd 1967: 468; Zeidler & Rowe 1986: 122.
Range: SW Australia to Victoria, 0-20 m.

ocellata Perrier, 1875
Perrier 1875: 189 [1876: 4]; H.L. Clark 1928: 378; 1946: 85; A.M. Clark 1966: 313; Shepherd 1967: 465; Zeidler & Rowe 1986: 119 (lectotype).
Range: Tasmania and Bass Strait*, W end of Great Australian Bight to northern N.S.W., 0-230 m.

ocellifera (Lamarck, 1816) (with possible synonym *Chaetaster munitus* Möbius, 1859 according to Sladen (1889) but see under *N. multispina*).
Lamarck 1816: 553 (as *Asterias).*
[?]Müller & Troschel, 1842: 60 (as *Goniodiscus).*
Gray 1866: 15; H.L. Clark 1946: 86; A.M. Clark 1966: 310; Shepherd 1967: 464; Zeidler & Rowe 1986: 119 (as *Nectria).*
Range: W and SW Australia, 45-180 m.

ocellifera: Sladen, 1889: 319, 752, non *Nectria ocellifera* (Lamarck, 1816), = *N. ocellata* Perrier, 1875 according to H.L. Clark (1946).

ocellifera: H.L. Clark 1909b: 529, non *Nectria ocellifera* (Lamarck, 1816), = *Asterodiscus* [i.e. *Asterodiscides*] *truncatus* Coleman, 1911 according to Coleman.

oculifera Gray, 1840: 287; lapsus for *ocellifera* Lamarck, 1816; conspecific with *Nectria ocellata* Perrier, 1875 according to Zeidler & Rowe (1986).

pedicelligera Mortensen, 1925
Mortensen 1925a: 291; Zeidler & Rowe 1986: 121.
Range: supposedly northern New Zealand but more likely from southern Australia, see Zeidler & Rowe (1986).

saoria Shepherd, 1967

Shepherd 1967: 475; Zeidler & Rowe 1986: 123.

Range: Spencer Gulf, South Australia* to Fremantle, W Australia, 0-25 m.

wilsoni Shepherd & Hodgkin, 1966

Shepherd & Hodgkin 1966: 119; Shepherd 1967: 474; Zeidler & Rowe 1986: 123.

Range: 'near Perth', western Australia*, SW and S Australia, 0-44 m.

NECTRIASTER H.L. Clark, 1946

H.L. Clark 1946: 84.

Type species: *Mediaster monacanthus* H.L. Clark, 1916.

monacanthus (H.L. Clark, 1916)

H.L. Clark 1916: 41 (as *Mediaster*).

Fisher 1917d: 167 (as *Nectria*).

H.L. Clark 1946: 84 (as *Nectriaster*).

Range: N.S.W., Australia, 85-91 m.

NIDORELLIA Gray, 1840

Gray 1840: 277 (as subgenus or division of *Pentaceros*).

Verrill 1870: 280; Döderlein 1916: 418 (as genus).

Type species: *Pentaceros (Nidorellia) armata* Gray, 1840.

armata (Gray, 1840) (with synonyms *Goniodiscus conifer* Möbius, 1859 and *G. michelini* Perrier, 1869)

Gray 1840: 277 (as *Pentaceros (Nidorellia)*).

Müller & Troschel 1842: 52; Bell 1884b: 59, 79 (as *Oreaster*).

Lütken 1859: 59 (as *Goniodiscus*).

Verrill 1870: 280; H.L. Clark 1910: 332; 1920: 86; Ziesenhenne 1937: 216; H. L. Clark 1940: 333; Caso 1961a: 63; Downey 1975: 86.

Range: Ecuador*, Peru and Galapagos Is N to the Gulf of California, 18-73 m.

horrida: Perrier, 1875, see *Paulia* (Asterodiscididae)

michelini (Perrier, 1869) Sladen, 1889: 764. A synonym of *Nidorellia armata* (Gray, 1840) according to Döderlein (1936).

OREASTER Müller & Troschel, 1842

Müller & Troschel 1842: 44; Lütken 1865: 148; Bell 1884b: 57 (sensu lato); Döderlein 1916: 409 (restricted).

Type species: *Asterias reticulatus* Linnaeus, 1758, by subsequent designation by Döderlein (1916).

aculeatus (Gray, 1840) Müller & Troschel 1842: 50. A synonym of *Oreaster reticulatus* (Linnaeus, 1758) according to Bell (1884b).

affinis Müller & Troschel, 1842, see *Pentaceraster*

alveolatus: Bell, 1884, see *Pentaceraster*

alveolatus: Livingstone, 1932a: 250, non *Pentaceros alveolatus* Perrier, 1875, = *Pentaceraster regulus* (Müller & Troschel, 1842) according to Döderlein (1936).

306

armatus: Müller & Troschel, 1842, see *Nidorellia*

australis Lütken, 1871: 252; also Livingstone 1932a: 247. A synonym of *Pentaceraster regulus* (Müller & Troschel, 1842) according to Döderlein (1936).

bermudensis H.L. Clark, 1942: 372. A variety of *Oreaster reticulatus* (Linnaeus, 1758).

*carinatus Müller & Troschel, 1842: 49. An uncertain species according to Bell (1884b). A possible synonym of *Oreaster clavatus* Müller & Troschel, 1842 according to Döderlein (1936, index).

castellum Grube in Sladen 1889: 761, undated. Cited only as a synonym of *Pentaceros muricatus* Gray, 1840, itself a synonym of *Protoreaster lincki* (de Blainville, 1830) according to Sladen (1889). [Not traced in any of Grube's papers found.]

*chinensis Müller & Troschel, 1842: 46. Untenable, being both a homonym and a synonym. [Apparently intended as a species distinct from *Pentaceros chinensis* Gray, 1840 of which it would be a secondary homonym following rejection of the pre-linnaean name *Pentaceros* and treated as conspecific by Bell (1884b), giving priority to *Pentaceros chinensis* Gray. But Döderlein (1936), ignoring Gray's priority, treated both *chinensis* Gray, 1840 and *chinensis* Müller & Troschel, 1842 as junior synonyms of *Pentaceraster orientalis* (Müller & Troschel, 1842), itself treated as a probable junior synonym of *Pentaceraster regulus* (Müller & Troschel, 1842) by Hayashi (1938b)!].

clavatus Müller & Troschel, 1842 (with synonym *Pentaceros dorsata* Perrier, 1875, possibly also *Oreaster carinatus* Müller & Troschel, 1842) Müller & Troschel, 1842: 49; Döderlein 1936: 321; Madsen 1950: 212; 1958: 94; A. M. Clark 1962b: 174; Opinion 707 1964: 206; Alvarado & Alvares 1964: 279; Nataf & Cherbonnier 1975: 822.

Range: Cape Verde Is, SE to Cameroon and São Thomé, 0-15 m.

A synonym of *Asterias nodosa* Linnaeus, 1758 according to Madsen (1959) but that name restricted by the ICZN to conform with *Protoreaster nodosus* sensu [e.g.] Döderlein (1936), following proposal by A.M.C. (1962b), simultaneously validating *O. clavatus* Müller & Troschel, 1842.

clouei Perrier, 1869: 271. A synonym of *Oreaster* [i.e. *Protoreaster*] *nodosus* Linnaeus, 1758) according to Bell (1884).

decipiens Bell, 1884, see *Pentaceraster*

desjardinsi Michelin, 1844: 173; 1845: 23. A synonym of *Nardoa* [i.e. *Gomophia*] *egyptiaca* (Gray, 1840) (Ophidiasteridae) according to Sladen (1889).

doederleini Goto, 1914: 451. A synonym of *Pentaceraster regulus* (Müller & Troschel, 1842) according to Hayashi (1938b).

dorsatus Lütken, 1865: 161 (from Linnaeus, 1753); see Madsen 1959b: 164. A synonym of *Oreaster clavatus* Müller & Troschel, 1842 according to Döderlein (1916).

307

*forcipulosus (Lütken, 1865)
Lütken 1865: 156 (as *Pentaceros*).
Bell, 1884b: 62 (as *Oreaster*).
Range: West Africa.
A doubtful species according to Döderlein (1936).
franklini (Gray, 1840) Bell, 1884b: 62. A synonym of *Protoraster nodosus* (Linnaeus, 1758) according to Döderlein (1936).
gigas Lütken, 1859: 44; 1965: 161 (from Linnaeus, 1753). A synonym of *Oreaster reticulatus* (Linnaeus, 1758) according to Döderlein (1936) and Madsen (1959b).
gracilis Lütken, 1871, see *Pentaceraster*
granulosus (Gray, 1840) Bell, 1884b: 67. A synonym of *Gonidiscaster seriatus* (Müller & Troschel, 1843) according to Döderlein (1935).
grayi Bell, 1884b: 83 (nom. nov. for *Pentaceros nodosus:* Gray, 1840, non *Asterias nodosus* Linnaeus, 1758). A synonym of *Pentaceraster multispinus* (Martens, 1866) according to Döderlein (1936). [But Gray's type locality Mauritius, beyond the known range of *P. multispinus*.]
grayi: Sluiter, 1889: 304; 1895: 55, non *Oreaster grayi* Bell, 1884b, = *Pentaster obtusatus* (Bory de St Vincent, 1827) according to Engel (1938).
gustavianum: Rowe, 1974: 214. [MS name cited as if a synonym of *Anthenea pentagonula* (Lamarck, 1816), so a NOMEN NUDUM.
hawaiiensis (Fisher, 1906) Döderlein, 1916: 433. A synonym of *Pentaceraster cumingi* (Gray, 1840) according to Döderlein (1936).
hedemanni Lütken, 1871: 255. A synonym of *Pentaceraster multispinus* (Martens, 1866) according to Döderlein (1936).
hiulcus Müller & Troschel, 1842: 48. A synonym of *Protoreaster nodosus* (Linnaeus, 1758) according to Döderlein (1936).
hiulcus: Perrier, 1875: 243 [1876: 59], non *Oreaster hiulcus* Müller & Troschel, 1842, = *Pentaceraster mammillatus* (Audouin, 1826) according to Döderlein (1916).
hondurae Domantay & Roxas, 1938: 211. A variety of *Oreaster* [i.e. *Protoreaster*] *nodosus* (Linnaeus, 1758).
intermedia Martens, 1866: 80 (as variety of *Oreaster muricatus* (Gray, 1840)). A synonym of *Protoreaster nodosus* (Linnaeus, 1758) according to Döderlein (1936).
lapidarius Grube, 1857: 342. A synonym of *Oreaster reticulatus* Linnaeus, 1758 according to Perrier (1875).
lepidosus Grube in Verrill, 1915, lapsus for *lapidarius*
lincki: Lütken, 1864, see *Protoreaster*
luetkeni Bell, 1884b: 75. A synonym of *Pentaceraster regulus* (Müller & Troschel, 1842) according to A.M.C. in Clark & Rowe (1971).
magnificus Goto, 1914, see *Pentaceraster*

mammillatus: Müller & Troschel, 1842, see *Pentaceraster*

mammosus Perrier, 1869: 78 [270]. A synonym of *Protoreaster nodosus* (Linnaeus, 1758) according to Döderlein (1936).

modestus: Goto, 1914: 444, non *Pentaceros modestus* Gray, 1866, = *Protoreaster gotoi* Döderlein, 1936 (nom. nov.), though probably only a forma of *Protoreaster lincki* de Blainville according to Döderlein but a synonym of *P. nodosus* (Linnaeus, 1758) according to Hayashi (1938b).

muelleri Bell, 1884b: 86. A synonym of *Pentaceraster affinis* (Müller & Troschel, 1842) according to A.M.C. in Clark & Rowe (1971).

muricatus (Gray, 1840) Dujardin & Hupé, 1862: 383. A synonym of *Protoreaster lincki* (de Blainville, 1830) according to Döderlein (1936).

mutica Martens, 1866: 80 (as variety of *Oreaster muricatus* (Gray, 1840)). A synonym of *Protoreaster nodosus* (Linnaeus, 1758) according to Döderlein (1936).

nahensis Goto 1914: 463. A synonym of *Protoreaster nodosus* (Linnaeus, 1758) according to Hayashi (1938b).

nodosus: Bell, 1884b, see *Protoreaster*

nodosus: Michelin, 1845: 23; also Lütken 1865: 152, non *Asterias nodosus* Linnaeus, 1758, = *Pentaceraster horridus* (Gray, 1840) according to Döderlein (1936).

nodulosus: Bell, 1884, see *Protoreaster*

obtusangulus: Müller & Troschel, 1842, see *Pseudoreaster*

obtusatus: Müller & Troschel, 1842, see *Pentaster*

*occidentalis Verrill, 1870

Verrill 1870: 278; 1871: 574; Bell 1884b: 76; Ziesenhenne 1937: 215; H.L. Clark 1940: 333; Caso 1961: 55.

Range: Panama*, Lower California to Equador and Galapagos Is, 0-90 m.

Thought to be a synonym of *Pentaceraster cumingi* (Gray, 1840) by Döderlein (1936) but treated as a valid species of *Oreaster* by H.L. Clark (1940) (though without reference to Döderlein), followed by Caso (1961).

*orientalis Müller & Troschel, 1842, see under *Pentaceraster*, also notes above under *Oreaster chinensis*

productus Bell, 1884b: 74. Designated as type species of *Poraster* by Döderlein (1916) but reduced to a variety of *Poraster superbus* (Möbius, 1859) by Döderlein (1936).

regulus Müller & Troschel, 1842, see *Pentaceraster*

reinhardti Lütken, 1865: 159. A synonym of *Protoreaster lincki* (de Blainville, 1830) according to Döderlein (1916).

reticulatus (Linnaeus, 1758) (with synonyms *Pentaceros aculeatus, P. gibbus* and *P. grandis* all of Gray, 1840, *Oreaster lapidarius* Grube, 1857, *P. tuberosus* Behn in Möbius, 1859 and *O. gigas* Lütken, 1859; also variety *bermudensis* H. L. Clark, 1942).

Linnaeus 1758: 661 (as *Asterias*).

Gray 1840: 276; Perrier 1875: 246 [1876: 62]; A. Agassiz 1877: 108; Sladen 1889: 345 (as *Pentaceros*).

Müller & Troschel 1842: 45; Döderlein 1916: 418; H.L. Clark 1933: 53; Boone 1933: 80; Döderlein 1936: 319; H.L. Clark 1942: 372 (variety *bermudensis*); Madsen 1959b: 163 (holotype); Ummels 1963: 73; Tommasi 1970: 10; Walenkamp 1976: 72 (as *Oreaster*).

Range: South Carolina to Brazil (c.24°S), 0-69 m; var. *bermudensis* from Bermuda.

rouxi (Koehler, 1910b) Goto 1914: 463. A synonym of *Pentaceraster gracilis* (Lütken, 1871) according to Döderlein (1936).

superbus Möbius, 1859, see *Poraster*

*thurstoni Bell, 1888

Bell 1888: 385; Döderlein 1936: 310.

Range: SE India.

Position indeterminable according to Döderlein (1936). [Holotype redetermined A.M.C. (MS) and conspecific with *Pentaceraster affinis* (Müller & Troschel, 1842)].

troscheli Bell, 1884b: 85. A synonym of *Pentaceraster alveolatus* (Perrier, 1875) according to A.M.C. in Clark & Rowe (1971).

tuberculatus Müller & Troschel, 1842, see *Pentaceraster*

tuberosus Behn in Möbius, 1859: 5. A synonym of *Oreaster reticulatus* (Linnaeus, 1758) according to Sladen (1889).

turritus (Gray, 1840) Müller & Troschel, 1842: 47. A synonym of *Protoreaster nodosus* (Linnaeus, 1758) according to Döderlein (1936).

valvulatus Müller & Troschel, 1843, see *Anthaster*

*verrucosus Müller & Troschel, 1842

Müller & Troschel 1842: 49.

Range: India.

A possible synonym of *Pentaceraster mammillatus* (Audouin, 1826) according to Döderlein (1936, index).

westermanni Lütken, 1871, see *Pentaceraster*

PENTACERASTER Döderlein, 1916

Döderlein 1916: 424; 1936: 331; A.M.C. in Clark & Rowe 1971: 55.

Type species: *Asterias mammillata* Audouin, 1826.

affinis (Müller & Troschel, 1842) (with synonym *Oreaster muelleri* Bell, 1884)

Müller & Troschel 1842: 46 (as *Oreaster*).

Koehler 1910a: 92 (as *Pentaceros*).

Döderlein 1916: 432; 1936: 348 (as *Pentaceraster*).

Range: Sri Lanka area and Bay of Bengal.

alveolatus (Perrier, 1875) (with synonyms *Oreaster troscheli* Bell, 1884 and

Pentaceros bedoti Koehler, 1911).

Perrier 1875: 243 [1876: 59]; Koehler 1910a: 95 (as *Pentaceros*).

Bell 1884b: 73; Domantay & Roxas 1938: 212 (as *Oreaster*).

Döderlein 1916: 428; Jangoux 1986a: 126 (as *Pentaceraster*).

Range: New Caledonia*, Melanesia, Indonesia and Philippine Is, 1-25 m.

australis (Lütken, 1871) Döderlein, 1926: 7. A synonym of *Pentaceraster regulus* (Müller & Troschel, 1842) according to Döderlein (1936).

australis: Döderlein, 1916: 433, non *Oreaster australis* Lütken, 1871, = *Pentaster hybridus* Döderlein, 1936 (nom. nov.) according to Döderlein (1936).

cebuana Döderlein, 1936: 352. A forma of *Pentaceraster regulus* (Müller & Troschel, 1842).

chinensis (Gray, 1840) (with probable synonym *Oreaster orientalis* Müller & Troschel, 1842).

Gray 1840: 276 (as *Pentaceros*).

Clark & Rowe 1971: 34,56 (as *Pentaceraster*).

Range: South China Sea.

Inadmissably treated as a junior synonym of *Oreaster orientalis* Müller & Troschel, 1842 by Döderlein (1936), as noted in Clark & Rowe (1971).

crassimana Döderlein, 1936: 339. A forma of *Pentaceraster multispinus* (Martens, 1866).

cumingi (Gray, 1840) (with synonym *Pentaceros hawaiiensis* Fisher, 1906 possibly also *Oreaster occidentalis* Verrill, 1870)

Gray 1840: 276 (as *Pentaceros*).

Döderlein 1916: 433; 1936: 355 (as *Pentaceraster*).

Range: Gulf of California to Peru and Galapagos Is, also Hawaiian Is, 55-73 m.

decipiens (Bell, 1884)

Bell 1884b: 69; Goto 1914: 463 (as *Oreaster*).

A.M.C. in Clark & Rowe 1971: 55 (as *Pentaceraster*).

Range: Indonesia.

gracilis (Lütken, 1871) (with synonyms *Pentaceros rouxi* Koehler, 1910 and *P. mertoni* Koehler, 1910)

Lütken 1871: 260; Bell 1884b: 82 (as *Oreaster*).

Sladen 1889: 760 (as *Pentaceros*).

Döderlein 1916: 437; 1936: 357 (as *Pentaceraster*).

Range: Queensland to W Burma, ?Mozambique.

horridus (Gray, 1840) (with synonyms *Pentaceros belli, P. grayi* and *P. sladeni* all of de Loriol, 1885)

Gray 1840: 276 (as *Pentaceros*).

Döderlein 1916: 432; 1936: 345; Ebert 1979: 71 (as *Pentaceraster*).

Range: Mauritius, Seychelles.

*japonicus Döderlein, 1916

Döderlein 1916: 429; 1936: 340; Hayashi 1938b: 201.
Range: southern Japan.
Possibly a synonym of *Pentaceraster regulus* (Müller & Troschel, 1842).
according to Hayashi (1938b).
*magnificus (Goto, 1914)
  Goto 1914: 457 (as *Oreaster*).
  Döderlein 1936: 359 (as *Pentaceraster*).
  Range: southern Japan and S China.
  Possibly a synonym of *Pentaceraster regulus* (Müller & Troschel, 1842)
  according to Hayashi (1938b) but treated as valid by A. M. Clark (1982b).
  [Though the status of the several nominal species of *Pentaceraster* from
  this area badly need reassessment.]
mamillatus Döderlein, 1916, lapsus for *mammillatus*
mammillatus (Audouin, 1826) (with possible synonym *Oreaster verrucosus*
  Müller & Troschel, 1842)
  Audouin 1826: 209 (as *Asterias*).
  Müller & Troschel 1842: 48; Bell 1884b: 67; H.L. Clark 1923a: 273 (as
  Oreaster).
  Perrier 1875: 246 [1876: 62] (as *Pentaceros*).
  Döderlein 1916: 430; 1936: 341; Mortensen 1940: 66; Tortonese 1953:
  27; Jangoux 1973: 20; A.M. Clark & Courtman-Stock 1976: 68; Price
  1982: 42 (as *Pentaceraster*).
  Range: northern Red Sea*, Arabian (Iranian) Gulf to Natal.
mammillatus: Tortonese, 1960 (pt), non *Pentaceraster mammillatus*
  (Audouin, 1826), = *Asteropsis carinifera* (Lamarck, 1816) (Asteropsei-
  dae) according to A.M. Clark (1967a).
multispinus (Martens, 1866) (with synonym *Oreaster hedemanni* Lütken,
  1871, possibly also *O. grayi Bell*, 1884)
  Martens 1866: 79 (as *Oreaster muricatus* var. *multispina*).
  Döderlein 1916: 430; 1936: 336; A.M.C. in Clark & Rowe 1971: 56;
  Marsh & Marshall 1983: 675 (as *Pentaceraster*).
  Range: Indonesia, NW Australia.
odhneri Döderlein 1926: 8. A synonym of *Pentaceraster regulus* (Müller &
  Troschel, 1842) according to Döderlein (1936).
*orientalis (Müller & Troschel, 1842)
  Müller & Troschel 1842: 128 (as *Oreaster*).
  Sladen 1889: 345 (as *Pentaceros*).
  Döderlein, 1916: 433; 1936: 356; Jangoux & de Ridder 1987: 89.
  Range: China.
  A probable synonym of *Pentaceraster chinensis* (Gray, 1840) according to
  A.M.C. herein (inadmissably treated as a junior synonym by Döderlein).
regulus (Müller & Troschel, 1842) (with synonyms *Oreaster australis*
  Lütken, 1871, *O. luetkeni* Bell, 1884, *O. doederleini* Goto, 1914 and

Pentaceraster odhneri Döderlein, 1926, also possibly *O. magnificus* Goto, 1914 and *P. japonicus* Döderlein, 1916; forma *cebuana* Döderlein, 1936).
Müller & Troschel 1842: 51 (as *Oreaster*).
Sladen 1889: 762; Koehler 1910a: 99 (as *Pentaceros*).
Döderlein 1916: 433, 352; 1936: 350 (forma *cebuana*); Hayashi 1938b: 198, 201; A.M.C. in Clark & Rowe 1971: 55; Marsh & Marshall 1983: 675; Jangoux 1984: 280.
Range: SE India*, to southern Japan, Melanesia, New Caledonia, and northern Australia.

sibogae Döderlein, 1916
Döderlein 1916: 432; 1936: 353.
Range: Indonesia.

tuberculatus (Müller & Troschel, 1842)
Müller & Troschel 1842: 46 (as *Oreaster*).
Sluiter 1895: 55 (as *Pentaceros*).
Döderlein 1916: 431; A.M.C. in Clark & Rowe 1971: 55.
Range: Red Sea, East Africa.
Possibly a forma of *Pentaceraster mammillatus* (Audouin, 1826) according to A.M.C. in Clark & Rowe (1971) also Price (1982).

westermanni (Lütken, 1871)
Lütken 1871: 237 (as *Oreaster*).
Sladen 1889: 764 (as *Pentaceros*).
Döderlein 1936: 354 (as *Pentaceraster*).
Range: Bay of Bengal.

PENTACEROPSIS Sladen, 1889
Sladen 1889: 350.
Type species: *Asterias obtusata* Bory de St Vincent, 1827.
A junior homonym of *Pentaceropsis* Steindachner & Döderlein, 1884 (Pisces) and replaced by *Pentaster* Döderlein (1935) nom. nov.

*euphues Sluiter, 1895
Sluiter 1895: 56.
Range: Molucca Is.
Doubtful validity according to Fisher (1919), holotype immature.

mindorensis Domantay & Roxas, 1937: 213. A variety of *Pentaceropsis tyloderma* Fisher and presumably also synonymous with *Pentaster obtusatus* (Bory de St Vincent, 1827).

obtusata: Sladen, 1889, see *Pentaster*

tyloderma Fisher, 1913c: 209; 1919a: 350. A synonym of *Pentaster obtusatus* (Bory de St Vincent, 1827) according to Döderlein (1936). (Also presumably variety *mindorensis* Domantay & Roxas, 1937: 213, though variety *ilocosensis* Domantay & Acosta, 1970 invalid under date limitation for infra-specific names.)

PENTACEROS Gray, 1840 (after Linck, 1733; *Pentaceros* Schultze, 1760

invalid, a non-binomial work officially rejected by the ICZN.)
Gray 1840: 276; Sladen 1889: 343.

Type species: None designated.

A junior homonym of *Pentaceros* Cuvier & Valenciennes, 1829 (Pisces),
= *Oreaster* Müller & Troschel, 1842.

aculeatus Gray, 1840: 277. A synonym of *Oreaster reticulatus* (Linnaeus,
1758) according to Sladen (1889).

affinis: Sladen, 1889; also Koehler 1910a, see *Pentaceraster*

alveolatus Perrier, 1875, see *Pentaceraster*

armatus Gray, 1840, see *Nidorellia*

australis (Lütken, 1871) Sladen, 1889: 345; also Koehler 1910a: 93. A
synonym of *Pentaceraster regulus* (Müller & Troschel, 1842) according
to Döderlein (1936).

bedoti Koehler, 1911b: 1. A synonym of *Pentaceraster alveolatus* (Perrier,
1875) according to Döderlein (1916).

belli de Loriol, 1885: 53. A synonym of *Pentaceraster horridus* (Gray, 1840)
according to Döderlein (1936).

caledonicus Perrier 1878: 83; Sladen 1889: 345, 760. A NOMEN NUDUM,
undescribed.

callimorphus Sladen, 1889: 347. A synonym of *Pentaceraster gracilis* (Lütken, 1871) according to Döderlein (1936).

*carinatus (Müller & Troschel, 1842) Sladen, 1889: 760. A possible syno-
nym of *Oreaster clavatus* Müller & Troschel, 1842 according to Döderlein (1936, index).

chinensis Gray, 1840, see *Pentaceraster,* also *Oreaster*

clouei (Perrier, 1869) Sladen, 1889: 345. A synonym of *Protoreaster nodosus* (Linnaeus, 1758) according to Döderlein (1936, index).

cumingi Gray, 1840, see *Pentaceraster*

decipiens: Sladen, 1889, see *Pentaceraster*

dorsata Perrier, 1875: 245 [1876: 61], from *Asterias dorsata* Linnaeus, 1753
(invalid). A synonym of *Oreaster clavatus* Müller & Troschel, 1842
according to Döderlein (1936).

*forcipulosus Lütken, 1865
Lütken 1865: 156.
Range: Guinea, West Africa.
A doubtful species according to Döderlein (1936).

franklini Gray, 1840: 277. A synonym of *Protoreaster nodosus* (Linnaeus,
1758) according to Fisher (1919a).

gibbus Gray, 1840: 277. A synonym of *Oreaster reticulatus* (Linnaeus, 1758)
according to Müller & Troschel (1842).

gracilis: Sladen, 1889, see *Pentaceraster*

grandis Gray, 1840: 277. A synonym of *Oreaster reticulatus* (Linnaeus,
1758) according to Müller & Troschel (1842).

314

granulosus Gray, 1847: 75. A synonym of *Goniodiscaster seriatus* (Müller & Troschel, 1843) (Goniasteridae) according to Döderlein (1935).

grayi (Bell, 1884b) Sladen, 1889: 760. A synonym of *Pentaceraster multispinus* (Martens, 1866) according to Döderlein (1936). [But see note under *Oreaster grayi*.]

grayi de Loriol, 1885: 60, non *Oreaster grayi* Bell, 1884. A synonym of *Pentaceraster horridus* (Gray, 1840) according to Döderlein (1936).

hawaiiensis Fisher, 1906a: 1072. A synonym of *Pentaceraster cumingi* (Gray, 1840) according to Döderlein (1936).

hedemanni (Lütken, 1871) Sladen, 1889: 345. A synonym of *Pentaceraster multispinus* (Martens, 1866) according to Döderlein (1936).

hiuculus Gray, 1840: 276. A synonym of *Protoreaster nodosus* (Linnaeus, 1758) according to Döderlein (1936).

hiulcus (Müller & Troschel, 1842) Studer, 1884: 37. A synonym of *Protoreaster nodosus* (Linnaeus, 1758) according to Döderlein (1936).

hiulcus: Perrier 1875: 243 [1876: 59], non *Oreaster hiulcus* Müller & Troschel, 1842, = *Pentaceraster mammillatus* (Audouin, 1826) according to Döderlein (1936).

horridus Gray, 1840, see *Pentaceraster*

indicus Koehler, 1910a: 110. A variety of *Poraster superbus* (Möbius, 1859) according to Döderlein (1936).

lincki: Brown, 1910, see *Protoreaster*

luetkeni (Bell, 1884) Sladen, 1889: 345, 760. A synonym of *Pentaceraster regulus* (Müller & Troschel, 1842) according to A.M.C. in Clark & Rowe (1971).

mammillatus: Perrier, 1875, see *Pentaceraster*

margaritifer Döderlein, 1926: 7. Conspecific with *Pentaceraster australis:* Döderlein, 1926, according to Döderlein (1936), presumably also *Oreaster australis* Lütken, 1871 so a synonym of *P. regulus* (Müller & Troschel, 1842). [But omitted from synonymy of *P. regulus* and bracketted in index, as if a lapsus, by Döderlein (1936)].

mertoni Koehler, 1910b: 275. A synonym of *Oreaster* [i.e. *Pentaceraster*] *gracilis* (Lütken, 1871) according to Döderlein (1916), also a lapsus for *Pentaceros rouxi* Koehler, 1910, according to H.L. Clark (1921).

modestus Gray, 1866: 6. A synonym of *Protoreaster nodosus* (Linnaeus, 1758) according to Döderlein (1936).

muelleri (Bell, 1884) Sladen, 1889: 345, 762. A synonym of *Pentaceraster affinis* (Müller & Troschel, 1842) according to A.M.C. in Clark & Rowe (1971).

muricatus Gray, 1840: 277. A synonym of *Protoreaster lincki* (de Blainville, 1830) according to Döderlein (1936).

muricatus: Studer, 1884: 40, non *Pentaceros muricatus* Gray, 1840, = *Pentaceraster alveolatus* (Perrier, 1875) according to Döderlein (1936).

nodosus (Linnaeus, 1758) Gray, 1866: 6 (species no. 5, not 11), see *Proto-reaster*

nodosus: Gray, 1840: 277; 1866: 6 (species no. 11, not 5), non *Asterias nodosa* Linnaeus, 1758, = *Pentaceraster multispinus* (Martens, 1866) according to Döderlein (1936).

nodulosus Perrier, 1875, see *Protoreaster*

obtusatus: Perrier, 1875, see *Pentaster*

occidentalis (Verrill, 1870) Sladen, 1889: 345, 762. A synonym of *Pentace-raster cumingi* (Gray, 1840) according to Döderlein (1936), though treated as valid by H. L. Clark (1940).

orientalis: Sladen, 1889, see under *Pentaceraster*

productus (Bell, 1884) Sladen, 1889: 762, with variety *tuberata* Sladen, 1889: 347. A variety of *Poraster superbus* (Möbius, 1859) according to Döderlein (1936).

regulus: Sladen, 1889, see *Pentaceraster*

reinhardti (Lütken, 1864) Sladen, 1889: 345, 762; also Koehler, 1910a: 101. A synonym of *Protoreaster lincki* (de Blainville, 1830) according to Döderlein (1936).

reticulatus: Gray, 1840, see *Oreaster*

*rugosus Hutton, 1872
    Hutton 1872: 812.
    Range: no locality, presumably New Zealand.
    [Validity doubtful.]

rouxi Koehler, 1910b: 272. A synonym of *Pentaceraster gracilis* (Lütken, 1871) according to Döderlein (1916).

sladeni de Loriol, 1885: 57. A synonym of *Pentaceraster horridus* (Gray, 1840) according to Döderlein (1936).

superbus: Sladen, 1889, see *Poraster*

thurstoni: Sladen, 1889: 762, see under *Oreaster*

troscheli (Bell, 1884) Sladen, 1889: 345, 762. A synonym of *Pentaceraster alveolatus* (Perrier, 1875) according to A. M.C. in Clark & Rowe (1971).

tuberata Sladen, 1889: 347 (as variety of *Pentaceros productus* (Bell, 1884)). [Presumably also a synonym of *Poraster superbus* (Möbius, 1859); not mentioned by Döderlein (1936) among the varieties of *P. superbus*.]

tuberculatus: Sladen, 1889, see *Pentaceraster*

turritus Gray, 1840: 276; also Perrier 1875: 240. A synonym of *Protoreaster nodosus* (Linnaeus, 1758) according to Döderlein (1936).

valvulatus: Sladen, 1889, see *Anthaster*

verrucosus: Sladen, 1889: 764, see under *Oreaster*

westermanni Sladen, 1889, see *Pentaceraster*

PENTASTER Döderlein, 1935 (nom. nov. for *Pentaceropsis* Sladen, 1889, preoccupied)
    Döderlein 1935: 110; 1936: 360; A.M.C. in Clark & Rowe 1971: 54.

Type species: *Asterias obtusata* Bory de St Vincent, 1827.

hybridus Döderlein, 1936

Döderlein 1936: 364.

Range: New Britain, Melanesia.

obtusatus (Bory de St Vincent, 1827) (with synonym *Pentaceropsis tyloderma* Fisher, 1913 and forma *spinosa* Döderlein, 1936)

Bory de St Vincent 1827: 140 (as *Asterias*).

Müller & Troschel 1842: 50 (as *Oreaster*).

Perrier 1875: 249 [1876: 65] (as *Pentaceros*).

Sladen 1889: 351; Döderlein 1916: 434 (as *Pentaceropsis*).

Döderlein 1935: 110; 1936: 360; Engel 1938a: 8; Guille & Jangoux 1978: 53.

Range: Indonesia, New Britain and Philippine Is.

PORASTER Döderlein, 1916

Döderlein 1916: 438; 1936: 364.

Type species: [*Oreaster*] *productus* Bell, 1884, a synonym (variety) of *Oreaster superbus* Möbius, 1859.

bengalensis Döderlein, 1916: 438. A variety of *Poraster productus* (Bell, 1884) [i.e. *P. superbus* (Möbius, 1859).]

indicus: Döderlein, 1916: 440; also 1926: 10. A variety of *Poraster superbus* (Möbius, 1859) according to Döderlein (1936).

productus: Döderlein, 1916: 438. A variety of *Poraster superbus* (Möbius, 1859) according to Döderlein (1936).

superbus (Möbius, 1859) (with presumed synonym *tuberata* Sladen [as variety of *Pentaceros productus]* and varieties *productus* Bell, 1884 [as *Oreaster*], *indicus* Koehler, 1910 and *bengalensis* Döderlein, 1936).

Möbius 1859: 5; Bell 1884: 81 (as *Oreaster*).

Sladen 1889: 345; Simpson & Brown 1910: 51; Brown 1910: 33 (as *Pentaceros*).

Döderlein 1916: 440; 1936: 364; Jangoux 1984: 280 (as *Poraster*).

Range: Sumatra*, Indonesia, New Caledonia W to Sri Lanka, 25-55 m.

PROTOREASTER Döderlein, 1916

Döderlein 1916: 420; 1936: 323.

Type species: *Asterias nodosus* Linnaeus, 1758.

*****gotoi** Döderlein, 1936 (a nom. nov. for *Oreaster modestus* Goto, 1914; supposedly invalidated by *Pentaceros modestus* Gray, 1866, but that name simultaneously treated as a synonym of *Protoreaster nodosus*.)

Döderlein 1936: 330.

Range: Ryu Kyu Is, southern Japan.

But possibly only a forma of *Protoreaster lincki* (de Blainville, 1830) according to Döderlein (1936).

lincki (de Blainville) (with synonyms *Pentaceros muricatus* Gray, 1840, *Oreaster castellum* Grube, [date?] and *O. reinhardti* Lütken, 1864, also possible forma *gotoi* Döderlein, 1936).

de Blainville 1830: 238; 1834: 219 (as *Asterias*).

Lütken 1864: 156; Bell 1884: 72; H.L. Clark 1923a: 273 (as *Oreaster*).

Döderlein 1916: 423; 1936: 328; Jangoux 1973: 23; A.M. Clark & Courtman-Stock 1976: 68; Marsh 1978: 222; Ebert 1979: 72 (as *Protoreaster*).

Range: Mozambique to NW Australia.

nodosus (Linnaeus, 1758) (with synonyms *Pentaceros franklini* Gray, 1840, *P. turritus* Gray, 1840, *P. hiuculus* Gray, 1840, *Oreaster hiulcus* Müller & Troschel, 1842, *P. modestus* Gray, 1866, varieties *intermedia* and *mutica* of *O. muricatus* Martens, 1866 [but not *muricatus* itself], *O. mammosus* Perrier, 1869, *O. clouei* Perrier, 1869 and *O. nahensis* Goto, 1914, also variety [?valid] *hondurae* Domantay & Roxas, 1938).

Linnaeus 1758: 661 (pt: Linck, 1733, pl.3, fig.3)(as *Asterias*).

Gray 1866: 6 (sp. no. 5, not 11)(as *Pentaceros*).

Bell 1884b: 62 (no.15); H.L. Clark 1908: 280 (as *Oreaster*).

Döderlein 1916: 420; 1936: 324; Hayashi 1938b: 200; Engel 1938b: 5; A.H. Clark 1949a: 74; A.M. Clark 1962b: 175 (proposed lectotype); Opinion 707 1964: 206 (validated, added to Official List of Species); Chang & Liao 1964: 60; Jangoux 1984: 280.

Range: Ceram, Bona or Amboina*, E to New Caledonia, W to the Seychelles, N to southern China, Japan and Palao, S to northern Australia, 1-30 m.

nodulosus (Perrier, 1875)

Perrier 1875: 237 [1876: 53] (as *Pentaceros*).

Bell 1884b: 66; H.L. Clark 1914: 143 (as *Oreaster*).

Döderlein 1936: 323; H.L. Clark 1938: 130 (as *Protoreaster*).

Range: NW Australia.

PSEUDANTHENEA Döderlein, 1915

Döderlein 1915: 26.

Type species: *Anthenea grayi* Perrier, 1875.

grayi (Perrier, 1875)

1875: 278 [1876: 94] (as *Anthenea*).

Döderlein 1915: 26 (as *Pseudanthenea*).

Range: Philippine Is.

PSEUDOREASTER Verrill, 1899

Verrill 1899: 148; H.L. Clark 1938: 103.

Type species: *Asterias obtusangulus* Lamarck, 1816.

obtusangulus (Lamarck, 1816)

Lamarck 1816: 556 (as *Asterias*).

Müller & Troschel 1842: 51 (as *Oreaster*).

Bell 1884: 62 (as *Goniaster*).

Verrill 1899: 148; Fisher 1911c: 174; H.L. Clark 1938: 104; 1946: 99 (as *Pseudoreaster*).

Döderlein 1915: 48 (as *Anthenea*).

Range: NW Australia.

RANDASIA Gray, 1840

Gray 1840: 278.

Type species: *Randasia luzonica* Gray, 1840.

A synonym of *Culcita* L. Agassiz, 1836 according to Fisher (1919a: 362).

granulata Gray 1847: 196. A synonym of *Culcita novaeguineae* Müller & Troschel, 1842 according to Perrier (1875).

*luzonica Gray, 1840: 278. Type (not traced) an immature *Culcita* according to Fisher (1919a) (by inference). [From locality could be *C. novaeguineae* Müller & Troschel, 1842 so better suppressed by the ICZN.]]

Family ASTEROPSEIDAE Hotchkiss & Clark

Asteropidae Fisher 1908a: 88; 1908b: 356; 1911c: 247-248; Verrill 1915: 86.

Asteropseidae Hotchkiss & Clark 1976: 266; Blake 1980: 179; 1981: 391; A. M. Clark 1984: 19-20.

Echinasteridae (Valvasterinae) Viguier 1878: 131.

Valvasteridae: Fisher 1911c: 252; Blake 1980: 167 (a synonym of Asteropseidae).

Nomenclatural note: As remarked in 1976 (Hotchkiss & Clark p.266, footnote), the earlier family name Asteropsidae, used once by Perrier (1884: 154), is invalidated by his mistaken concept of the genus *Asteropsis* Müller & Troschel, 1842, omitting the type species, *Asterias carinifera* Lamarck, 1816. The transfer of *Valvaster* to this family by Blake (1980) means that the name Valvasteridae Viguier, 1878 should take precedence over Asteropseidae but there is a good case for the ICZN to set aside this priority to avoid further confusion.

Genus-group names: Asterope, **Asteropsis**, **Dermasterias**, Gymnasteria, **Petricia**, **Poraniella**, **Valvaster**.

ASTEROPE Müller & Troschel, 1840

Müller & Troschel 1840a: 104; Fisher 1908a: 88; 1911c: 248; H.L. Clark 1946: 109.

Type species: *Asterias carinifera* Lamarck, 1816.

A junior primary homonym of *Asterope* Hübner, 1819 according to Wright in Spencer & Wright (1966), replaced by *Asteropsis* Müller & Troschel, 1840b nom. nov.

carinifera: Müller & Troschel 1840a, see *Asteropsis*

ASTEROPSIS Müller & Troschel, 1840 (with synonym *Gymnasteria* Gray, 1840)

Müller & Troschel 1840b: 322; 1842: 63; A.M. Clark 1967a: 37; Hotchkiss & Clark 1976: 263 (pt); A.M. Clark 1984: 19. (Presumably a replacement name for *Asterope* Müller & Troschel, 1840a, preoccupied.) Type species: *Asterias carinifera* Lamarck, 1816.

ASTEROPSIS: Perrier, 1875: 281 [1876: 96], non *Asteropsis* Müller & Troschel, 1840b, = *Petricia* Gray, 1847 according to Fisher (1908).

capreensis Gasco, 1876, see *Marginaster* (Poraniidae).

carinifera (Lamarck, 1816) (with synonyms *Gymnasteria inermis* Gray, 1840, *G. spinosa* Gray, 1840, *G. biserrata* Martens, 1866 and probably *G. valvulata* Perrier, 1875).

Lamarck 1816: 556 (as *Asterias*).

Müller & Troschel 1840a: 104; Fisher 1908a: 90; Ely 1942: 24; H.L. Clark 1946: 109; Endean 1956: 124; Chang & Liao 1964: 61 (as *Asterope*).

Müller & Troschel 1840b: 322; 1842: 63; A.M. Clark 1967a: 37; Downey 1975: 89; Tortonese 1979: 317; Liao 1980: 155; Price 1983: 46 (as *Asteropsis*).

Martens 1866: 74; Goto 1914: 610; Domantay & Roxas 1938: 218 (as *Gymnasteria*).

Range: tropical Indo-West and East Pacific.

ctenacantha Müller & Troschel, 1842: 63. A synonym of *Porania pulvillus* (O. F. Müller, 1776) (Poraniidae) according to Perrier (1875).

imbricata Grube, 1857, see *Dermasterias*

imperialis Farquhar, 1897, see *Petricia*

*lissoterga (Benham, 1911)

Benham 1911: 145 (as *Gymnasteria*).

H.E.S. Clark 1970b: 5 (as *?Asteropsis*).

Range: Kermadec Is, Tasman Sea.

[Validity doubtful.]

pulvillus: Müller & Troschel, 1842, see *Porania* (Poraniidae)

vernicina Müller & Troschel, 1842, see *Petricia*

DERMASTERIAS Perrier, 1875 Perrier 1875: 282 [1876: 98]; Fisher 1911c: 248.

Type species: *Dermasterias inermis* Perrier, 1875, a synonym of *Asteropsis* [i.e. *Dermasterias*] *imbricata* Grube, 1857.

imbricata (Grube, 1857) (with synonym *Dermasterias inermis* Perrier, 1875 and variety *valvulifera* Verrill, 1914)

Grube 1857: 340; A. Agassiz 1877: 106 (as *Asteropsis*).

Sladen 1889: 766; Fisher 1911c: 249; Verrill 1914a: 308 (variety *valvulifera*) (as *Dermasterias*).

Range: Alaska to mid-California (var. *valvulifera* from Alaska).

inermis Perrier, 1875: 282 [1876: 98]. A synonym of *Dermasterias imbricata* (Grube, 1857) according to Fisher (1911c).

GYMNASTERIA Gray, 1840

Gray 1840: 278.

Type species: None designated but both included nominal species synonymous with *Asteropsis carinifera* (Lamarck, 1816).

A synonym of *Asteropsis* Müller & Troschel, 1840b according to Müller & Troschel, (1842).

biserrata Martens, 1866: 74. A synonym of *Gymnasteria* [i.e. *Asteropsis*] *carinifera* (Lamarck, 1816) according to de Loriol (1885).

carinifera: Martens, 1866, see *Asteropsis*

inermis Gray, 1840: 278. A synonym of *Gymnasteria* [i.e. *Asteropsis*] *carinifera* (Lamarck, 1816) according to Perrier (1875).

lissotergum Benham, 1911, see *Asteropsis*

spinosa Gray, 1840: 278; also Verrill 1869: 384; 1871: 574. A synonym of *Gymnasteria* [i.e. *Asteropsis*] *carinifera* (Lamarck, 1816) according to Perrier (1875).

valvulata Perrier, 1875: 283 [1876: 99]; Bell 1893b: 28. No subsequent records. [Both syntypes in BM(NH) collection very small and almost certainly young *Asteropsis carinifera* (Lamarck, 1816), so a probable synonym.]

GYMNASTERIAS Bell, 1893b: 28, lapsus for *Gymnasteria*

PETRICIA Gray, 1840

Gray 1847: 81.

Type species: *Petricia punctata* Gray, 1847, a synonym of *Asterias vernicina* Lamarck, 1816.

imperialis (Farquhar, 1897)

Farquhar 1897: 193; 1907: 127; Benham 1911: 141 (as *Asteropsis*).

H.E.S. Clark 1970b: 5 (as *Petricia*).

Range: Kermadec Is, Tasman Sea.

obesa H.L. Clark, 1923b: 241; also 1938: 142; 946: 110. A possible synonym of *Petricia vernicina* (Lamarck, 1816) according to A.M. Clark (1966), confirmed by Shepherd (1968).

punctata Gray, 1847: 81. A synonym of *Asterias* [i.e. *Petricia*] *vernicina* Lamarck, 1816 according to Perrier (1875).

vernicina (Lamarck, 1816) (with synonyms *Petricia punctata* Gray, 1847 and *P. obesa* H.L. Clark, 1923)

Lamarck 1816: 554 (as *Asterias*).

Müller & Troschel 1842: 64 (as *Asteropsis*).

Fisher 1911c: 247; H.L. Clark 1928: 389; A.M. Clark 1966: 318; Shepherd 1968: 743 (as *Petricia*).

Range: SE and South Australia, Abrolhos Is, W Australia, 0-60 m.

PORANIELLA Verrill, 1914

Verrill 1914b: 19; 1915: 68; Downey 1973: 81; A.M. Clark 1984: 19.
Type species: *Poraniella regularis* Verrill, 1914, a synonym of *Marginaster echinulatus* Perrier, 1881.

echinulata (Perrier, 1881) (with synonym *Poraniella regularis* Verrill, 1914)

Perrier, 1881: 17; 1884: 230; 1894: 169 (as *Marginaster*).
Verrill 1914b: 20; 1915: 73; Fisher 1928b: 490 (as *Poraniella*).
Range: Barbados*, N and W to the Bahamas and Gulf of Mexico, 20-310 m.

regularis Verrill, 1914b: 19. A probable synonym of *Poraniella echinulata* (Perrier, 1881) according to Fisher (1928b). [Now confirmed on the basis of new material from the Windward Is, the type locality.]

VALVASTER Perrier, 1875

Perrier 1875: 112 [376]; Blake 1980: 178.
Type species: *Asterias striata* Lamarck, 1816.

*spinifera H.L. Clark, 1921

H.L. Clark 1921: 102; A.M. Clark & Rowe 1971: 71; Marsh 1974: 94.
Range: Torres Strait.
Possibly a synonym of *Valvaster striatus* (Lamarck, 1816) according to A.M.C. in Clark & Rowe (1971), also Marsh (1974).

striatus (Lamarck, 1816) (with possible synonym *Valvaster spinifera* H.L. Clark, 1921)

Lamarck, 1816: 564 (as *Asterias*).
Müller & Troschel, 1842: 18 (as *Asteracanthion*).
Perrier 1875: 112 [376]; de Loriol 1885: 11; Fisher 1906a: 1093; Koehler 1910: 175; A.M. Clark & Rowe 1971: 71; Marsh 1974: 94; Blake 1980: 165 (as *Valvaster*).
Range: Mauritius*, Andaman Is, Philippine Is, Polynesia and Hawaiian Is.

Family ACANTHASTERIDAE Sladen

Echinasteridae (Acanthasterinae) Sladen 1889: 536.
Acanthasteridae Fisher 1911c: 252; 1919a: 441; Caso 1962: 315; Blake 1979: 313.
Genus-group names: **Acanthaster**, Echinites.

ACANTHASTER Gervais, 1841 (including *Echinites* Müller & Troschel, 1844, invalid)

Gervais 1841: 474 (nom. nov. for *Echinaster* Gray, 1840, non *Echinaster* Müller & Troschel, 1840); Viguier 1878: 140; Sladen 1889: 536; Fisher 1911c: 252; Caso 1962: 315; Blake 1979: 303.
Type species: '*Asterias echinus* Gervais, 1841' [? a lapsus for *A. echinites*

Ellis & Solander, 1786] but usually cited as a synonym of *Asterias planci* Linnaeus, 1758.

brevispinus Fisher, 1917 (with subspecies *seychellesensis* Jangoux & Aziz, 1984)

brevispinus brevispinus Fisher, 1917

Fisher 1917c: 92; 1919a: 442; Madsen 1955b: 179; Lucas & Jones 1976: 409.

Range: S Philippine Is*, coast of Queensland and Great Barrier Reef, c. 18 m.

brevispinus seychellesensis Jangoux & Aziz, 1984

Jangoux & Aziz 1984: 868.

Range: Seychelles, 52 m.

echinites (Ellis & Solander, 1786) Gervais, 1841: 474. A synonym of *Acanthaster planci* (Linnaeus, 1758) according to Verrill (1914b).

echinus Gervais, 1841: 474. A synonym of *Acanthaster planci* (Linnaeus, 1758) according to Fisher (1919).

ellisi (Gray, 1840) Gervais, 1841: 474; also Verrill 1869: 385; Ziesenhenne 1937: 219; Caso 1962: 316. A synonym of *Acanthaster planci* (Linnaeus, 1758) according to Glynn and Turner in Glynn (1974), while Lucas & Jones (1976) note that the status is 'disputed' and Nishida & Lucas (1988) cite it as a possible synonym.

mauritiensis de Loriol, 1885: 6. A synonym of *Acanthaster planci* (Linnaeus, 1758) according to Madsen (1955b).

planci (Linnaeus, 1758) (with synonyms *Asterias echinites* Ellis & Solander, 1786, *A. solaris* Schreber, 1793, *Echinaster ellisi* Gray, 1840, *Acanthaster echinus* Gervais, 1841, *A. mauritiensis* de Loriol, 1885 and almost certainly *A. ellisi* pseudoplanci Caso, 1962).

Linnaeus 1758: 823 (as *Asterias*).

Verrill 1914a: 373; Fisher 1919a: 441; H.L. Clark 1921: 101; Hayashi 1939a: 442; Madsen 1955b: 161; Caso 1972b: 64; McKnight 1978: 17; Blake 1979: 306, 311; Liao 1980: 164; Nishida & Lucas 1988: 359.

Range: Goa*, W India, entire coral reef area of Indo-West Pacific, also East Pacific from Galapagos Is to Gulf of California if *ellisi* is synonymous.

pseudoplanci Caso 1962: 322 (as subspecies of *Acanthaster ellisi* Gray, 1840). [Since holotype evidently resembles *A. planci* and E Pacific specimens have been recognised as conspecific with *A. planci,* this too must be a synonym.]

solaris (S[chreber] 1793) Dujardin & Hupé 1862: 352. A synonym of *Acanthaster planci* (Linnaeus, 1758) according to Madsen (1955b), the locality Magellan Strait a mistake.

ECHINITES Müller & Troschel, 1844

Müller & Troschel 1844: 180.

Type species: *'Echinaster solaris'* Schreber, 1793.
A junior homonym of *Echinites* Gesner, 1758; Leske, 1778 (Echinoidea),
= *Acanthaster* Gervais, 1841 according to Sladen (1889).

Family MITHRODIIDAE Viguier

Mithrodiae Viguier 1878: 682.
Echinasteridae (Mithrodinae) Viguier 1879: 128; Perrier 1884: 164.
Echinasteridae (Mithrodiinae): Sladen 1889: xxxviii, 538.
Mithrodidae: Perrier 1893: 849; 1894: 4, 141.
Mithrodiidae: Fisher 1906a: 1094; 1911c: 252; Verrill 1914a: 204; Spencer
    & Wright 1966: U71; A.M. Clark & Courtman-Stock 1976: 87; Caso
    1977: 2; Blake 1987: 520.
Ophidiasteridae (Mithrodiinae): Blake 1980: 178; 1981: 391.
    Genus-group names: Heresaster, **Mithrodia, Thromidia**.

HERESASTER Michelin, 1844
    Michelin 1844: 173.
    Type species: *Heresaster papillosus* Michelin, 1844.
    A synonym of *Mithrodia* Gray, 1840 according to Sladen (1889).
papillosus Michelin, 1844: 173. A synonym of *Mithrodia clavigera*
    (Lamarck, 1816) according to Sladen (1889).
MITHRODIA Gray, 1840 (with synonym *Heresaster* Michelin, 1844)
    Gray 1840: 287; Viguier 1879: 129; Sladen 1889: 538; Engel et al. 1948:
    3; Spencer & Wright 1966: U71; Pope & Rowe 1977: 213; Blake 1980:
    167, 176, 180.
    Type species: *Mithrodia spinulosa* Gray, 1840, a synonym of *Asterias
    clavigera* Lamarck, 1816.
bailleui Perrier (MS) in Engel et al., 1948. A NOMEN NUDUM and =
    Mithrodia fisheri Holly, 1932 according to Engel et al. (1948).
bradleyi Verrill, 1870
    Verrill 1870: 288; 1871: 575; H.L. Clark 1910: 327; Caso 1944: 253;
    1953: 214; Downey 1975: 87; Caso 1976: 3.
    Range: Lower California to Colombia and Galapagos Is, 0-14 m.
bradleyi: Fisher 1906a: 1094; also 1928b: 491; Ely 1942: 27, non *Mithrodia
    bradleyi* Verrill, 1870, = *Mithrodia* fisheri Holly, 1932 according to Engel
    et al. (1948).
clavigera (Lamarck, 1816) (with synonyms *Mithrodia spinulosa* Gray, 1840,
    Ophidiaster echinulatus Müller & Troschel, 1842 and *M. victoriae* Bell,
    1882).
    Lamarck 1816: 562 (as *Asterias*).
    Perrier 1875: 114 [378]; Hayashi 1938b: 216; 1938c: 287; Engel et al.
    1948: 5; Marsh 1977: 276; Liao 1980: 165; Jangoux 1986: 149 (as
    Mithrodia).

324

Range: tropical Indo-West Pacific except Hawaiian Is, also E of Brazil [and unpublished records from the western Caribbean, identified by A.M.C.]

echinulatus (Müller & Troschel, 1842) Lütken, 1871: 266. A synonym of *Mithrodia clavigera* (Lamarck, 1816) according to Sladen (1889).

*enriquecasoi Caso, 1976
   Caso 1976: 2.
   Range: Gulf of California.
   [Holotype differentiated from *Mithrodia bradleyi* only by variable characters and artefacts of preservation.]

fisheri Holly, 1932
   Holly 1932: 6; Ely 1942: 27.
   Range: Hawaiian Is. [Philippines and New Ireland, see Pope & Rowe]

gigas Mortensen, 1935, see *Thromidia*

spinulosa Gray, 1840: 288. A synonym of *Mithrodia clavigera* (Lamarck, 1816) according to Sladen (1889).

victoriae Bell, 1882: 123. A possible synonym of *Mithrodia clavigera* (Lamarck, 1816) according to Engel et al. (1948), confirmed by Pope & Rowe (1977).

THROMIDIA Pope & Rowe, 1977
   Pope & Rowe 1977: 202.
   Type species: *Thromidia catalai* Pope & Rowe, 1977.

catalai Pope & Rowe, 1977
   Pope & Rowe 1977: 203; De Celis 1980: 50; Jangoux 1986a: 150.
   Range: New Caledonia*, Hawaiian Is, Bonin Is, E of Japan and Philippine Is, 10-45 m.

gigas (Mortensen, 1935)
   Mortensen 1935: 1; Cherbonnier 1975: 631; A.M. Clark & Courtman-Stock 1976: 87 (as *Mithrodia*).
   Pope & Rowe 1977: 210 (as *Thromidia*).
   Range: SE South Africa, Madagascar, 45-55 m.

seychellesensis Pope & Rowe, 1977
   Pope & Rowe 1977: 207.
   Range: Seychelles, 12 m.

Family OPHIDIASTERIDAE Verrill

Ophidiasteridae Verrill 1870: 344; H.L. Clark 1921: 36-94 (review); Spencer & Wright 1966: U64; Downey 1970: 80; 1973: 61-62; A.M. Clark & Courtman-Stock 1976: 69; Blake 1980: 180; Liao 1980: 155; Miller 1984: 205.

Linckiidae Viguier 1878: 144; Sladen 1889: xxxv, 397; Perrier 1894: 327; Fisher 1911c: 240.

The family Ophidiasteridae was fully reviewed by H.L. Clark (1921) with keys to the genera and species of the major genera then included.

Genus-group names: Acalia, **Andora, Austrofromia, Bunaster,** Calliophidiaster, Catantes, **Celerina, Certonardoa,** Chione, **Cistina, Copidaster,** [Cribrella], **Dactylosaster, Devania, Dissogenes, Dorana, Drachmaster, Ferdina, Fromia, Gomophia, Hacelia, Heteronardoa, Leiaster,** Lepidaster, **Linckia,** Melia, **Narcissia, Nardoa, Neoferdina, Oneria, Ophidiaster, Paraferdina, Pharia, Phataria, Plenardoa,** Pseudolinckia, **Pseudophidiaster,** Scytaster, **Sinoferdina, Tamaria,** Undina.

*ACALIA Gray, 1840

Gray 1840: 285 (as subgenus of *Linckia* Nardo, 1834).

Type species: None designated.

Unrecognisable according to H.L. Clark (1921). [Types of all three included species untraced in the collections of the BM(NH) and only *Linckia (Acalia) erythraea* with a type locality (Red Sea). Diagnoses inadequate for recognition. Better treated as a NOMEN NUDUM since no type species designated.]

*erythraea Gray, 1840: 286. Unrecognisable according to H.L. Clark (1921).

*intermedia Gray, 1840: 286. Unrecognisable according to H.L. Clark (1921).

*pulchella Gray, 1840: 285. Unrecognisable according to H.L. Clark (1921).

ANDORA A.M. Clark. 1967 (with subgenus *Dorana* Rowe, 1977)

ANDORA (ANDORA) A.M. Clark, 1967

A.M. Clark 1967a: 187 (as subgenus of *Nardoa* Gray, 1840).

Rowe 1977: 237 (as genus).

Type species: *Nardoa faouzii* Macan, 1938.

bruuni Rowe, 1977

Rowe 1977: 238.

Range: Mozambique Channel.

A possible synonym of *Andora faouzii* (Macan, 1938) according to Jangoux & Aziz (1984).

faouzii (Macan, 1938)

Macan 1938: 407 (as *Nardoa*).

A.M. Clark 1967a: 186; James 1973: 553 (as *Nardoa (Andora)*).

Rowe 1977: 237; Jangoux & Aziz 1984: 865 (as *Andora*).

Range: SE Arabia*, southern India, Seychelles.

ANDORA (DORANA) Rowe, 1977

Rowe 1977: 236.

Type species: *Andora (Dorana) wilsoni* Rowe, 1977.

popei Rowe, 1977

326

Rowe 1977: 240.
Range: Great Barrier Reef, Queensland.
wilsoni Rowe, 1977
Rowe 1977: 236.
Range: Philippine Is.
AUSTROFROMIA H.L. Clark, 1921
H.L. Clark 1921: 48.
Type species: *Fromia polypora* H.L. Clark, 1916.
polypora (H.L. Clark, 1916)
H.L. Clark 1916: 51 (as *Fromia*).
H.L. Clark 1921: 48; 1928: 387; A.M. Clark 1966: 318; Shepherd 1968: 744 (as *Austrofromia*).
Range: southern half of Australia, 5-160 m.
schultzei (Döderlein, 1910)
Döderlein 1910: 249 (as *Fromia*).
H.L. Clark 1923a: 276; Mortensen 1933a: 247; A.M. Clark & Courtman-Stock 1976: 69 (as *Austrofromia*).
Range: False Bay* and south coast of South Africa, c.15 m.
BUNASTER Döerlein, 1896
Döderlein 1896: 317; H.L. Clark 1921: 69.
Type species: *Bunaster ritteri* Döderlein, 1896.
lithodes Fisher, 1917
Fisher 1917c: 91; 1919a: 398; H.L. Clark 1923b: 241; Liao 1980: 157.
Range: Philippine Is* and Paracel (Xisha) Is, S China Sea.
ritteri Döderlein, 1896
Döderlein 1896: 317.
Range: Amboina, S Molucca Is.
uniserialis H.L. Clark, 1921
H.L. Clark 1921: 69.
Range: Torres Strait.
variegatus H.L. Clark, 1938
H.L. Clark 1938: 134.
Range: SW and W coasts of Australia.
CALLIOPHIDIASTER Tommasi, 1970
Tommasi 1970: 9.
Type species: *Calliophidiaster psicodelica* Tommasi, 1970.
Holotype of type species in poor condition, conspecific with *Linckia nodosa* Perrier, 1875 according to Clark & Downey (1992).
psicodelica Tommasi, 1970
Tommasi 1970: 10.
Range: southern Brazil (c.31.5°S).
[See comment under genus.]
CATANTES Gistel, 1848 [?]

Gistel: 1848: p?

Fisher 1919a: 400 (footnote).

Established for '*Linckia* Agassiz', presumably not *Linckia* Nardo, according to Fisher who nevertheless rejects the name as a synonym. [I cannot trace either *Catantes* or *Undina* (mentioned at the same time by Fisher) on p.176 of the copy of Gistel's 'Naturgeschichte des Thierreichs' in the library of the British Museum (Nat. Hist.) with the other Stellerides, only *Chione* and *Melia* of new names by Gistel, so can do nothing but concur that *Catantes* and *Undina* should be rejected.]

CELERINA A.M. Clark, 1967

A.M. Clark 1967b: 193.

Type species: *Ferdina heffernani* Livingstone, 1931.

heffernani (Livingstone, 1931)

Livingstone 1931: 306 (as *Ferdina*).

A.M. Clark 1967b: 193; Marsh 1977: 255; Jangoux 1978b: 294; Guille & Jangoux 1978: 55; Oguro 1983: 221; Jangoux 1986a: 128 (as *Celerina*).

Range: Santa Cruz Is*, New Caledonia, Caroline Is, Philippines and Macclesfield Bank, South China Sea, 3-55 m.

CERTONARDOA H.L. Clark, 1921

H.L. Clark 1921: 56; Hayashi 1973b: 61.

Type species: *Scytaster semiregularis* Müller & Troschel, 1842.

Thought to be a synonym of *Nardoa* Gray, 1840 according to Tortonese (1955), but a valid genus according to Hayashi (1973b).

carinata: H.L. Clark, 1921, see *Heteronardoa*

semiregularis (Müller & Troschel, 1842)

Müller & Troschel 1842: 36 (as *Scytaster*).

Martens 1866: 351 (as *Linckia* with variety *japonica*).

Sladen 1889: 412; Tortonese 1955: 679 (as *Nardoa*).

H.L. Clark 1921: 56; Hayashi 1938a: 115; Rho & Kim 1966: 284; Hayashi 1973b: 59;

Jangoux & de Ridder 1987: 91 (as *Certonardoa*).

Range: Java*, N to Japan and S Korea, E to New Guinea.

squamulosa (Koehler, 1910) H.L. Clark 1921: 57. A synonym of *Heteronardoa carinata* (Koehler, 1910) according to Rowe (1976).

CHIONE Gistel, 1848

Gistel 1848: 176.

Type species: *Asterias ophidiana* Lamarck, 1816 (by monotypy).

A synonym of *Ophidiaster* according to Fisher (1919a).

CISTINA Gray, 1840

Gray 1840: 283; A.M.C. in Clark & Rowe 1971: 72; Blake 1978: 234.

Type species: *Cistina columbiae* Gray, 1840.

columbiae Gray, 1840 (with synonym *Echinaster sladeni* de Loriol, 1893)

Gray 1840: 283; A.M.C. in Clark & Rowe 1971: 72; Blake 1978: 239;

328

Jangoux 1986a: 128; James 1989: 106.

Range: Mauritius[?*], Lakshadweep (Laccadive) Is, Chagos Archipelago [MS record det. A.M.C.], New Caledonia.

COPIDASTER A.H. Clark, 1948

A.H. Clark 1948b: 55; Miller 1984: 194, 200.

Type species: *Copidaster lymani* A.H. Clark, 1948.

lymani A.H. Clark, 1948

A.H. Clark 1948b: 56; Miller 1984: 194.

Downey 1973: 62 (as *Leiaster*).

Nataf & Cherbonnier 1975: 818 (as *Ophidiaster*).

Range: Florida keys, Ascension Id. and Annobon Id, Gulf of Guinea.

schismochilus (H.L. Clark, 1922)

H.L. Clark 1922: 355.

Miller 1984: 200 (as *Copidaster*).

Range: Bermuda, 55 m.

[**CRIBRELLA** L. Agassiz, 1835

L. Agassiz 1836: 191.

Cited only as a synonym of *Linckia* Nardo, 1834.]

DACTYLOSASTER Gray, 1840

Gray 1840: 283; H.L. Clark 1921: 85.

Type species: *Asterias cylindrica* Lamarck, 1816.

cylindricus (Lamarck, 1816) (with synonym *Ophidiaster asperulus* Lütken, 1871 and subspecies *pacificus* Fisher, 1925)

cylindricus cylindricus (Lamarck, 1816)

Lamarck 1816: 567 (as *Asterias*).

Gray 1840: 283; Michelin 1845: 21; H.L. Clark 1921: 85; A.H. Clark 1952: 286; Sloan et al. 1979: 96; Liao 1980: 155 (as *Dactylosaster*).

Müller & Troschel 1842: 29; de Loriol 1885: 20; Simpson & Brown 1910: 55 (as *Ophidiaster*).

Martens 1866: 85 (as *Linckia*).

Range: Type locality 'les mers australes,' Mauritius, Mozambique to S China Sea, Amboina, Marshall Is and ?Fiji.

cylindricus pacificus Fisher, 1925

Fisher 1925; Ely 1942: 22.

Range: Hawaiian Is.

*gracilis Gray, 1840

1840: 283; H.L. Clark 1921: 85.

Müller & Troschel 1842: 33 (as *Ophidiaster*).

Range: 'West Coast of Columbia'.

A 'lost species' according to H.L. Clark (1921). [Although there are two specimens in the collection of the BM(NH) labelled as types of *Dactylosaster gracilis,* they agree better with the data given by Gray for *Cistina columbiae,* so H.L.C.'s comment was doubly correct.]

DEVANIA Marsh, 1974
  Marsh 1974: 82.
  Type species: *Devania naviculiforma* Marsh, 1974.
naviculiforma Marsh, 1974
  Marsh 1974: 83.
  Range: SE Polynesia, 38-40 m.
DISSOGENES Fisher, 1913
  Fisher 1913c: 212; 1919a: 367; Jangoux 1981b: 712.
  Type species: *Dissogenes styracia* Fisher 1913.
petersi Jangoux 1981
  Jangoux 1981b: 709.
  Range: New Caledonia.
styracia Fisher, 1913
  Fisher 1913c: 212; 1919a: 368; Jangoux 1981b: 712.
  Range: Moluccas, 240 m.
DORANA Rowe, 1977. A subgenus of *Andora* A.M. Clark, 1967.
DRACHMASTER Downey, 1970.
  Downey 1970b: 77.
  Type species: *Drachmaster bullisi* Downey 1970.
bullisi Downey, 1970
  Downey 1970: 78.
  Range: Windward Is, 62-73 m.
FERDINA Gray, 1840
  Gray 1840: 282; Fisher 1919a: 370; H.L. Clark 1921: 58; Livingstone 1931: 305 (restricted); A.M. Clark 1967b: 191 (further restricted).
  Type species: *Ferdina flavescens* Gray, 1840, by subsequent designation by Fisher (1919a).
cancellata (Grube, 1857) Sladen 1889: 780. A synonym of *Neoferdina cumingi* (Gray, 1840) according to Jangoux (1973b).
cumingi Gray, 1840, see *Neoferdina*
cumingi: H.L. Clark, 1921, non *Ferdina cumingi* Gray, 1840, according to Livingstone (1931) but true identity not surmised.
flavescens Gray, 1840
  Gray 1840: 282; de Loriol 1885: 47; Livingstone 1931: 305; A.M. Clark 1967b: 192.
  Range: Mauritius.
glyptodisca Fisher, 1913, see *Neoferdina*
heffernani Livingstone, 1931, see *Celerina*
intermedia Djakonov, 1930: 248. A synonym of *Neoferdina offreti* (Koehler, 1910) according to Jangoux (1973).
kuhli: Sladen, 1889, see *Neoferdina*
ocellata H.L. Clark, 1921, see *Neoferdina*
offreti Koehler, 1910, see *Neoferdina*

FROMIA Gray, 1840

Gray 1840: 286; Fisher 1919a: 373; H.L. Clark 1921: 38.

Type species: *Asterias milleporella* Lamarck, 1816.

andamanensis Koehler, 1909b: 105; also H.L. Clark 1923b: 239. A synonym of *Fromia indica* (Perrier, 1869) according to Hayashi (1938c).

armata Koehler, 1910

Koehler 1910a: 141; Julka & Sumita Das 1978: 346.

Range: Andaman and Nicobar Is.

balansae Perrier, 1875

Perrier 1875: 178 [442]; Koehler 1910a: 140; A.H. Clark 1952: 285; Tortonese 1955: 677.

Range: New Caledonia, Marshall Is, Samoa.

***elegans** H.L. Clark, 1921

H.L. Clark 1921: 43; 1923b: 240; A.H. Clark 1949a: 74; A.M. Clark 1967b: 189; A.M.C. in Clark & Rowe 1971: 62; Marsh 1977: 257.

Range: Torres Strait*, N and W Australia, Indonesia, Philippine Is.

A synonym of *Fromia indica* (Perrier, 1869) according to Hayashi (1938c) but maintained as a valid species by A.M.C. (1971), though status again doubted by Marsh (1977).

elegans: Engel, 1938a: 11, non *Fromia elegans* H.L. Clark, 1921, = *Fromia indica* (Perrier, 1869) according to A.M.C. in Clark & Rowe (1971).

eusticha Fisher, 1913

Fisher 1913c: 213; 1919a: 375; Domantay & Roxas 1938: 220; A.H. Clark 1952: 286; Jangoux 1978b: 294.

Range: Philippine Is*, Marshall Is, E Indonesia, 0-55 m.

ghardaqana Mortensen, 1938

Mortensen 1938: 37; A.M. Clark 1952c: 205; Tortonese 1960: 18; A.M. Clark 1967a: 37; James & Pearse 1971: 83.

Range: northern Red Sea and Gulf of Aqaba (Elat).

ghardaqana: Tortonese 1979: 317, non *Fromia ghardaqana* Mortensen, 1938, = *F. monilis* (Perrier, 1869) (q.v.).

hadracantha H.L. Clark, 1921

H.L. Clark 1921: 45; Hayashi 1938b: 207.

Range: Philippine Is*, Ryu Kyu Is.

hemiopla Fisher, 1913

Fisher 1913c: 214; 1919a: 377; A.H. Clark 1952: 286.

Range: Philippine Is*, Marshall Is, 0-45 m.

indica (Perrier, 1869) (with synonyms *Fromia tumida* Bell, 1882 and *F. andamanensis* Koehler, 1909, possibly also *F. elegans* H.L. Clark, 1921)

Perrier 1869: 63 [255] (as *Scytaster*).

Perrier 1875: 177 [441]; Koehler 1910a: 140; H.L. Clark 1921: 42; Hayashi 1938b: 207; 1938d: 59; 1938c: 279; A.M. Clark 1967b: 188; Jangoux 1978b: 295; 1986a: 130.

Range: 'Mers des Indes', Bay of Bengal, Indonesia, S Japan, Fiji and New Caledonia, 3-5 m.

japonica Perrier, 1881: 14; 1884: 227; also Fisher 1919a: 373. A synonym of *Fromia monilis* (Perrier, 1869) according to H.L. Clark (1921).

major Koehler, 1895a: 339; 1910a: 140; 1910b: 283. A synonym of *Fromia japonica* Perrier, 1881 [i.e. *F. monilis* (Perrier, 1869)] according to Fisher (1919a).

mexicana Perrier, 1884: 172; 1894: 37. A NOMEN NUDUM according to H.L. Clark (1921), undescribed.

milleporella (Lamarck, 1816) (with synonym *Scytaster pistoria* Müller & Troschel, 1842)

Lamarck 1816: 564 (as *Asterias*).

Müller & Troschel, 1840: 103 (as *Linckia*).

Gray 1840: 286; Fisher 1919a: 378; H.L. Clark 1921: 40; Hayashi 1938b: 205; Marsh 1977: 257; Liao 1980: 159 (as *Fromia*).

Michelin 1845: 22 (as *Scytaster*).

Range: tropical Indo-West Pacific except Hawaiian Is.

monilis (Perrier, 1869) (with synonyms *Fromia japonica* Perrier, 1881 and *F. major* Koehler, 1895)

Perrier 1869: 62 [254] (as *Scytaster*).

H.L. Clark 1921: 46; Döderlein 1926: 19; Hayashi 1938d: 425; 1973b: 58; Marsh 1977: 258; 1978: 222; Jangoux 1978b: 296; Julka & Sumita Das 1978: 347.

Range: Indonesia, New Caledonia, Caroline Is, S Japan, NW Australia, Andaman Is [also Red Sea, Fridman, Birtles, Jangoux (pers. comms) and personal identifications and record of Tortonese (1979) as *F. ghardaqana* from colour figure], 1-35 m.

narcissae Perrier 1885c: 28; 1894: 331. A synonym of *Narcissia canariensis* (d'Orbigny, 1839) according to Koehler (1909a).

nodosa A.M. Clark, 1967

A.M. Clark 1967b: 189; Oguro 1984: 101.

Range: Amirante, Maldive and Marshall Is.

pacifica H.L. Clark, 1921

H.L. Clark 1921: 42; Jangoux 1986a: 130.

Range: Hawaiian Is*, W Polynesia, New Caledonia, 15-30 m.

pistoria (Müller & Troschel, 1842) Perrier, 1881: 15. A synonym of *Fromia milleporella* (Lamarck, 1816) according to H.L. Clark (1921).

polypora H.L. Clark, 1916, see *Austrofromia*

schultzei Döderlein, 1910, see *Austrofromia*

subtilis: Perrier, 1878: 78 (as *Fromia (Metrodira)),* see *Metrodira* (Echinasteridae).

subulata: Perrier, 1878: 78 (as *Fromia (Metrodira)),* see *Metrodira* (Echinasteridae).

tumida Bell, 1882a: 124. A synonym of *Fromia indica* (Perrier, 1869) according to A. M. Clark (1967b).

GOMOPHIA Gray, 1840

Gray 1840: 286; H.L. Clark 1921: 55.

Type species: *Gomophia egyptiaca* Gray, 1840.

aegyptiaca Tortonese, 1979: 316, correct spelling according to Tortonese for *egyptiaca* Gray, 1840. [But since Gray used 'e' not 'æ' (diphthong) I consider that his spelling should be retained.]

egeriae A.M. Clark, 1967b: 169. A subspecies of *Gomophia egyptiaca* Gray, 1840.

egyptiaca Gray, 1840 (with synonyms *Scytaster zodiacalis* Müller & Troschel, 1842, *Oreaster desjardinsi* Michelin, 1844, and probably *Ophidiaster watsoni* Livingstone, 1936, also subspecies *egeriae* A.M. Clark, 1967).

egyptiaca egyptiaca Gray, 1840

Gray 1840: 286; Koehler 1910a: 157; H.L. Clark 1921: 55; A.M. Clark 1952: 206; Endean 1965: 230; A.M. Clark 1967b: 176; Marsh 1974: 85; Guille & Jangoux 1978: 56; Jangoux 1986a: 134.

Perrier 1875: 164 [428] (as *Scytaster*).

Sladen 1889: 788 (as *Nardoa*).

Range: Red Sea*, East Africa, Mauritius, Sri Lanka, Christmas Id, Indian Ocean, S Moluccas, Queensland, New Caledonia, southern Polynesia, 15-50 m.

egyptiaca egeriae A.M. Clark, 1967

A.M. Clark 1967b: 169.

Range: Macclesfield Bank, S China Sea.

*watsoni (Livingstone, 1936)

Livingstone 1936: 386; Endean 1956: 125 (as *Ophidiaster*).

[?]Jangoux 1986a: 134 (as *Gomophia*).

Range: Queensland*, [?]New Caledonia, 10-55 m.

Probably a synonym of *Gomophia egyptiaca* Gray, 1840 according to A. M. Clark (1967b), which Endean's (1965) Queensland record of *G. egyptiaca* supports. But a valid species extending to New Caledonia according to Jangoux (1986a). [However, the illustration given by Jangoux of '*G. watsoni*' shows very slender arms unlike the holotype of *O. watsoni*.]

HACELIA Gray, 1840

Gray 1840: 284 (as subgenus of *Ophidiaster* L. Agassiz, 1836).

Ludwig 1897: 209; H.L. Clark 1921: 36 (as genus). Type species: *Ophidiaster (Hacelia) attenuatus* Gray, 1840.

attenuata (Gray, 1840) (with synonyms *Asterias coriacea* Grube, 1840 and *Ophidiaster lessonae* Gasco, 1876).

Gray 1840: 284 (as *Ophidiaster (Hacelia)*).

Müller & Troschel 1842: 29; Perrier 1875: 119 [1876: 133]; Marenzeller 1895b: 190 (as *Ophidiaster*).

Ludwig 1897: 272; Koehler 1909a: 90; H.L. Clark 1921: 86; Koehler 1924: 164; Tortonese 1965: 164; Nataf & Cherbonnier 1975: 813 (as *Hacelia*).

Range: Mediterranean, Cape Verde area, Azores, 98-128 m.

capensis Mortensen, 1925

Mortensen 1925b: 152 (as variety of *H. superba* H.L. Clark, 1921).

A.M. Clark 1974: 435; A.M. Clark & Courtman-Stock 1976: 70 (as species).

Range: Natal, 70 m.

floridae: A.H. Clark 1948, see *Tamaria*

helicosticha: H.L. Clark 1909a, see *Ophidiaster*

inarmata (Koehler, 1895)

Koehler 1895: 400 (as variety of *Ophidiaster helicostichus*).

H.L. Clark 1921: 87; A.M.C. in Clark & Rowe 1971: 59 (as species of *Hacelia*).

Range: Indonesia.

superba H.L. Clark, 1921

H.L. Clark 1921: 87; Mortensen 1933b: 426; Downey 1973: 61; Miller 1984: 202.

Range: Barbados* to Gulf of Mexico and North Carolina, also St Helena. 40-100 m.

Thought to be a synonym of *Ophidiaster floridae* Perrier, 1881 by A.H. Clark (1948) but a valid species of *Hacelia* according to Downey (1973).

tuberculata Liao, 1985

Liao 1985: 30.

Range: E China Sea, 100-200 m.

HETERONARDOA Hayashi, 1973

Hayashi 1973a: 6; 1973b: 65; Rowe 1976: 85.

Type species: *Heteronardoa sagamina* Hayashi, 1973, a synonym of *Nardoa carinata* Koehler, 1910.

carinata (Koehler, 1910) (with synonyms *Nardoa squamulosa* Koehler, 1910, *Narcissia mohamedi* Macan, 1938 and *Heteronardoa sagamina* Hayashi, 1973)

Koehler 1910a: 165 (as *Nardoa*).

H.L. Clark 1921: 56 (as *Certonardoa*).

Rowe 1976: 86; Jangoux 1986a: 134 (as *Heteronardoa*).

Range: Gulf of Aden, Maldive Is, Bay of Bengal, W Australia, Philippine Is, Indonesia and New Caledonia, 34-230 m.

diamantinae Rowe, 1976

Rowe 1976: 94.

Range: NW Australia, Mozambique Channel, 88-128 m.

334

sagamina Hayashi, 1973a: 6; 1973b: 65. A synonym of *Heteronardoa carinata* (Koehler, 1910) according to Rowe (1976).

LEIASTER Peters, 1852

Peters 1852: 177 (as subgenus of *Ophidiaster,* but treated as genus in headings).

Fisher 1919a: 396; H.L. Clark 1921: 71; 1946: 119; Jangoux 1980: 88 (as genus).

Type species: *Ophidiaster (Leiaster) coriaceus* Peters, 1852, by subsequent designation by Fisher (1919a).

analogus Fisher, 1913c: 215; 1919a: 396. A synonym of *Leiaster coriaceus* Peters, 1852 according to Döderlein (1926).

brevispinus H.L. Clark, 1921: 74; also Ely 1942: 21. A synonym of *Leiaster leachi* (Gray, 1840) according to Jangoux (1980).

callipeplus Fisher, 1906a: 1083; also Koehler 1910a: 153; H.L. Clark 1921: 73; Fisher 1925: 77; Downey 1975: 86. A synonym of *Leiaster glaber* (Peters, 1852) according to Döderlein (1926) and Jangoux (1980), though maintained as valid by H. L. Clark (1921), Fisher (1925) and Downey (1975), with *L. glaber* referred to the synonymy of *L. leachi* (Gray, 1840).

coriaceus (Peters, 1852) (with synonym *Leiaster analogus* Fisher, 1913)

Peters 1852: 177 (as [*Ophidiaster*] *(Leiaster)*.

de Loriol 1885: 37; Fisher 1919a: 397; Döderlein 1926: 18; Jangoux 1980: 96; 1986a: 139 (as *Leiaster*).

Range: Mozambique*, Mauritius, Seychelles area, New Caledonia, also Lower California and Panama, 10-25 m.

glaber (Peters, 1852) (with synonym *Leiaster callipeplus* Fisher, 1913)

Peters 1852: 177 (as [*Ophidiaster*] *(Leiaster))*.

Döderlein 1926: 17; A.M.C. in Clark & Rowe 1971: 57; Jangoux 1980: 100; [?]Applegate 1984: 97 (as *Leiaster*).

Range: Mozambique* and Hawaiian Is; [?]Taiwan and East Pacific off Colombia.

Thought to be a synonym of *Leiaster leachi* Gray, 1840 by H.L. Clark (1921) but a valid species according to A.M.C (1971). [Geographic range needs confirmation.]

*grandis Hayashi, 1938

Hayashi 1938c: 283; 1973b: 69.

Range: W Japan.

A possible synonym of *Leiaster leachi* (Gray, 1840) according to Hayashi (1973), from suggestion by H.L. Clark.

hawaiiensis Fisher, 1925: 77, as subspecies of *Leiaster leachi* (Gray, 1840). A probable synonym of *L. leachi* according to Tortonese (1979), confirmed by Jangoux (1980).

leachi (Gray, 1840) (with synonyms *Leiaster brevispinus* H.L. Clark, 1921, *L. hawaiiensis* Fisher, 1925 [as subspecies] and possibly *L. grandis*

335

Hayashi, 1938).

Gray 1840: 284; Perrier 1875: 121 (as *Ophidiaster*).

de Loriol 1885: 40; H.L. Clark 1921: 73; Fisher 1925: 77; H.L. Clark 1938: 136; Hayashi 1938b: 211; A.M. Clark 1967a: 36; Hayashi 1973b: 69; Marsh 1974: 80; Tortonese 1979: 317; Jangoux 1980: 91; Price 1983: 43; Jangoux 1986a: 139 (as *Leiaster*).

Range: tropical Indo-West Pacific, 10-30 m.

leachi: Tortonese 1953: 29, 1956: 193, non *Leiaster leachi* (Gray, 1840), = *Leiaster coriaceus* (Peters, 1852) according to Tortonese (1979).

lymani: Downey, 1973, see *Copidaster*

speciosus Martens, 1866

Martens 1866: 70; H.L. Clark 1921: 74; Hayashi 1938b: 211; Jangoux 1980: 94; 1986a: 139.

Range: Indonesia, Philippine Is, Ryu Kyu Is, Lord Howe Id, E Australia and New Caledonia, 10-30 m.

Thought to be a synonym of *Leiaster leachi* (Gray, 1840) by Döderlein (1926), followed by H.L. Clark (1938), but a doubtful synonym according to H. L. Clark (1946) and a valid species according to Jangoux (1980).

speciosus: Sladen, 1889: 408, non *Leiaster speciosus* Martens, 1866, = *Leiaster leachi* (Gray, 1840) according to A.M.C. (1967a).

teres (Verrill, 1871)

Verrill 1871: 578 (as *Lepidaster*).

Sladen 1889: 408; Ziesenhenne 1937: 217; Blake 1978: 235; Jangoux 1980: 101 (as *Leiaster*).

Range: Lower California, 37-55 m.

LEPIDASTER Verrill, 1871

Verrill 1871: 577.

Type species: *Lepidaster teres* Verrill, 1871.

A junior homonym of †*Lepidaster* Forbes, 1850 and = *Leiaster* Peters, 1852 according to Sladen (1889).

teres Verrill, 1871, see *Leiaster*

LINCKIA Nardo, 1834 (as *Linkia* incorrect spelling; with synonyms *Catantes* and *Undina* both of Gistel, 1848 according to Fisher (1919a) [unconfirmed].

Nardo 1834: 717 *(Linkia)*.

Gray 1840: 284 (name corrected); Verrill 1870: 285; Perrier 1875: 135 [399]; Fisher 1906a: 990; 1919a: 400; H.L. Clark 1921: 62; Downey 1968: 144; 1973: 66 (as *Linckia*).

Type species: *Linkia typus* Nardo, 1834, a synonym of *Asterias laevigata* Linnaeus, 1758.

aegyptiaca Martens, 1866, lapsus for *egyptiaca* Gray, 1840, see *Gomophia*

bifascialis Gray, 1840: 285. A synonym of *Phataria unifascialis* (Gray, 1840) according to H.L. Clark (1921).

336

bouvieri Perrier, 1875: 150 [414]; 1876: 69; Studer 1884: 28; H.L. Clark 1941: 50; Madsen 1950: 214; Gray, Downey & Cerame-Vivas 1968: 148; Downey 1968: 42; 1973: 67; Nataf & Cherbonnier 1975: 815.
   Range: Cape Verde Is*, Gulf of Guinea, North Carolina, Cuba, 10-380 m.
browni Gray, 1840: 285. A synonym of *Linckia laevigata* (Linnaeus, 1758) according to H.L. Clark (1921).
bullisi Moore, 1960: 414. A synonym of *Linckia nodosa* Perrier, 1875 according to Downey (1973).
colombiae: H.L. Clark 1940: 334, lapsus for
columbiae Gray, 1840 (with synonyms *Phataria fascialis* Monks, 1904 and *bifascialis* Monks, 1904 [as variety of *P. unifascialis* (Gray)])
   Gray 1840: 285; Fisher 1911c: 242; Ziesenhenne 1937: 216; Caso 1961: 68; 1979: 206.
   Müller & Troschel 1842: 33; Dujardin & Hupé 1862: 364 (as *Ophidiaster*).
   Range: W Colombia, Peru and Galapagos Is, N to Southern California, 0-100 m.
costae Russo, 1894: 163. A probable synonym of *Linckia multifora* (Lamarck, 1816) according to H.L. Clark (1921).
crassa Gray, 1840: 284. A short-armed form of *Linckia laevigata* (Linnaeus, 1758) according to H.L. Clark (1921) [holotype extant and conspecific with *laevigata*].
cylindrica: Martens, 1866, see *Dactylosaster*
desjardinsi (Michelin, 1845) Hoffman, 1874: 54. A synonym of *Nardoa* [i.e. *Gomophia*] *egyptiaca* Gray, 1840 according to Sladen (1889).
diplax (Müller & Troschel, 1842) Lütken, 1871: 269. A synonym of *Linckia guildingi* Gray, 1840 according to Fisher (1919); of doubtful validity as distinct from *guildingi* according to H.L. Clark (1921) and A.M.C. in Clark & Rowe (1971), though meantime treated as valid by Hayashi (1939).
diplax: Perrier, 1875: 144 [408], non *Ophidiaster diplax* Müller & Troschel, 1842, = *Linckia columbiae* Gray, 1840 according to Fisher (1911c).
dubiosa Koehler, 1910a, see *Tamaria*
ehrenbergi (Müller & Troschel, 1842) de Loriol 1885: 31. A synonym of *Linckia guildingi* Gray, 1840 according to H.L. Clark (1921).
*erythraea Gray, 1840, see under *Acalia*
formosa Mortensen, 1933b: 430. A synonym of *Linckia bouvieri* Perrier, 1875 according to Madsen (1950) but of *L. nodosa* Perrier, 1875 according to Downey (1973).
franciscanus: Sladen, 1889, lapsus for
*franciscus Nardo, 1834
   Nardo 1834: 717; Gray 1840: 284. Unidentifiable but probably an *Ophi-*

diaster according to H.L. Clark (1921).

gracilis Liao, 1985

Liao 1985: 32.

Range: E China Sea, 100-200 m.

guildingi Gray, 1840 (with synonyms *Linckia pacifica* Gray, 1840, *Ophidiaster diplax, O. ornithopus* and *O. ehrenbergi* all of Müller & Troschel, 1842, *Scytaster stella* Duchassaing, 1850, *O. flaccidus* Lütken, 1859, *O. irregularis* Perrier, 1869 and *Linckia nicobarica* Lütken, 1871)

Gray 1840: 285; A. Agassiz 1877: 105; H.L. Clark 1921: 67; Hayashi 1938c: 284; Caso 1941: 155; Ely 1942: 18; Ummels 1963: 81; A.M. Clark & Davies 1966: 608 (lectotype); Downey 1968: 41; Tommasi 1970: 9; Nataf & Cherbonnier 1975: 816. Müller & Troschel 1842: 33 (as *Ophidiaster*).

Range: 'Tropicopolitan'.

hemprichi: Martens, 1866, see *Ophidiaster*

hondurae Domantay & Roxas, 1938. A variety of *Linckia laevigata* (Linnaeus, 1758).

*intermedia Gray, 1840, see under *Acalia*

kuhli: Martens, 1866, see *Neoferdina*

laevigata (Linnaeus, 1758) (with synonyms *Linckia typus* Nardo, 1834, *Ophidiaster miliaris* Müller & Troschel, 1842, *Linckia browni* Gray, 1840, *L. crassa* Gray, 1840 (or forma), *O. clathrata* Grube, 1865, *L. rosenbergi* Martens, 1866 and *O. propinquus* Livingstone, 1932, also variety *hondurae* Domantay & Roxas, 1938).

Linnaeus 1758: 662 (as *Asterias*).

Nardo 1834: 717; Lütken 1871: 265; H.L. Clark 1908: 282; Fisher 1919a: 400; H. L. Clark 1921: 64; Engel 1938a: 15; Domantay & Roxas 1938: 221 (variety *hondurae);* Hayashi 1938d: 434; 1973b: 68; Jangoux 1973: 29; Marsh 1974: 86; 1977: 260.

Müller & Troschel 1842: 30 (as *Ophidiaster*).

Range: tropical Indo-West Pacific except Red Sea and Hawaiian Is, 0-25 m.

leachi Gray, 1840: 285. A synonym of *Linckia multifora* (Lamarck, 1816) according to H.L. Clark (1921).

leviuscula Stimpson, 1857, see *Henricia* (Echinasteridae)

mammillata (Müller & Troschel, 1842) Philippi, 1857: 132, see under *Tosia* (Goniasteridae).

marmorata: Martens in von der Decken, 1869, see *Tamaria*

megaloplax Bell, 1884b, see *Tamaria*

miliaris (Müller & Troschel, 1840) Martens, 1866: 66. A synonym of *Linckia laevigata* (Linnaeus, 1758) according to H.L. Clark (1921).

milleporella: Müller & Troschel, 1840, see *Fromia*

milleporella: Martens, 1866: 69, non *Asterias milleporella* Lamarck, 1816, =

338

Fromia monilis (Perrier, 1875 according to H.L. Clark (1921).

multiflora: [e.g.] McKnight, 1972: 43, multiforas: Gray, 1840 and multiforis: Martens, 1866, lapsi for

multifora (Lamarck, 1816) (with synonyms *Linckia leachi* Gray, 1840, and probably *L. costae* Russo, 1894)
Lamarck 1816: 565 (as *Asterias*).
Müller & Troschel 1842: 31 (as *Ophidiaster*).
Lütken, 1871: 267; Fisher 1919a: 401; H.L. Clark 1921: 66; Engel 1938a: 16; Hayashi 1938d: 435; Ely 1942: 19; Jangoux 1973: 32; Liao 1980: 157; Price 1983: 45 (as *Linckia*).
Range: tropical Indo-West Pacific, 0-40 m.

nicobarica Lütken, 1871: 265. A synonym of *Linckia guildingi* Gray, 1840 according to H.L. Clark (1921).

nodosa Perrier, 1875 (with synonyms *Linckia bullisi* Moore, 1960 and *L. formosa* Mortensen, 1933b)
Perrier 1875: 153 [417]; Verrill 1915: 93; Caso 1961: 72; Brito 1971: 263; Downey 1968: 42; 1973: 67; Jangoux 1978a: 97.
Range: North Carolina to southern Brazil and St Helena, 35-475 m.

oculata: Forbes, 1841, see *Henricia* (Echinasteridae)

ornithopus (Müller & Troschel, 1842) Verrill, 1868: 367. A synonym of *Linckia guildingi* Gray, 1840 according to Sladen (1889).

pacifica Gray, 1840: 285. A synonym of *Linckia guildingi* Gray, 1840 according to Fisher (1919).

pauciforis Martens, 1866, see *Nardoa*

pertusa: Stimpson, 1853, ?non *Asterias pertusa* O. F. Müller, 1776, probably = *Henricia sanguinolenta* (O. F. Müller, 1776) (Echinasteridae) according to Fisher (1911c).

pistoria (Müller & Troschel, 1842) Martens, 1869: 130. A synonym of *Fromia milleporella* (Lamarck, 1816) according to Sladen (1889).

pulchella Gray, 1840, see under *Acalia*

pusilla: Martens, 1866, see *Tamaria*

pustulata Martens, 1866: 62. A synonym of *Ophidiaster hemprichi* Müller & Troschel, 1842 according to Döderlein (1926).

rosea: Thompson, 1840 (as *Linkia*), see *Stichastrella* (Asteriidae)

rosenbergi Martens, 1866: 63. A synonym of *Linckia laevigata* (Linnaeus, 1758) according to Engel (1942).

semiregularis: Martens, 1866, see *Nardoa*

semiseriata Martens, 1866, see *Plenardoa*

subulata: Martens, 1867, see *Metrodira* (Echinasteridae)

suturalis: Martens, 1866: 85, non *Ophidiaster suturalis* Müller & Troschel, 1842, = *Linckia laevigata* (Linnaeus, 1758) according to H.L. Clark (1921).

tuberculata: Martens, 1866, see *Nardoa*

tyloplax H.L. Clark, 1914
  H.L. Clark 1914: 147.
  Range: Western Australia, 146-220 m.
typus Nardo, 1834: 717. A synonym of *Linckia laevigata* (Linnaeus, 1758) according to H.L. Clark (1921).
typus: Gray, 1840: 284, non *Linckia typus* Nardo, 1834, = *Linckia multifora* (Lamarck, 1816) according to Sladen (1889).
unifascialis Gray, 1840, see *Phataria*
variolosa Nardo, 1834: 717, lapsus for *variolata,* see *Nardoa*
variolata: L. Agassiz, 1836, see *Nardoa*
LINCKIA (ACALIA) Gray, a NOMEN NUDUM, see under *Acalia*
erythraea Gray, 1840, see under *Acalia*
intermedia Gray, 1840, see under *Acalia*
pulchella Gray, 1840 see under *Acalia*
LINCKIA (PHATARIA) Gray, 1840, a genus according to Sladen (1889).
bifascialis Gray, 1840, see *Phataria*
unifascialis Gray, 1840, see *Phataria*
MELIA Gistel, 1848
  Gistel 1848: 176.
  Type species: *Asterias variolata* Lamarck, 1816.
  A synonym of *Nardoa* Gray, 1840 according to Fisher (1919a).
variolata: Gistel, 1848, see *Nardoa*
NARCISSIA Gray, 1840
  Gray 1840: 287; Sladen 1889: 414; Perrier 1894: 329; H.L. Clark 1921: 57.
  Type species: *Narcissia teneriffae* Gray, 1840, a synonym of *Asterias canariensis* d'Orbigny, 1839.
canariensis (d'Orbigny, 1839) (with synonyms *Narcissia teneriffae* Gray, 1840 and *Fromia narcissae* Perrier, 1885c).
  d'Orbigny 1839: 148 (as *Asterias*).
  Dujardin & Hupé 1862: 368 (as *Scytaster*).
  Perrier 1875: 170 [434]; 1885c: 28 (as *Scytaster (Narcissia)*).
  Sladen 1889: 413; Perrier 1894: 330; Koehler 1909a: 91; H.L. Clark 1921: 57; Madsen 1950: 216; A.M. Clark 1955: 33; Nataf & Cherbonnier 1975: 817 (as *Narcissia*).
  Range: Canary and Cape Verde Is to Congo, 37-155 m.
gracilis A.H. Clark, 1916 (with subspecies *malpeloensis* Downey, 1975)
gracilis gracilis A.H. Clark 1916
  A.H. Clark 1916a: 58; Ziesenhenne 1937: 216; Downey 1975: 87.
  Range: Lower California, 56-90 m.
gracilis malpeloensis Downey, 1975
  Downey 1975: 87.
  Range: Malpelo Id, off W coast of Colombia.

340

helenae Mortensen, 1933b: 429. A variety of *Narcissia trigonaria* Sladen, 1889.

malpeloensis Downey, 1975: 87. A subspecies of *Narcissia gracilis* A. H. Clark, 1916.

mohamedi Macan, 1938: 408. A synonym of *Heteronardoa carinata* (Koehler, 1910a) according to Rowe (1976).

teneriffae Gray, 1840: 287. A synonym of *Narcissia canariensis* (d'Orbigny, 1839) according to Sladen (1889).

trigonaria Sladen, 1889 (with variety *helenae* Mortensen, 1933b)
Sladen 1889: 414; Verrill 1915: 97; H.L. Clark 1921: 58; Mortensen 1933b: 429 (variety *helenae*); Gray, Downey & Cerame-Vivas 1968: 147; Tommasi 1970: 9; Downey 1973: 64; Walenkamp 1976: 74; Jangoux 1978a: 97.
Range: NE Brazil*, to North Carolina, 37-210 m.; var. *helenae* from St Helena.

NARDOA Gray, 1840 (with synonym *Melia* Gistel, 1848)
Gray 1840: 268; Fisher 1919a: 378; H.L. Clark 1921: 49; A.M. Clark 1967b: 175 (reviewed).
Type species: *Asterias variolata* Retzius, 1805, by subsequent designation by H. L. Clark (1921).

ægyptiaca: Sladen, 1889, lapsus or alternative spelling for *egyptiaca,* see *Gomophia*

agassizi Gray, 1840: 287. A synonym of *Nardoa variolata* (Retzius, 1805) according to Sladen (1889).

bellonae Koehler 1910a: 164, lapsus for *Nardoa mollis* de Loriol, 1891 [a synonym of *N. novaecaledoniae* (Perrier, 1875)] according to H.L. Clark (1921).

carinata Koehler, 1910a, see *Heteronardoa*

faouzii Macan, 1938, see *Andora*

finschi de Loriol, 1891: 28. A synonym of *Nardoa pauciforis* (Martens, 1866) according to H.L. Clark (1921), itself a synonym (or forma) of *N. tuberculata* Gray, 1840.

frianti Koehler, 1910 (with possible synonyms *Nardoa tumulosa* Fisher, 1917 and *N. mamillifera* Livingstone, 1930)
Koehler 1910a: 158; Fisher 1919a: 385; Hayashi 1938c: 280; A.M. Clark 1967b: 178; Liao 1980: 159; Jangoux 1986a: 140.
Range: Andaman Is to Paracel Is, S China Sea, Philippines, Caroline Is and New Caledonia, 3-45 m.

galatheae (Lütken, 1865)
Lütken 1865: 167 (as *Scytaster*).
Sladen 1889: 412; H.L. Clark 1921: 53; Jangoux in Guille & Jangoux 1978: 56 (*?galatheae*).
Range: Nicobar Is, ?Moluccas.

gamophia: Fisher 1919a: 380, lapsus for *gomophia* (Perrier, 1875), but not the same species and = *Nardoa novaecaledoniae* (Perrier, 1875) according to A. M. Clark (1967b).

gomophia (Perrier, 1875)

Perrier 1875: 167 (as *Scytaster*).

Sladen, 1889: 412; A.M. Clark 1967b: 172; Jangoux 1984: 281 (as *Nardoa*).

Range: New Caledonia, 1-40 m.

Thought to be a synonym of *Nardoa novaecaledoniae* (Perrier, 1875) by H. L. Clark (1921) but a valid species according to A.M. Clark (1967b).

indica Koehler, 1910, see *Fromia*

lemonnieri Koehler, 1910 (with subspecies *platyspina* Jangoux & Aziz, 1984)

lemonnieri lemonnieri Koehler, 1910

Koehler 1910a: 161; Fisher 1919a: 382; A.M. Clark 1967b: 183; A.M.C. in Clark & Rowe 1971: 41; Julka & Sumita Das 1978: 347; Jangoux & Aziz 1984: 866.

Range: Andaman and Philippine Is, ?Maldives.

Thought to be a synonym of *Nardoa mollis* de Loriol, 1891 according to H. L. Clark (1921), a valid species according to A.M. Clark (1967b), a synonym of *N. galatheae* (Lütken, 1864) according to Jangoux in Guille & Jangoux (1978) but again valid according to Jangoux & Aziz (1984).

lemonnieri platyspina Jangoux & Aziz, 1984

Jangoux & Aziz 1984: 866.

Range: Seychelles. 48 m.

*mamillifera Livingstone, 1930

Livingstone 1930: 20; H.L. Clark 1946: 116.

Range: Torres Strait.

A possible synonym of *Nardoa frianti* Koehler, 1910 according to A. M. Clark (1967b).

mollis de Loriol, 1891: 26; also H.L. Clark 1921: 52; A.M. Clark 1967b: 184. A synonym of *Nardoa novaecaledoniae* (Perrier, 1875) according to Jangoux (1985).

novaecaledoniae (Perrier, 1875) (with synonym *Nardoa mollis* de Loriol, 1891)

Perrier 1875: 162 [426] (as *Scytaster*).

Sladen 1889: 412; Fisher 1919a: 380; A.M. Clark 1967b: 184; Jangoux 1985b: 27 (as *Nardoa*).

Range: New Caledonia*, Queensland, Philippine Is, 1-5 m.

obtusa (Perrier, 1875) Sladen, 1889: 412; also Djakonov 1930: 247. A synonym of *Nardoa tuberculata* Gray, 1840 according to A.M. Clark (1967b).

pauciforis (Martens, 1866) Sladen 1889: 412; also Hayashi 1938d: 432; A.

M. Clark 1967b: 186; Marsh 1977: 262. A forma of *Nardoa tuberculata* Gray, 1840 according to Hayashi (1938d) and probably so according to Marsh (1977).

platyspina Jangoux & Aziz, 1984. A subspecies of *Nardoa lemonnieri* Koehler, 1910.

rosea H.L. Clark 1921

H.L. Clark 1921: 53; 1946: 115; A.M. Clark 1967b: 179, 181.

Range: N Queensland.

semiregularis: Sladen, 1889, see *Certonardoa*

semiregularis: Fisher, 1919a: 383, non *Certonardoa semiregularis* (Müller & Troschel, 1842), = *Heteronardoa diamantinae* Rowe, 1976 according to Rowe.

semiseriata: Sladen, 1889, see *Plenardoa*

sphenisci A.M. Clark, 1967

A.M. Clark 1967b: 173; Jangoux & Aziz 1984: 867.

Range: NW Australia*, Seychelles, 22-51 m.

squamulosa Koehler, 1910a: 168; Fisher 1919a: 383. A synonym of *Heteronardoa carinata* (Koehler, 1910) according to Rowe (1976).

tuberculata Gray, 1840 (with synonyms *Scytaster obtusus* Perrier, 1875 and *Nardoa finschi* de Loriol, 1891, also forma *pauciforis* (Martens, 1866))

Gray, 1840: 287; Sladen 1889: 788; Koehler 1910a: 160; Fisher 1919a: 384; Engel 1938b: 2; Hayashi 1939a: 430; 1952: 153; A.M. Clark 1967b: 180; Marsh 1977: 262; Jangoux in Guille & Jangoux 1978: 57.

Müller & Troschel 1842: 32 (as *Ophidiaster*).

Perrier 1875: 157 [421] (as *Scytaster*).

Range: Philippine Is*, S Japan, Caroline Is.

tumulosa Fisher, 1917

Fisher 1917c: 90; 1919a: 386; Hayashi 1938b: 210; 1939a: 432; Hayasaka 1949: 15; A. M. Clark 1967b: 180.

Range: Philippine Is*, Caroline and Ryu Kyu Is, 21-63 m.

A possible synonym of *Nardoa frianti* Koehler, 1910 according to A. M. Clark (1967b).

variolata (Retzius, 1805) (with synonym *Nardoa agassizi* Gray, 1840)

Retzius 1805: 19; Lamarck 1816: 565 (as *Asterias*).

Gray 1840: 286; Fisher 1919a: 379; H.L. Clark 1921: 51; A.M. Clark 1967b: 171, 183 (as *Nardoa*).

Müller & Troschel 1842: 34 (as *Scytaster*).

Gistel 1848: 176 (as *Melia*).

Range: Mauritius, Seychelles, East Africa.

variolata: A.M. Clark in Clark & Davies, 1966: 609, non *Nardoa variolata* (Retzius, 1805), possibly = *Nardoa lemonnieri* Koehler, 1910 according to A. M. Clark (1967b).

variolata: Sluiter, 1895, also Domantay & Roxas, 1938: 224, non *Nardoa variolata* (Retzius, 1805), probably = *Nardoa novaecaledoniae* (Perrier, 1875) according to H.L. Clark (1921) and A.M. Clark (1967b) respectively.

NEOFERDINA Livingstone, 1931

Livingstone 1931: 307; Jangoux 1973b: 776.

Type species: *Ferdina cumingi* Gray, 1840.

cancellata (Grube, 1857) Livingstone, 1931: 307. A synonym of *Neoferdina cumingi* (Gray, 1840) according to Jangoux (1973b).

cumingi (Gray, 1840) (with synonym *Scytaster cancellata* Grube, 1857 with subspecies *tylota* Fisher, 1925 [?still distinct])

Gray 1840: 283; H.L. Clark 1921: 59 (as *Ferdina*).

Livingstone 1931: 307; Fisher 1935: 74; Jangoux 1973b: 788; Marsh 1974: 89; Liao 1980: 160; Jangoux 1986a: 132 (as *Neoferdina*).

Range: Type locality 'W coast of Columbia' but more likely Philippine Is according to H.L. Clark (1921), Christmas Id (Indian Ocean) E to Queensland, New Caledonia and SE Polynesia, N to S China Sea, 5-30 m; *tylota* [?subspecies] from Wake Id.

glyptodisca (Fisher, 1913)

Fisher 1913c: 213; 1919a: 370 (as *Ferdina*).

Livingstone 1931: 307; Jangoux 1973b: 783.

Range: Sulawesi (Celebes), 44 m.

*insolita Livingstone, 1936

Livingstone 1936: 384.

Range: Papua-New Guinea.

Status doubtful according to Jangoux (1973b).

intermedia (Djakonov, 1930) Livingstone, 1936: 385. A synonym of *Neoferdina offreti* (Koehler, 1910) according to Jangoux (1973b).

*kuhli (Müller & Troschel, 1842)

Müller & Troschel 1842: 36 (as *Scytaster*).

Martens 1866: 85 (as *Linckia (Scytaster)*).

Sladen 1889: 780 (as *Ferdina*).

Livingstone 1931: 307 (as *Neoferdina*).

Range: Java.

Status doubtful according to Jangoux (1973b). Type untraced by Jangoux & de Ridder (1987).

mahei Jangoux 1973a: 784. A synonym of *Neoferdina offreti* (Koehler, 1910) according to Jangoux & Aziz (1984) but only possibly so according to Jangoux & Massin (1986).

ocellata (H.L. Clark, 1921) Livingstone, 1931: 307; also A.H. Clark 1952: 285; Endean 1953: 54. A synonym of *Neoferdina cumingi* (Gray, 1840) according to Jangoux (1973b).

offreti (Koehler, 1910) (with synonyms *Ferdina intermedia* Djakonov, 1930 and *Neoferdina mahei* Jangoux, 1973)

Koehler 1910a: 143 (as *Ferdina*).

Livingstone 1931: 307; A.H. Clark 1954: 255; Jangoux 1973b: 778; Marsh 1977: 263; Jangoux & Aziz 1984: 867 (as *Neoferdina*).

Range: Andaman Is*, W to Seychelles, E to Caroline Is and New Caledonia, 0-62 m.

tylota Fisher, 1925. A subspecies of *Neoferdina cancellata* (Grube, 1857) so presumably also conspecific with *N. cumingi* (Gray, 1840).

ONERIA Rowe, 1981

Rowe 1981: 89.

Type species: *Oneria tasmanensis* Rowe, 1981.

tasmanensis Rowe, 1981

Rowe 1981: 91.

Range: Lord Howe Id, E Australia, c.100-180 m.

OPHIDIASTER L. Agassiz, 1836 (with synonym *Chione* Gistel, 1848)

L. Agassiz 1836: 191; Lütken 1859: 79; Perrier 1875: 384 [1876: 259]; Verrill 1915: 89; H.L. Clark 1921: 76; Downey 1973: 68.

Type species: *Asterias ophidiana* Lamarck, 1816.

aequalis: Dujardin & Hupé, 1862, see *Mediaster* (Goniasteridae).

agassizi Perrier, 1881

Perrier 1881: 10; H.L. Clark 1921: 83; Lieberkind 1924: 387; Codoceo 1977: 94.

Range: Juan Fernandez group, SE Pacific, 0-36 m.

alexandri Verrill, 1915 (with synonym *Ophidiaster pinguis* H.L. Clark, 1941)

Verrill 1915: 91; H.L. Clark 1941: 50; Tommasi 1970: 10; Downey 1973: 61.

Range: Georgia, SE U.S.A.* to S Brazil.

Thought to be a synonym of *Hacelia* [i.e. *Tamaria*] *floridae* (Perrier, 1881) by A.H. Clark (1948) but a valid species according to Downey (1973).

*arenatus (Lamarck, 1816)

Lamarck 1816: 566 (as *Asterias*).

Dujardin & Hupé 1862: 365 (as *Ophidiaster*).

Range: Unknown.

Unidentifiable according to H.L. Clark (1921).

armatus Koehler, 1910

Koehler 1910b: 277; 1910a: 148; H.L. Clark 1938: 137; A.M. Clark 1982b: 490; Aziz & Jangoux 1984: 135.

Range: Aru Is, Indonesia*, N to Hong Kong, S to Queensland and W to the Andaman Is, 5-33 m.

asperulus Lütken, 1871: 274. A synonym of *Dactylosaster cylindricus* (Lamarck, 1816) according to H.L. Clark (1921).

astridae Engel, 1938a: 12. A probable synonym of *Ophidiaster helicostichus* Sladen, 1889 according to A.M. Clark (1967b), confirmed by Engel (pers.

comm.) and Jangoux & Massin (1986).

attenuatus Gray, 1840 (as *Ophidiaster (Hacelia))*, see *Hacelia*

attenuatus Perrier, 1869: 60 [252]. (Locality Zanzibar). 'Quite unknown, abandoned by its author without a word of explanation, possibly a *Nardoa*' according to H. L. Clark (1921). [Invalid as a junior primary homonym of *O. attenuatus* Gray, 1840.]

aurantius Gray, 1840: 284. A synonym of *Ophidiaster ophidianus* (Lamarck, 1816) according to Müller & Troschel (1842).

bayeri A. H. Clark 1948

A. H. Clark 1948: 58.

Range: Florida keys.

*bicolor (Lamarck, 1816)

Lamarck 1816: 566 (as *Asterias)*.

Dujardin & Hupé 1862: 364 (as *Ophidiaster*, but citing Müller & Troschel, 1842: 33, not Lamarck).

Unidentifiable according to H.L. Clark (1921).

campbelli Perrier in Filhol 1885b: 573, a NOMEN NUDUM, undescribed.

canariensis Greeff, 1872: 104. A synonym of *Ophidiaster ophidianus* (Lamarck, 1816) according to Ludwig (1897).

*chinensis Perrier, 1875

Perrier 1875: 123 [387]; H.L. Clark 1921: 80.

Range: 'Canton'.

'Probably valid' according to H.L. Clark (1921) but also implies resemblance to *Ophidiaster ophidianus* and *O. guildingi* from Atlantic!

clathratus Grube, 1865: 51. A synonym of *Linckia laevigata* (Linnaeus, 1758) according to H.L. Clark (1921).

colombiae: Müller & Troschel, 1842: 33, see *Linckia*

columbiae: Müller & Troschel, 1842: 34, see *Cistina*

confertus H.L. Clark, 1916

H.L. Clark 1916: 53; Livingstone 1930: 19; H.L. Clark 1938: 138; McKnight 1967b: 321.

Range: Lord Howe Island, Queensland, N.S.W. and Norfolk Id.

crassa (Gray, 1840) Dujardin & Hupé, 1862: 365. A synonym of *Linckia laevigata* (Linnaeus, 1758) according to H.L. Clark (1921).

cribrarius Lütken, 1871 (with synonym *Ophidiaster germani* Perrier, 1875)

Lütken 1871: 277; H.L. Clark 1921: 84; A.M.C. in Clark & Rowe 1971: 61; [?]Rho & Kim 1966: 285.

Range: Tonga*, New Caledonia, [?]S Korea.

cylindricus: Müller & Troschel, 1842, see *Dactylosaster*

diplax Müller & Troschel, 1842: 30. A synonym of *Linckia guildingi* Gray, 1840 according to Fisher (1919a).

dubiosus: Fisher, 1919, see *Tamaria*

duncani de Loriol, 1885 (with synonym *Ophidiaster lioderma* H.L. Clark, 1921)

de Loriol 1885: 15; A.M.C. in Clark & Rowe 1971: 60; Jangoux 1985b: 21 (neotype).

Range: Mauritius.

easterensis Ziesenhenne, 1964

Ziesenhenne 1964: 461.

Range: Easter Id, Pacific.

echinulatus Müller & Troschel, 1842: 32. A synonym of *Mithrodia clavigera* (Lamarck, 1816) (Mithrodiidae) according to H.L. Clark (1921).

ehrenbergi Müller & Troschel, 1842: 31. A synonym of *Linckia guildingi* Gray, 1840 according to H.L. Clark (1921).

flaccidus Lütken, 1859: 86. A synonym of *Ophidiaster guildingi* Gray, 1840 according to Lütken (1871) but of *Linckia guildingi* Gray, 1840 according to H. L. Clark (1921) [?lapsus].

floridae Perrier, 1881, see *Tamaria*

fuscus: Müller & Troschel, 1842, see *Tamaria*

germani Perrier, 1875: 130 [394]. A synonym of *Ophidiaster cribrarius* (Lütken, 1871) according to H.L. Clark (1921).

gracilis: Müller & Troschel, 1842, see under *Dactylosaster*

granifer Lütken, 1871 (with synonym *Ophidiaster trychnus* Fisher, 1913)

Lütken 1871: 276; H.L. Clark 1921: 81; Hayashi 1939: 437; H.L. Clark 1946: 121; A.H. Clark 1949a: 76; 1952: 286; Endean 1956: 125; Marsh 1978: 222; Liao 1980: 156.

Range: Indonesia N to Paracel (Xisha) Is, S China Sea, E and S to Tonga, New Caledonia and northern Australia.

Thought to be a synonym of *Ophidiaster pusillus* Müller & Troschel, 1842 by Döderlein (1926) but a valid species according to H.L. Clark (1946).

guildingi Gray, 1840 (with synonym *Scytaster muelleri* Duchassaing, 1850 and probably *Ophidiaster flaccidus* Lütken, 1859)

Gray 1840: 284; Verrill 1915: 90; H.L. Clark 1933: 23; Fisher 1940: 269; Ummels 1963: 86; Tommasi 1970: 10; Downey 1973: 68; Pawson 1978: 10.

Range: Lesser Antilles*, N to Bermuda and S to southern Brazil (c.31.5°S), also Ascension Id, 0-330 m.

guildingi: Müller & Troschel, 1842: 33, non *Ophidiaster guildingi* Gray, 1840, = *Linckia guildingi* Gray, 1840.

helicostichus Sladen, 1889 (with synonym *Ophidiaster astridae* Engel, 1938 and variety *inarmata* Koehler, 1895)

Sladen 1889: 405; Koehler 1895a: 400 (variety *inarmata*); A.M. Clark 1967b: 195; A.M.C. in Clark & Rowe 1971: 59, 60; Jangoux 1984: 281; Jangoux & Massin 1986: 90.

H.L. Clark 1909: 111; 1921: 86; 1938: 139; 1946: 122 (as *Hacelia*).

Range: Torres Strait*, northern Australia, Indonesia (Enöe) and New Caledonia.

hemprichi Müller & Troschel, 1842 (with synonyms *Linckia pustulata* Martens, 1866, *Ophidiaster purpureus* Perrier, 1869, probably also *O. squameus* Fisher, 1906).

Müller & Troschel 1842: 29; Döderlein 1926: 14; A.M.C. in Clark & Rowe 1971: 61; Julka & Sumita Das 1978: 348; Sloan et al., 1979: 98; Liao 1980: 157.

Range: Red Sea* S to East Africa and Mauritius, E to Maldive Is and tropical W Pacific except Hawaiian Is.

Thought to be a synonym of *Dactylosaster cylindricus* (Lamarck, 1816) by H. L. Clark (1921) but a valid species according to Döderlein (1926).

hirsutus Koehler, 1910, see *Tamaria*

inarmata Koehler, 1895a: 400. A variety of *Ophidiaster helicostichus* Sladen, 1889.

irregularis Perrier, 1869. A probable synonym of *Linckia pacifica* var. *diplax* (Müller & Troschel, 1842) [i.e. *Linckia guildingi* Gray, 1840] according to Sladen (1889).

kermadecensis Benham, 1911

Benham 1911: 148; Mortensen 1925a: 294; Farquhar 1927: 239; McKnight 1968: 510.   ·

Range: Kermadec Is, N of New Zealand.

laevigatus: Müller & Troschel, 1842, see *Linckia*

leachi Gray, 1840, see *Leiaster*

leachi: Perrier, 1878: 80, non *Leiaster leachi* (Gray, 1840), = *Leiaster coriaceus* (Peters, 1852) according to Sladen (1889).

lessonae Gasco, 1876: 8. A synonym of *Hacelia attenuata* (Gray, 1840) according to Ludwig (1897).

linearis Perrier, 1869: 254. A synonym of *Metrodira subulata* Gray, 1840 (Echinasteridae) according to H.L. Clark (1921).

lioderma H.L. Clark, 1921: 80. A synonym of *Ophidiaster duncani* de Loriol, 1885 according to Jangoux (1985).

lorioli Fisher, 1906

Fisher 1906a: 1077; H.L. Clark 1921: 84; ?Hayashi 1938c: 281; A.M.C. in Clark & Rowe 1971: 36, 61; Marsh 1977: 269; 1978: 78.

Range: Hawaiian Is*, Samoa, SE Polynesia, ?W Japan.

lorioli: A.H. Clark, 1952: 287, non *Ophidiaster lorioli* Fisher, 1906, = *Ophidiaster robillardi* de Loriol, 1885 according to Marsh (1977).

ludwigi de Loriol, 1900

de Loriol 1900: 78; H.L. Clark 1910: 336.

Range: 'Perou' (but doubted by H.L. Clark (1910)).

lymani: Nataf & Cherbonnier, 1975, see under *Copidaster* but misidentified according to Downey (pers. comm.).

macknighti H.E.S. Clark, 1962

H.E.S. Clark 1962b: 2.

Range: NE New Zealand, 153-204 m.

marmoratus Michelin, 1844, see *Tamaria*

miliaris Müller & Troschel, 1842: 30. A synonym of *Linckia laevigata* (Linnaeus, 1758) according to H.L. Clark (1921).

multiforis: Müller & Troschel, 1842, see *Linckia*

ophidianus (Lamarck, 1816) (with synonyms *Ophidiaster aurantius* Gray, 1840 and *O. canariensis* Greeff, 1872)

Lamarck 1816: 567 (as *Asterias*).

L. Agassiz 1836: 191; d'Orbigny, 1839: 148; Perrier 1894: 330; Ludwig 1897: 300; Nobre 1930: 47; Madsen 1950: 218; Alvarado & Alvarez 1964: 1063 (as *Ophidiaster*).

Range: W Mediterranean, W to Azores, S to Gulf of Guinea and St Helena, 0-100 m.

ornatus Koehler, 1910, see *Tamaria*

ornithopus Müller & Troschel, 1842: 31; also Lütken 1859: 80. A synonym of *Linckia guildingi* Gray, 1840 according to Sladen (1889).

pacifica (Gray, 1840) Müller & Troschel, 1842: 33. A synonym of *Linckia guildingi* Gray, 1840 according to H.L. Clark (1921).

*perplexus A.H. Clark, 1954

A.H. Clark 1954: 256; A.M.C. in Clark & Rowe 1971: 61.

Range: S Polynesia.

Probably a *Linckia* according to A.M.C. (1971).

perrieri de Loriol, 1885

de Loriol 1885: 17; A.M.C. in Clark & Rowe 1971: 60.

Range: Mauritius* and Zanzibar.

pinguis H.L. Clark, 1941: 51. A synonym of *Ophidiaster alexandri* Verrill, 1915 according to Downey in Miller (1984).

porosissimus Lütken, 1859: 33, 87. A synonym of *Pharia pyramidata* (Gray, 1840) according to Sladen (1889).

propinquus Livingstone, 1932a: 255. A synonym of *Linckia laevigata* (Linnaeus, 1758) according to Rowe & Pawson (1977).

purpureus Perrier, 1869: 61 [253]; 1875: 127 [391]. A synonym of *Ophidiaster hemprichi* Müller & Troschel, 1842 according to Döderlein (1926) and A.M.C. in Clark & Rowe (1971), though meantime treated as valid by Macan (1938).

pusillus Müller & Troschel, 1844, see *Tamaria*

pustulatus (Martens, 1866) Döderlein, 1896: 317. A synonym of *Ophidiaster hemprichi* Müller & Troschel, 1842 according to Döderlein (1926) and A.M.C. in Clark & Rowe (1971), though meantime treated as valid by Hayashi (1938b) and A.H. Clark (1949a).

pyramidatus Gray, 1840, see *Pharia*

reyssi Sibuet, 1977

Sibuet 1977: 1085.

Range: Azores, 350 m.

rhabdotus Fisher, 1906

Fisher 1906a: 1082.

Range: Hawaiian Is, c.426 m.

robillardi de Loriol, 1885

de Loriol 1885: 24; Marsh 1977: 266; Jangoux 1985b: 29 (lectotype).

Range: Mauritius*, NW Australia, Micronesia.

schismochilus H.L. Clark, 1922, see *Copidaster*

sclerodermus Fisher, 1906, see *Tamaria*

*squameus Fisher, 1906

Fisher 1906a: 1079; H.L. Clark 1921: 83; Domantay & Roxas 1938: 223; Ely 1942: 21; A.H. Clark 1952: 287.

Range: Hawaiian Is*, [?]Torres Strait, Marshall and Philippine Is.

A probable synonym of *Ophidiaster hemprichi* Müller & Troschel, 1842 according to A.M.C. in Clark & Rowe (1971).

superba: Nataf & Cherbonnier, 1975: 820, non *Hacelia superba* H.L. Clark, 1921, probably = *H. attenuata* (Gray, 1840) according to Miller (1984).

suturalis Müller & Troschel, 1842: 30. A synonym of *Phataria unifascialis* (Gray, 1840) according to Verrill (1870).

tenellus Fisher, 1906, see *Tamaria*

triseriatus Fisher, 1906, see *Tamaria*

trychnus Fisher, 1913c: 215; 1919a: 388. A synonym of *Ophidiaster granifer* Lütken, 1871 according to H.L. Clark (1921).

tuberculatus: Müller & Troschel, 1842, see *Nardoa*

tuberifer Sladen, 1889: 404; also Koehler 1910a: 148. A synonym of *Tamaria megaloplax* (Bell, 1884) according to H.L. Clark (1946).

tumescens Koehler, 1910b, see *Tamaria*

vestitus Perrier, 1869, see *Echinaster* (Echinasteridae)

watsoni Livingstone, 1936, see *Gomophia*

OPHIDIASTER (HACELIA) Gray, 1840, raised to generic rank by Ludwig (1897)

OPHIDIASTER (LEIASTER) Peters, 1852, raised to generic rank, or at least treated as a genus by Martens (1866), positively by Fisher (1919a).

OPHIDIASTER (PHARIA) Gray, 1840, raised to generic rank by Sladen (1889)

pyramidatus Gray, 1840, see *Pharia*

PARAFERDINA James, 1976

James 1976: 556.

Type species: *Paraferdina laccadivensis* James, 1976.

laccadivensis James, 1976

James 1976: 556.

Range: Laccadive Is, 2 m.

PHARIA Gray, 1840

Gray 1840: 284 (as subgenus of *Ophidiaster*).

Sladen 1889: 398; H.L. Clark 1921: 75 (as genus).

Type species: *Ophidiaster (Pharia) pyramidatus* Gray, 1840.

pyramidatus (Gray, 1840) (with synonym *Ophidiaster porosissimus* Lütken, 1859)

Gray 1840: 284 (as *Ophidiaster (Pharia)*).

Verrill 1870: 287 (as *Ophidiaster*).

Sladen 1889: 784; H.L. Clark 1910: 335; 1920: 86; Ziesenhenne 1937: 217; H. L. Clark 1940: 334; Caso 1961: 74; 1979: 206 (as *Pharia*).

Range: W Colombia*, Peru, Galapagos Is, N to Gulf of California, 0-18 m.

PHATARIA Gray, 1840

Gray 1840: 285 (as subgenus of *Linckia*).

Sladen 1889: 398; H.L. Clark 1921: 68 (as genus).

Type species: *Linckia (Phataria) unifascialis* Gray, 1840.

bifascialis (Gray 1840) Monks, 1904 (as variety of *Phataria unifascialis*) and

fascialis Monks, 1904: 351. Synonyms of *Linckia columbiae* Gray, 1840 according to Fisher (1911c).

mionactis Ziesenhenne, 1942

Ziesenhenne 1942: 206.

Range: Equador, 82-91 m.

unifascialis (Gray, 1840) (with synonym *Ophidiaster suturalis* Müller & Troschel, 1842)

Gray 1840: 285 (as *Linckia (Phataria)*).

Verrill 1870: 285 (as *Linckia*).

Sladen 1889: 786; H.L. Clark 1910: 334; Ziesenhenne 1937: 217; H.L. Clark 1940: 334; Caso 1961: 77; 1979: 206 (as *Phataria*).

Range: W Colombia*, Peru, Galapagos Is, N to Gulf of California, 0-18 m.

PLENARDOA H.L. Clark, 1921

H.L. Clark 1921: 57.

Type species: *Linckia semiseriata* Martens, 1866.

semiseriata (Martens, 1866)

Martens 1866: 355 (as *Linckia*).

Sladen 1889: 412 (as *Nardoa*).

H.L. Clark 1921: 57 (as *Plenardoa*).

Range: S China Sea, 73 m.

PSEUDOLINCKIA in combination with

rhysa H.L. Clark, 1916: 5, lapsi for *Pseudophidiaster* and *P. rhysus* H. L. Clark (1916).

PSEUDOPHIDIASTER H.L. Clark, 1916

  H.L. Clark 1916: 54.

  Type species: *Pseudophidiaster rhysus* H.L. Clark, 1916.

rhysus H.L. Clark, 1916

  H.L. Clark 1916: 55; 1921: 94.

  Range: Tasmania and South Australia, 110-366 m.

SCYTASTER Müller & Troschel, 1842

  Müller & Troschel, 1842: 34.

  Type species: None designated but *Asterias variolata* Retzius, 1805 implied by Verrill (1870) and accordingly given as a synonym of *Nardoa* Gray, 1840 by Verrill.

aegyptiaca: Perrier, 1875, lapsus [or alternative] for *egyptiaca* Gray, see *Gomophia*

canariensis: Dujardin & Hupé, 1862, see *Narcissia*

cancellatus Grube, 1857: 340. A synonym of *Neoferdina cumingi* (Gray, 1840) according to Jangoux (1973b).

crucellatis Grube (MS) in Sladen, 1889, A NOMEN NUDUM, = *Fromia monilis* (Perrier, 1869) according to Sladen (1889).

erythraea: Müller & Troschel, 1842: 37, see under *Acalia*

galatheae Lütken, 1864, see *Nardoa*

gomophia Perrier, 1875, see *Nardoa*

indicus Perrier, 1869, see *Fromia*

intermedia: Müller & Troschel, 1842: 37, see under *Acalia*

kuhli Müller & Troschel, 1842, see *Neoferdina*

milleporellus: Müller & Troschel, 1842, non *Asterias milleporella* Lamarck, 1816,

  = *Fromia ghardaqana* Mortensen, 1938 according to A.M. Clark (1952c) [but might alternatively be *F. monilis* (Perrier, 1869), subsequently identified in the Red Sea].

milleporellus: Michelin, 1845, see *Fromia*

monilis Perrier, 1869, see *Fromia*

muelleri Duchassaing, 1850: 4. A synonym of *Ophidiaster guildingi* Gray, 1840 according to Sladen (1889).

novaecaledoniae Perrier, 1875, see *Nardoa*

obtusus Perrier, 1875: 169 [435]. A synonym of *Nardoa tuberculata* Gray, 1840 according to A.M. Clark (1967b).

pistorius Müller & Troschel, 1842: 35. A synonym of *Fromia milleporella* (Lamarck, 1816) according to Sladen (1889).

pulchella: Müller & Troschel, 1842: 37, see under *Acalia*

semiregularis Müller & Troschel, 1842, see *Nardoa*

stella Duchassaing, 1850: 4. A synonym of *Linckia guildingi* Gray, 1840 according to Perrier (1875).

subulatus: Müller & Troschel, 1842, see *Metrodira* (Echinasteridae) *tuberculatus:* Perrier, 1875, see *Nardoa*

variolatus: Müller & Troschel, 1842, see *Nardoa*

zodiacalis Müller & Troschel, 1842: 35. A synonym of *Scytaster aegyptiacus* [i.e. *Gomophia egyptiaca*] (Gray, 1840) according to Perrier, 1875.

SINOFERDINA Liao, 1982

    Liao 1982: 93.

    Type species: *Sinoferdina gigantea* Liao, 1982.

gigantea Liao, 1982

    Liao 1982: 93.

    Range: E China Sea, 131-162 m.

TAMARIA Gray, 1840

    Gray 1840: 283; H.L. Clark 1921: 88; Livingstone 1932c: 368; H.L. Clark 1946: 123; Downey 1971b: 43; 1973: 64.

    Type species: *Tamaria fusca* Gray, 1840.

ajax Livingstone, 1932c: 371. A synonym of *Tamaria tumescens* (Koehler, 1910) according to A.M. Clark (1967b).

dubiosa (Koehler, 1910)

    Koehler 1910a: 155 (as *Linckia*).

    ?Fisher 1919a: 394 (as *Ophidiaster*).

    H.L. Clark 1921: 94; Jangoux 1981a: 475 (as *Tamaria*).

    Range: Andaman Is* and Philippines, 129-685 m.

floridae (Perrier, 1881)

    Perrier 1881: 9; 1884: 221 (as *Ophidiaster*).

    H.L. Clark 1921: 91; Downey 1971b: 44; 1973: 64 (as *Tamaria*).

    A.H. Clark 1948: 60; 1954: 376 (as *Hacelia*).

    Range: Florida Strait* N to Georgia, 250-600 m.

fusca Gray, 1840

    Gray 1840: 283; H.L. Clark 1921: 89; Livingstone 1932a: 257; Jangoux 1984: 281.

    Müller & Troschel 1842: 34; Fisher 1919a: 388 (as *Ophidiaster*).

    Range: Philippine Is*, Queensland and New Caledonia, 25-45 m.

halperni Downey, 1971

    Downey 1971b: 46; 1973: 65.

    Range: E of Cuba* to Florida Strait, 180-510 m.

hirsuta (Koehler, 1910)

    Koehler 1910a: 149 (as *Ophidiaster*).

    Livingstone 1932a: 260; 1936: 387; A.M.C. in Clark & Rowe 1971: 58 (as *Tamaria*).

    Range: Andaman Is* and NW Australia.

lithosora H.L. Clark, 1921

    H.L. Clark 1921: 90; A.M.C. in Clark & Rowe 1971: 36, 59; Sloan et al. 1979: 98.

    Range: Zanzibar* and Seychelles, 98m.

marmorata (Michelin, 1844)

Michelin 1844: 173; 1845: 21 (as *Ophidiaster*).

Martens in von der Decken 1869: 130 (as *Linckia*).

H. L. Clark 1921: 92; A. M. C. in Clark & Rowe 1971: 58.

Range: Mauritius* and E Africa.

Thought to be a synonym of *Tamaria pusilla* (Müller & Troschel, 1844) by Engel (1938a) but treated as valid by A. M. C. (1971). [Engel's material probably misidentified.]

megaloplax (Bell, 1884) (with synonym *Ophidiaster tuberifer* Sladen, 1889)

Bell 1884b: 126 (as *Linckia*).

Livingstone 1932a: 259; 1932c: 369; H. L. Clark 1946: 124; A. M. C. in Clark & Rowe 1971: 58; Julka & Sumita Das 1978: 348 (as *Tamaria*).

Range: Queensland, NW Australia and Andaman and Nicobar Is, 15-95 m.

obstipa Ziesenhenne, 1942

Ziesenhenne 1942: 208.

Range: pacific coast of Costa Rica, Cocos and Galapagos Is, 85-100 m.

ornata (Koehler, 1910)

Koehler 1910a: 151 (as *Ophidiaster*).

Livingstone 1932a: 260; A. M. C. in Clark & Rowe 1971: 58.

Range: S Sri Lanka*, ?Maldive Is, NW Australia, 56 m.

passiflora Downey 1971

Downey 1971b: 49; 1973: 65.

Range: Bahamas*, Florida Strait, 198-278 m.

propetumescens Livingstone, 1932c: 369. A synonym of *Tamaria tumescens* (Koehler, 1910) according to H. L. Clark (1938).

pusilla (Müller & Troschel, 1844)

Müller & Troschel 1844: 180 (as *Ophidiaster*).

H. L. Clark 1921: 92; Döderlein 1926: 16; A. M. C. in Clark & Rowe 1971: 58; Jangoux in Guille & Jangoux 1978: 57 (as *Tamaria*).

Range: Philippine Is, Indonesia and New Caledonia.

scleroderma (Fisher, 1906)

Fisher 1906a: 1081 (as *Ophidiaster*).

H. L. Clark 1921: 92 (as *Tamaria*).

Range: Hawaiian Is, 180-194 m.

stria Downey, 1975

Downey 1975: 87.

Range: Malpelo Id, W of Colombia, 36-49 m.

tenella (Fisher, 1906)

Fisher 1906a: 1082 (as *Ophidiaster*).

H. L. Clark 1921: 91 (as *Tamaria*).

Range: Hawaiian Is, 240-285 m.

triseriata (Fisher, 1906)
  Fisher 1906a: 1080 (as *Ophidiaster*).
  H.L. Clark 1921: 94 (as *Tamaria*).
  Range: Hawaiian Is, 125-165 m.
tuberifera (Sladen, 1889) H.L. Clark, 1921: 90; also 1938: 139. A synonym
  of *Tamaria megaloplax* (Bell, 1884) according to H.L. Clark (1946).
tumescens (Koehler, 1910) (with synonyms *Tamaria propetumescens* and *T.
  ajax* both of Livingstone, 1932)
  Koehler 1910b: 277 (as *Ophidiaster*).
  H.L. Clark 1921: 94; 1938: 141; 1946: 123; A.M. Clark 1967b: 194.
  Range: Aru Is, Indonesia*, Banda Sea and N Australia, 18-28 m.
UNDINA Gistel, 1848 [?]
  Gistel 1848: [p.?].
  Fisher 1919a: 400 (footnote).
  See comments under *Catantes*. Should be rejected.

Family LEILASTERIDAE Jangoux & Aziz

  Leilasteridae Jangoux & Aziz 1988: 646.

LEILASTER A.H. Clark, 1938
  A.H. Clark 1938: 1; Fisher 1940: 152; Downey 1973: 80; A.M. Clark
  1983: 361.
  Type species: *Korethraster radians* Perrier, 1881.
  Not a ganeriid, family uncertain according to A.M. Clark (1983).
radians (Perrier, 1881)
  Perrier 1881: 12; Sladen 1889: 459 (as *Korethraster*).
  Perrier 1884: 167, 169 (as *Lophaster*).
  Perrier 1884: 275 (as *Solaster*).
  A.H. Clark 1938: 2; Downey 1973: 80; A.M. Clark 1983: 361 (as
  Leilaster).
  Range: Barbados* E to NW Brazil and W to Puerto Rico, Cuba and the
  Yucatan Channel (unpublished, identified by A.M.C.), 102-274 m.
spinulosus Aziz & Jangoux, 1985
  Aziz & Jangoux 1985: 287 (as subspecies of *L. radians*).
  Jangoux & Aziz 1988: 645 (as species).
  Range: Philippine Is, Réunion Id, 137-192 (?220) m.
MIRASTRELLA Fisher, 1940
  Fisher 1940: 152; A.M. Clark 1983: 362.
  Type species: *Mirastrella biradialis* Fisher, 1940.
  Not an asterinid, family uncertain according to A.M. Clark (1983).
biradialis Fisher, 1940
  Fisher 1940: 155.
  Range: Clarence Id and Shag Rocks, Weddell Quadrant, 177-342 m.

REFERENCES (supplementary to those for part 1)

Opinion 707. 1964. *Asterias nodosa* Linnaeus 1758 (Asteroidea): added to the Official List of Specific Names. *Bull. zool. Nom.* 21: 206-207.

Achituv, Y. 1969. Studies on the reproduction and distribution of *Asterina burtoni* and *A. wega* in the Red Sea and eastern Mediterranean. *Israel J. Zool.* 18: 329-342.

Agassiz, L. 1836a. Notice sur les fossiles du terrain Cretacé du Jura Neuchatelois. *Mém. Soc. Sci. nat., Neuchatel* 1: 126-145. [Includes citation of type species for *Goniaster*].

Ayres, W.O. 1851. Remarks on new asteroids. *Proc. Boston Soc. nat. Hist.* 4: 1851-1854.

Aziz, A. & M. Jangoux 1984b. Descriptions de quatre nouvelles espèces d'astérides profonds (Echinodermata) de la région Indo-Malaise. *Indo-Malayan Zool.* 2: 187-194.

Aziz, A. & M. Jangoux 1984c. Note on the statute of the goniasterid genera *Aphroditaster* Sladen, 1889 and *Fisheraster* Halpern, 1970 (Echinodermata: Asteroidea). *Indo-Malayan Zool.* 2: 255-256.

Baker, A.N. & L.M. Marsh 1976. The rediscovery of *Halityle regularis* Fisher (Echinodermata: Asteroidea). *Rec. W. Aust. Mus.* 4(2): 107-116.

Bell, F.J. 1883. Descriptions of two new species of Asteroidea in the collection of the British Museum. *Ann. Mag. nat. Hist.* 12: 333-335.

Bell, F.J. 1884b. Contributions to the systematic arrangement of the Asteroidea. 2. The species of *Oreaster*. *Proc. zool. Soc. Lond.* 1884: 57-87.

Bell, F.J. 1887. Report on a collection of Echinodermata from the Andaman Islands. *Proc. zool. Soc. Lond.* 1887: 139-145.

Bell, F.J. 1888. Report on a collection of echinoderms made at Tuticorin, Madras by Mr Edgar Thurston. *Proc. zool. Soc. Lond.* 1888: 383-389.

Bell, F.J. 1891a. Stray notes on the nomenclature etc. of some british starfishes. *Ann. Mag. nat. Hist.* 7: 233-235.

Bell, F.J. 1893b. On the names or existence of three exotic starfishes. *Ann. Mag. nat. Hist.* 12: 25-29.

Bell, F.J. 1893c. On *Odontaster* and the allied or synonymous genera of asteroid echinoderms. *Proc. zool. Soc. Lond.* 1893: 259-262.

Belyaev, G.M. 1974. [A new family of abyssal starfishes.] *Zool. Zh.* 53(10): 1502-1508. [English summary].

Belyaev, G.M. & N.M. Litvinova 1977. The second finding of deep-sea starfishes of the family Caymanostellidae. *Zool. Zh.* 56(12): 1893-1896.

Bernasconi, I. 1956. Dos nuevos Equinodermos de la costa del Brasil. *Neotropica* 2: 33-36.

Bernasconi, I. 1957. Otra nueva especie de Asteroideo brasileyo. *Neotropica* 3: 33-34.

Bernasconi, I. 1961b. Una nueva especie de asteroideo. *Neotropica* 7(22): 1-2.

Bernasconi, I. 1962a. *Perknaster densus patagonicus* nueva subespecie de la Argentina (Echinoderma). *Physis, B. Aires* 23: 257-258.

Bernasconi, I. 1962b. Asteroideos argentinos. 3. Familia Odontasteridae. *Rev. Mus. argent. Inst. nac. Cienc. nat.* (Zool.) 8: 27-51.

Bernasconi, I. 1963a. *Ceramaster patagonicus fisheri* nueva subespecie de California. *An. Inst. biol. Univ. Mex.* 33: 287-291 [*Physis, B. Aires* 24: 287-291].

Bernasconi, I. 1963b. Asteroideos argentinos. 4. Familia Goniasteridae. *Rev. Mus. argent. Cienc. nat.* (Zool.) 9(1): 1-25.

Bernasconi, I. 1964a. Asteroideos argentinos. 5. Familia Ganeriidae. *Rev. Mus. argent. Cienc. nat.* (Zool.) 9(4): 59-89.

Bernasconi, I. 1965. Nuevo género y nueva especie abisal de Goniasteridae (Echinodermata, Asteroidea). *Physis, B. Aires* 25: 333-335.

Bernasconi, I. 1973b. Asteroideos Argentinos. 6. Familia Asterinidae. *Revta Mus. argent. Cienc. nat. Bernardino Rivadavia Inst. nac. Invest. Cienc. nat.* (Hidrol.) 3(4): 335-346.

Blake, D.B. 1978. The taxonomic position of the modern sea-star *Cistina* Gray, 1840. *Proc. biol. Soc. Wash.* 91(1): 234-241.

Blake, D.B. 1979. The affinities and origins of the crown-of-thorns sea star *Acanthaster* Gervais. *J. nat. Hist.* 13: 303-314.

Blake, D.B. 1980. On the affinities of three small sea-star families. *J. nat. Hist.* 14: 163-182.

Blake, D.B. 1981. A re-assessment of the sea-star orders Valvatida and Spinulosida. *J. nat. Hist.* 15: 375-394.

Blake, D.B. 1987. A classification and phylogeny of post-Palaeozoic sea stars (Asteroidea: Echinodermata). *J. nat. Hist.* 21: 481-528.

Brandt, J.F. 1835. *Prodromus descriptionis animalium ab H. Mertensio observatorum.* Fasc. 1: 1-77. [BM(NH) library copy sep. pag., original ?706-783.] Petropoli.

Caso, M.E. 1941. Contribucion al conocimento de los asteridos de Mexico. 1. La existencia de *Linckia guildingii* en la costa pacifica. *An. Inst. Biol. Mexico* 12: 155-160.

Caso, M.E. 1962. Estudios sobre Asteridos de Mexico. Observaciones sobre especies pacificos del genero *Acanthaster* y descripcion de una subespecie nueva *A. ellisii pseudoplanci. An. Inst. Biol. Univ. Mex.* 32: 313-331.

Caso, M.E. 1972b. Morfologia externa de *Acanthaster planci* (Linnaeus). *An. Inst. Biol. Univ. nac. Aut. Mex.* (Ser. Cienc. mar. Limnol.) 41: 63-78.

Caso, M.E. 1976. Contribucion al estudio de los Asterozoa de Mexico. La familia Mithrodiidae: descripcion de una nueva especie del genero *Mithrodia, Mithrodia enriquecasoi* sp. nov. *An. Centro Cienc. mar y Limnol. Univ. nac. Aut. Mex.* [1975]2: 1-28.

Caso, M.E. 1978. Especies de la familia Asterinidae en la costa pacifico de Mexico. Descripcion de una nueva especie del genero *Asterina: Asterina Agustincasoi* sp. nov. *An. Centro Cienc. mar y Limnol. Univ. nac. Aut. Mex.* [1977]4: 209-232.

Cherbonnier, G. 1970b. Etude critique de l'astérie *Anseropoda lobiancoi* (Ludwig). *Bull. Mus. natn. Hist. nat. Paris* 41 [1969]: 946-951.

Cherbonnier, G. 1970c. *Podosphaeraster thalassae*, nov. sp., espèce actuelle d'Asterie de la famille jurassique des Sphaerasteridae. *C. r. hebd. Seanc. Acad. Sci. Paris* (D)271: 203-206.

Cherbonnier, G. 1974. *Podosphaeraster crassus* nov. sp., espèce actuelle d'Asterie de la famille jurassique des Sphaerasteridae. *C. r. hebd. Seanc. Acad. Sci. Paris* (D)278: 1731-1733.

Cherbonnier, G. 1975. Sur la présence, à Madagascar, de l'astérie *Mithrodia gigas* Mortensen. *Bull. Mus. natn. Hist. nat. Paris* (Zool.) No. 210: 639-645.

Clark, A.H. 1916c. Seven new genera of echinoderms. *J. Wash. Acad. Sci.* 6: 115-122.

Clark, A.H. 1916d. A new starfish *(Lydiaster americanus)* from the Gulf of Mexico. *J. Wash. Acad. Sci* 6: 141-144.

Clark, A.H. 1917a. A new starfish from the magellanic region. *Proc. biol. Soc. Wash.* 30: 7.

Clark, A.H. 1917b. Four new echinoderms from the West Indies. *Proc. biol. Soc. Wash.* 30: 63-70.

Clark, A.H. 1917d. Two new astroradiate echinoderms from the Pacific coast of Colombia, and Equador. *Proc. biol. Soc. Wash.* 30: 171-174.

Clark, A.H. 1938. A new genus of starfishes from Puerto Rico. *Smithson. misc. Colls* 91(29): 1-7.

Clark, A.H. 1939b. A new genus of starfishes from the Aleutian Islands. *Proc. U.S. natn. Mus.* 86: 497-500.

Clark, A.H. 1948b. Two new starfishes and a new brittle-star from Florida and Alabama. *Proc. biol. Soc. Wash.* 61: 55-64.

Clark, A.H. 1949b. Echinoderms from the mid-Atlantic dredged by the 'Atlantis' in the summer of 1948. *J. Wash. Acad. Sci.* 39: 371-377.

Clark, A.M. 1956. A note on some species of the family Asterinidae (class Asteroidea). *Ann. Mag. nat. Hist.* 9: 374-383.

Clark, A.M. 1962b. *Asterias nodosa* Linnaeus, 1758 (Asteroidea); selection of a lectotype and addition to the Official List. *Bull. zool. Nom.* 19: 174-176.

Clark, A.M. 1963. A note on *Patiria ocellifera* Gray, 1847 (Asteroidea). *Doriana* 3(127): 1-9.

Clark, A.M. 1967b. Notes on asteroids in the British Museum (Nat. Hist.). 5. *Nardoa* and some other ophidiasterids. *Bull. Br. Mus. nat. Hist.* (Zool.) 15: 167-198.

Clark, A.M. 1967c. Variable symmetry in fissiparous Asterozoa. In N. Millott (ed.), *Echinoderm Biology.* Symp. zool. Soc. Lond. No. 20: 143-157.

Clark, A.M. 1970. The name of the starfish *Anthenea acuta* (Perrier), preoccupied. *Proc. Linn. Soc. N.S.W.* 95: 157.

Clark, A.M. 1976. Asterozoa from Amsterdam and St Paul Islands, southern Indian Ocean. *Bull. Br. Mus. nat. Hist.* (Zool.) 30: 247-261.

Clark, A.M. 1983. Notes on Atlantic Asteroidea. 3. Families Ganeriidae and Asterinidae. *Bull. Br. Mus. nat. Hist.* (Zool.) 45: 359-380.

Clark, A.M. 1984. Notes on Atlantic and other Asteroidea. 4. Families Poraniidae and Asteropseidae. *Bull. Br. Mus. nat. Hist.* (Zool.) 47: 19-51.

Clark, A.M. & P.S. Davies 1966. Echinoderms of the Maldive Islands. *Ann. Mag. nat. Hist.* [1965] 8: 597-612.

Clark, A.M. & M.E. Downey 1992. *Starfishes of the Atlantic.* London: Natural History Museum & Chapman Hall.

Clark, A.M. & C.W Wright 1962. A new genus and species of recent starfishes belonging to the aberrant family Sphaerasteridae, with notes on the possible origin and affinities of the family. *Ann. Mag. nat. Hist.* 5: 243-251.

Clark, H.E.S. 1962a. *Odinia* and *Ophidiaster* (Asteroidea) in New Zealand. *Zool. Publs Vict. Univ. N.Z.* No. 30: 1-10.

Clark, H.E.S. 1971. Pentopliidae, a new family of Asteroidea from the southern Atlantic Ocean. *Bull. mar. Sci.* 21: 545-551.

Clark, H.E.S. 1972. *Knightaster,* a new genus of asteroid from northern New Zealand. *J. R. Soc. N.Z.* 2(2): 147-150.

Clark, H.E.S. 1982. A new genus and two new species of seastars from north of New Zealand, with notes on *Rosaster* species. *Rec. natn. Mus. N.Z.* 2(5): 35-42.

Codoceo, R.M. & V.H. Andrade 1981. Nuevo asteroideo para Chile: *Criptopeltaster philippi* n. sp. (Goniasteridae, Hippasteriinae). *Rev. Biol. mar.* 17(3): 379-387. [Not seen.]

358

Colman, H.L. 1911. Supplement to the Echinodermata. *Mem. Aust. Mus.* 4: 699-701.

Crump, R.G. & R.H. Emson 1984. Comparative studies on the ecology of *Asterina gibbosa* Pennant and *Asterina phylactica* Emson & Crump at Lough Ine. *J. mar. biol. Ass. U.K.* 64: 35-53.

Dartnall, A.J. 1968. *Asterodiscus truncatus* (Coleman, 1911) – a new record for Tasmanian waters. *Pap. Proc. R. Soc. Tasm.* 102: 23.

Dartnall, A.J. 1969a. New Zealand sea stars in Tasmania. *Pap. Proc. R. Soc. Tasm.* 103: 53-55.

Dartnall, A.J. 1969b. A viviparous species of *Patiriella* (Asteroidea, Asterinidae) from Tasmania. *Proc. Linn. Soc. N.S.W.* 93: 294-296.

Dartnall, A.J. 1970a. The asterinid sea stars of Tasmania. *Pap. Proc. R. Soc. Tasm.* 104: 73-78.

Dartnall, A.J. 1970b. Some species of *Asterina* from Flinders, Victoria. *Victorian Nat.* 87: 19-22.

Dartnall, A.J. 1970d. A new species of *Marginaster* (Asteroidea: Poraniidae) from Tasmania. *Proc. Linn. Soc. N.S.W.* 94: 207-211.

Dartnall, A.J. 1971a. Australian sea stars of the genus *Patiriella* (Asteroidea, Asterinidae). *Proc. Linn. Soc. N.S.W.* 96: 39-49.

Desor, E. 1848. Zoological investigations among the shoals of Nantucket. *Proc. Boston Soc. nat. Hist.* 3: 11, 17, 65-68.

Döderlein, L. 1908. *Asterina lüderitziana*, eine neue Art aus Südwest-Afrika. *Jb. nassau. Ver. Naturk.* 61: 296-298.

Döderlein, L. 1915. Die Arten der Asteroidea-Gattung *Anthenea*. *Jb. Nassau. Ver. Naturk. Wiesbaden* 68: 21-55.

Döderlein, L. 1916. Über die Gattung *Oreaster* und Verwandte. *Zool. Jb.* (Syst. Abt.) 40: 409-440.

Döderlein, L. 1922. Über die Gattung *Calliaster* Gray. *Bijdr. Dierk.* 1922: 47-52.

Döderlein, L. 1924. Die Asteriden der Siboga-Exped. ii. Pentagonasteridae. *Siboga-Exped.* 46(2): 49-69.

Döderlein, L. 1935. Die Asteriden der Siboga-Expedition. iii. Oreasteridae. *Siboga-Exped.* 46(3): 71-110.

Döderlein, L. 1936. Die Asteriden der Siboga-Expedition. III. Die Unterfamilie Oreasterinae. *Siboga-Exped.* 46c: 295-369.

Dollfus, R. 1936. *Stellasteropsis fouadi* n.g., n.sp. stelleride commun dans le Golfe de Suez. *Bull. Soc. zool. Fr.* 61: 151-158.

Dons, C. 1929. Zoologiske Notizen. 7. *Tremaster mirabilis*. *K. norske Vidensk. Selsk. Forh.* 11(27): 98-101.

Dons, C. 1936a. Zoologiske Notizen. 26. *Pseudoporania stormi* n. gen., n. sp. *K. norske Vidensk. Selsk. Forh.* 8: 17-20.

Dons, C. 1938. Zoologiske Notizen. 34. *Hippasteria insignis* n. sp.; 35. *Sphaeraster Berthae* n. gen., n. sp. 36. *Sphaeraster Björlykkei* n. sp. *K. norske Vidensk. Selsk. Forh.* 10: 16-19; 161-164; 165-168.

Dons, C. 1939. Zoologiske Notizen. 37. Die Asteriden Gattung *Sphaeriaster*, nomen novum. *K. norske Vidensk. Selsk. Forh.* 11: 37.

Downey, M.E. 1968. A note on the Atlantic species of the starfish genus *Linckia*. *Proc. biol. Soc. Wash.* 81: 41-44.

Downey, M.E. 1970b. *Drachmaster bullisi*, new genus and species of Ophidiasteridae (Echinodermata: Asteroidea) with a key to the Caribbean species of the family. *Proc. biol. Soc. Wash.* 83: 77-82.

Downey, M.E. 1971b. Two new species of the genus *Tamaria* (Echinodermata: Asteroidea) from the tropical western Atlantic. *Proc. biol. Soc. Wash.* 84: 43-50.

Downey, M.E. 1975. Asteroidea from Malpelo Island, with a description of a new species of the genus *Tamaria. Smithson. Contr. Zool.* No. 176: 86-90.

Downey, M.E. 1980. *Floriaster maya,* new genus and species of the family Goniasteridae (Echinodermata: Asteroidea). *Proc. biol. Soc. Wash.* 93: 346-349.

Downey, M.E. 1981. A new goniasterid seastar, *Evoplosoma scorpio* (Echinodermata: Asteroidea), from the northeastern Atlantic. *Proc. biol. Soc. Wash.* 94: 561-563.

Downey, M.E. 1982. *Evoplosoma virgo,* a new goniasterid starfish (Echinodermata: Asteroidea) from the Gulf of Mexico. *Proc. biol. Soc. Wash.* 95: 772-773.

Ebert, T.A. 1979. Natural history notes on two Indian Ocean starfishes in Seychelles: *Protoreaster lincki* (de Blainville) and *Pentaceraster horridus* (Gray). *J. mar. biol. Ass. India* 18(1): 71-77.

Ellis, J. & D.C. Solander 1786. *The natural history of many curious and uncommon zoophytes.* xii + 208 pp. London.

Emson, R.G. & R.H. Crump 1979. Description of a new species of *Asterina* (Asteroidea), with an account of its ecology. *J. mar. biol. Ass. U.K.* 59: 77-94.

Engel, H. 1942. *Linckia rosenbergi* von Martens, a synonym of *L. laevigata* (L.). *Zool. Meded.* D1. 23: 273-274.

Engel, H., D.D. John & G. Cherbonnier 1948. The genus *Mithrodia* Gray, 1840. *Zool. Verh. Leiden* 2: 1-39.

Farquhar, H. 1894. Notes on New Zealand echinoderms. *Trans. N.Z. Inst.* 27: 194-208.

Farquhar, H. 1897. A contribution to the history of New Zealand echinoderms. *J. Linn. Soc.* 26: 186-198.

Farquhar, H. 1898. Notes on New Zealand starfishes. *Trans. N.Z. Inst.* 30: 187-191.

Farquhar, H. 1907. Notes on New Zealand echinoderms; with description of a new species. *Trans. N.Z. Inst.* 39: 123-130.

Farquhar, H. 1909. Further notes on New Zealand starfishes. *Trans. N.Z. Inst.* 41: 126-129.

Farquhar, H. 1913. Two new echinoderms. *Trans. N.Z. Inst., Wellington* 45: 212-215.

Farquhar, H. 1927. Notes on New Zealand seastars. *N.Z. J. Sci.* 9: 237-240.

Filhol. H. 1885b. Echinodermes. pp. 572-573. In: *Recueil de Mémoires – Passage de Vénus sur le Soleil.* 3(2). Mission Ile Campbell. Inst. Franc. Academie des Sciences. [Asteroid name by Perrier, nom. nud.].

Fisher, W.K. 1911b. *Hyalinothrix,* a new genus of starfishes from the Hawaiian Islands. *Proc. U.S. natn. Mus.* 39: 639-664.

Fisher, W.K. 1911c. Asteroidea of the North Pacific and adjacent waters. 1. Phanerozonia and Spinulosa. *Bull. US natn. Mus.* 76: XIII + 420.

Fisher, W.K. 1916b. New East Indian starfishes. *Proc. Biol. Soc. Wash.* 29: 27-36.

Fisher, W.K. 1917e. A new genus and subgenus of East-Indian sea-stars. *Ann. Mag. nat. Hist.* 20: 172-173.

Fisher, W.K. 1918. Notes on Asteroidea. *Ann. Mag. nat. Hist.* 1: 103-111.

Fisher, W.K. 1922. A new sea-star from Hong Kong. *Ann. Mag. nat. Hist.* 10: 415-418.

Fisher, W.K. 1931. Report on the South American sea stars collected by Waldo L. Schmitt. *Proc. U.S. natn. Mus.* 78: 1-10.

Fisher, W.K. 1935. Note on a starfish from Christmas Island, Indian Ocean. *Bull. Raffles Mus., Singapore* 9: 74.

Fisher, W.K. 1939. A new sea star of the genus *Poraniopsis* from Japan. *Proc. U.S. natn. Mus.* 86: 469-472.

Fisher, W.K. 1941. A new genus of sea stars *(Plazaster)* from Japan, with a note on the genus *Parasterina*. *Proc. U.S. natn. Mus.* 90: 447-456.

Forbes, E. 1843. On a new british starfish of the genus *Goniaster (G. abbensis). Ann. nat. Hist.* 11: 280-281.

Gallo, V.R. 1937. Sur le genre 'Culcitopsis' Verrill (Astéroides). *Compte Rendu XIIe Congrès Internationale Zoologique, Lisbon,* 1935. pp. 1664-1667.

Gasco, F. 1870. Intorno ad una nuova specie di *Asteriscus. Boll. Ass. Med. nat. Napoli* : 86-90. [Not seen.]

Gasco, F. 1876. Descrizione di alcune Echinodermi nuovi o per la prima volta trovato nel Mediterraneo. *Rend. Acad. Sci. Fis. Matem.* 15(2): 9-11 [?38-40].

Gervais, P. 1841. Astérie. *Asterias.* pp. 461-481. In: *Dictionnaire des Sciences Naturelles.* Suppl. 1. Paris. [Dated 1840 but includes taxa of Gray, Dec., 1840.]

Glynn, P.W. 1974. The impact of *Acanthaster* on corals and coral reefs in the Eastern Pacific. *Environmental Conserv.* 1(4): 295-304.

Gray, J.E. In: Johnston 1836. Illustrations in british Zoology. *Mag. nat. Hist.* 8: 147.

Grieg, J.A. 1905. *Goniaster nidarosiensis* Storm og dens synonymer. *Bergens Mus. Aarb.* 1905(3): 1-14.

Grube, A.E. 1865. Mehrere noch unbeschreibene oder doch nicht hinreichend bekannte Seesterne des Breslauer Museums. *Jahrsber. schles. Ges. vaterl. Cultur.* 42: 51-53.

Guzman, A. 1980. *Asteroideos chilenos* de la familia Odontasteridae. *Acta Zool. lilloana* 35(1): 111-117.

Halpern, J.A. 1969a. Biological investigations of the deep sea. 46. The genus *Litonotaster* (Echinodermata, Asteroidea). *Proc. biol. Soc. Wash.* 82: 129-142.

Halpern, J.A. 1969b. Biological investigations of the deep sea. 50. The validity and generic position of *Pentagonaster parvus* Perrier (Echinodermata, Asteroidea). *Proc. biol. Soc. Wash.* 82: 503-506.

Halpern, J.A. 1970. A monographic revision of the goniasterid sea stars of the North Atlantic. 253 pp. Unpublished thesis: University of Miami. 1970.

Halpern, J.A. 1970a. Biological investigations of the deep sea. 53. New species and genera of goniasterid sea stars. *Proc. biol. Soc. Wash.* 83: 1-12.

Halpern, J.A. 1970b. Goniasteridae (Echinodermata, Asteroidea) of the Straits of Florida. *Bull. mar. Sci.* 20: 193-286.

Halpern, J.A. 1972. Pseudarchasterinae (Echinodermata, Asteroidea) of the Atlantic. *Proc. biol. Soc. Wash.* 85: 359-384.

Hartlaub, C. 1893. Ueber die Arten und den Skelettbau von *Culcita. Notes Leyden Mus.* 14: 65-118.

Harvey, R., A.M. Clark et al. 1988. Echinoderms of the Rockall Trough and adjacent areas. 3. Additional records. *Bull. Br. Mus. nat. Hist.* (Zool.) 54(4): 153-198.

Hayashi, R. 1938b. Sea-stars of the Ryukyu Islands. *Bull. biogeograph. Soc. Japan* 8: 197-222.

Hayashi, R. 1939a. Sea-stars of the Caroline Islands. *Palao trop. biol. Stn Stud.* 3: 417-446.

Hayashi, R. 1940. Contributions to the classification of the sea-stars of Japan. 1. Spinulosa. *J. Fac. Sci. Hokkaido imp. Univ.* Zool. 7: 107-204.

Hayashi, R. 1947. The fauna of Akkeshi Bay. 16. Asteroidea. *J. Fac. Sci. Hokkaido imp. Univ. Zool.* 9: 227-280.

Hayashi, R. 1974. A new sea-star from Japan, *Asterina minor* sp. nov. *Proc. jap. Soc. syst. Zool.* No. 10: 41-44.

Hayashi, R. 1977. A new sea star of *Asterina* from Japan, *Asterina pseudoexigua pacifica* n. ssp. *Proc. jap. Soc. syst. Zool.* No. 13: 88-91.

Hernandez, D.A. & A. Tablado '1985' [?1987] Asteroidea de Puerto Deseado, Santa Cruz, Argentina. *Contrçoes Centro Nac. Patagonico Invest. Cienc. Téc.* No. 104: 1-16.

Hotchkiss, F.H.C. & A.M. Clark 1976. Restriction of the family Poraniidae, sensu Spencer & Wright. *Bull. Br. Mus. nat. Hist.* Zool. 30(6): 263-268.

Hupé, M.H. 1857. Zoophytes. pp. 97-100. In: Castelnau. *Voyage dans l'Amerique du Sud.* 3. Paris. [Date 1859 in BM(NH) library catalogue but generally assumed to be 1857.]

Hutton, F.W. 1872. Descriptions of some new starfishes from New Zealand. *Proc. zool. Soc. Lond.* 1872: 810-812.

Hutton, F.W. 1879. Notes on some New Zealand Echinodermata, with descriptions of new species. *Trans. Proc. N.Z. Inst.* 11: 305-308.

James, D.B. 1976. Studies on Indian echinoderms. 5. New and little-known starfishes from the Indian seas. *J. mar. Biol. Ass. India* [1973] 15(2): 556-559.

James, D.B. 1989. Echinoderms of Lakshadweep and their zoogeography. *Bull. cent. mar. Fish. Res. Inst.* 43: 97-144.

Jangoux, M. 1973b. Le genre *Neoferdina* Livingstone (Echinodermata, Asteroidea: Ophidiasteridae). *Rev. Zool. Bot. afr.* 87(4): 775-794.

Jangoux, M. 1980. Le genre *Leiaster* Peters. *Rev. Zool. afr.* 94(1): 87-108.

Jangoux, M. 1981b. Une nouvelle espèce d'astéride bathyale des eaux de Nouvelle-Caledonie (Echinodermata, Asteroidea). *Bull. Mus. natn. Hist. nat. Paris* 3[3]: 709-712.

Jangoux, M. 1982. On *Tremaster* Verrill, 1879, an odd genus of recent starfish (Echinodermata: Asteroidea). In J.M. Lawrence (ed.) *Echinoderms: Proc. int. Conf. Tampa Bay.* 155-163. Rotterdam: Balkema.

Jangoux, M. 1986b. La collection d'Echinodermes de Musée Zoologique de Strassbourg. *Bull. Ass. philomath. Alsace Lorr.* 22: 125-131.

Jangoux, M. & A. Aziz 1988. Les asterides (Echinodermata) récoltés autour de l'île de la Réunion par le N.O. 'Marion-Dufresne' en. 1982. *Bull. Mus. natn. Hist. nat. Paris* (4)10: 631-650.

Jangoux, M. & C. De Ridder 1987. Annotated catalogue of recent echinoderm type specimens in the collection of the Rijksmuseum van Natuurlijke Historie at Leiden. *Zool. Meded.* 61(6): 79-96.

Janssen, H.H. et al. 1984. Some recent findings on the seastar *Archaster typicus* M. et Tro., 1840. *Philipp. Scient.* 21: 51-74.

Keough, M.J. & A.J. Dartnall 1978. A new species of viviparous asterinid asteroid from Eyre Peninsula, South Australia. *Rec. S. Aust. Mus.* 17(28): 407-416.

Koehler, R. & C. Vaney 1906. Mission des pêcheries de la côte occidentale d'Afrique. 2. Echinodermes. *Act. Soc. linn. Bordeaux* 60: 58-66.

Leeling, B. 1985. Remarks on *Odontaster penicillatus*. In B. Keegan & B. O'Connor (eds). *Echinoderms: Proc. Fifth int. Echinoderm Conf. Galway, 1984.* 529-532. Rotterdam: Balkema.

Liao, Y.-L. 1982. A new genus of starfish from the continental shelf of the East China Sea. *Stud. mar. Sinica* 19(19): 93-97.

Liao, Y.-L. 1983. A new species of goniasterid sea-star from the Huanghai [Yellow] Sea, China. *Chin. J. oceanol. limnol.* 1(3): 367-369.

362

Liao, Y.-L. 1984. *Rosaster attenuatus,* a new species of the family Goniasteridae (Asteroidea) from the East China Sea. *Oceanol. Limnol. Sinica* 15(5): 478-481.

Liao, Y.-L. 1989. *Calliaster quadrispinus,* a new species of the family Goniasteridae (Asteroidea) from southern China. *Oceanol. Limnol. Sinica* 20(1): 23-27.

Liao, Y.-L. & A.M. Clark 1989. Two new species of the genus *Anthenoides* (Echinodermata: Asteroidea) from southern China. *Chin. J. oceanol. limnol.* 7(1): 37-42.

Livingstone, A.A. 1930. On some new and little known australian asteroids. *Rec. Aust. Mus.* 18: 15-24.

Livingstone, A.A. 1931a. On a new asteroid from Queensland. *Rec. Aust. Mus.* 18: 135-137.

Livingstone, A.A. 1931b. On the restriction of the genus *Ferdina* Gray. *Aust. Zool.* 6(4): 305-309.

Livingstone, A.A. 1932b. Notes on some representatives of the asteroid genus *Culcita. Aust. Zool.* 7(3): 265-273.

Livingstone, A.A. 1932c. Some further notes on species of *Tamaria. Rec. Aust. Mus.* 18: 368-372.

Livingstone, A.A. 1932d. The australian species of *Tosia. Rec. Aust. Mus.* 18: 373-382.

Livingstone, A.A. 1933. Some genera and species of Asterinidae. *Rec. Aust. Mus.* 19: 1-22.

Livingstone, A.A. 1934. Two new asteroids from Australia. *Rec. Aust. Mus.* 19: 177-180.

Livingstone, A.A. 1936. Descriptions of new Asteroidea from the Pacific. *Rec. Aust. Mus.* 19: 383-397.

Loriol, P.de 1906. Rectification de nomenclature. *Rev. paléozool., Paris* 10: 77-78.

Lucas, J. & M. Jones 1976. Hybrid crown-of-thorns starfish *(Acanthaster planci x A. brevispinus)* reared to maturity in the laboratory. *Nature, Lond.* No. 5576: 409-412.

Ludwig, H. 1912. Uber die J. E. Gray'schen Gattungen *Pentagonaster* und *Tosia. Zool. Jb.* Suppl. 15(1): 1-44.

Lütken, C. 1869. Ueber *Choriaster granulatus,* eine neue Gattung aus der Familie der Asteriden. p. xxxv. In J.D.E. Schmeltz (ed.) *Catalog der zum verkauf stehenden Doubletten aus den naturhistorichen Expeditionen der Herren J.C. Godeffroy & Sohn in Hamburg.* 4. Hamburg.

McKnight, D.G. 1967b. Some Asterozoans from Norfolk Island. *N.Z. Jl mar. freshwat. Res.* 1(3): 324-326.

McKnight, D.G. 1972. Echinoderms collected by the Cook Islands Eclipse Expedition, 1965. *N.Z.O.I. Rec.* 1(3): 37-45.

McKnight, D.G. 1973a. Additions to the asteroid fauna of New Zealand: family Goniasteridae. *N.Z.O.I. Rec.* 1(13): 171-195.

McKnight, D.G. 1978. *Acanthaster planci* (Linnaeus) (Asteroidea: Echinodermata) of the Kermadec Islands. *N.Z.O.I. Rec.* 4(3): 17-19.

Madsen, F.J. 1955b. A note on the sea star genus *Acanthaster. Vidensk. Meddr dansk. naturh. Foren.* 117: 179-192.

Madsen, F.J. 1958. On *Sphaeriodiscus placenta* and a few other sea-stars from W. Africa. *Bull. Inst. franç. Afr. N.* N.20a: 90-94.

Madsen, F.J. 1959a. On a new North Atlantic seastar *Chondraster hermanni* n. sp. with some remarks on related forms. *Vidensk. Meddr dansk. naturh. Foren.* 121: 153-160.

Marenzeller, E.von 1875. Revision adriatischer Seesterne. *Verh. zool.-bot. Wien* 25: 361-372.

Marenzeller, E.von 1895b. [Echinodermen gesammelt in östlichen Mittelmeer 1893 und 1894.] *Anz. Akad. wiss. Wien (Math.-naturw.* Cl.) 32: 189-191.

Miller, J.E. 1984. Systematics of the ophidiasterid sea stars *Copidaster lymani* A.H. Clark and *Hacelia superba* H.L. Clark (Echinodermata: Asteroidea) with a key to the species of Ophidiasteridae from the western Atlantic. *Proc. biol. Soc. Wash.* 97(1): 194-208.

Monks, S.P. 1904. Variability and autotomy of *Phataria. Proc. Acad. nat. Sci. Philad.* 56: 596-600.

Moore, D.R. 1960. *Linckia bullisi,* a new asteroid from the northeast coast of South America. *Bull. mar. Sci. Gulf Caribb.* 10: 414-416.

Mortensen, T. 1935. A new giant sea-star, *Mithrodia gigas* n. sp. from South Africa. *Ann. S. Afr. Mus.* 32: 1-4.

Mortensen, T. 1938. Contributions to the study of the development and larval forms of Echinoderms. 4. *K. danske Vidensk. selsk. Skr.* (naturv.-math.) (9)7(3): 1-59.

Nishida, M. & J.S. Lucas 1988. Genetic differences between geographic populations of the crown-of-thorns starfish throughout the Pacific region. *Mar. Biol., Berlin* 98: 359-368.

Oguro, C. 1983. Supplementary notes on the sea-stars from the Palau and Yap Islands. 1. *Annot. zool. Jap.* 56(3): 221-226.

Oguro, C. 1984. Occurrence of *Fromia nodosa* A.M. Clark (Asteroidea, Ophidiasteridae) from the Marshall Islands, the western Pacific. *Proc. jap. Soc. syst. Zool.* No. 27: 101-106.

Parelius, J.von der L. 1768. Beskrivelse over Nogle Korstrold. *K. norske Vidensk. Selsk. Skr.* 4: 423-428.

Pearse, J.S. 1965. Reproductive periodicities in several contrasting populations of *Odontaster validus* Koehler, a common antarctic asteroid. *Antarctic Res. Ser.* 5: 39-70.

Perrier, E. 1876. Les stellérides des Iles du Cap-Vert. *Bull. Soc. zool. Fr.* 1: 63-71.

Pope, E.C. & F.W.E. Rowe 1977. A new genus and two new species in the family Mithrodiidae (Echinodermata: Asteroidea) with comments on the status of species of *Mithrodia* Gray, 1840. *Austral. Zool.* 19(2): 201-216.

Powell, A.W.B. 1937. A starfish of the genus *Asterodiscus* new to New Zealand. *Trans. R. Soc. N.Z.* 67: 78-79.

Rowe, F.W.E. 1976. The occurrence of the genus *Heteronardoa* (Asteroidea: Ophidiasteridae) in the Indian Ocean, with the description of a new species. *Rec. W. Aust. Mus.* 4(1): 85-100.

Rowe, F.W.E. 1977a. A new family of Asteroidea (Echinodermata) with the description of five new species and one new subspecies of *Asterodiscides. Rec. Aust. Mus.* 31(5): 187-233.

Rowe, F.W.E. 1977b. The status of *Nardoa (Andora)* A.M. Clark, 1967 (Asteroidea: Ophidiasteridae) with the description of two new genera and three new species. *Rec. Aust. Mus.* 31(4-6): 235-245.

Rowe, F.W.E. 1981. A new genus and species in the family Ophidiasteridae (Echinodermata: Asteroidea) from the vicinity of Lord Howe Island, Tasman Sea. *Proc. linn. Soc. N.S.W.* 105(2): 89-94.

Rowe, F.W.E. 1985. On the genus *Podosphaeraster* A.M. Clark & Wright (Echinodermata: Asteroidea), with description of a new species from the North Atlantic. *Bull. Mus. natn. Hist. nat. Paris* (4e ser.) 7A(2): 309-325.

Rowe, F.W.E. 1989a. Nine new deep-water species of Echinodermata from Norfolk

Island and Wanganella Bank, northeastern Tasman Sea, with a checklist of the echinoderm fauna. *Proc. linn. Soc. N.S.W.* 111(4): 257-291.

Rowe, F.W.E. 1989b. A review of the family Caymanostellidae (Echinodermata: Asteroidea) with the description of a new species of *Caymanostella* Belyaev and a new genus. *Proc. linn. Soc. N.S.W.* 111(4): 293-307.

Rowe, F.W.E., A.N. Baker & H.E.S. Clark 1988. The morphology, development and taxonomic status of *Xyloplax* Baker, Rowe & Clark (1986)(Echinodermata: Concentricycloidea), with the description of a new species. *Proc. R. Soc. Lond.* B233: 431-459.

Rowe, F.W.E. & L.M. Marsh 1982. A revision of the asterinid genus *Nepanthia* Gray, 1840 (Echinodermata: Asteroidea), with the description of three new species. *Mem. Aust. Mus.* 16: 89-120.

Rowe, F.W.E. & D. Nichols 1980. A new species of *Podosphaeraster* Clark & Wright, 1962. (Echinodermata: Asteroidea). from the Pacific. *Micronesica* 16(2): 289-295.

Rowe, F.W.E., D. Nichols & M. Jangoux 1982. Anatomy of the spherical, valvatid starfish *Podosphaeraster* (Echinodermata: Asteroidea) with comments on the affinities of the genus. *Micronesica* 18(1): 83-95.

Sars, M. 1872. Tillaeg. In G.O. Sars Nye Echinodermer fra den Norske Kyst. *K. Vidensk. Selsk. Forh.* 1871: 27-31.

Schöndorf, F. 1906. Die organisation und systematische Stellung der Sphaeriden. *Arch. Biontol.* 1: 245-306.

S[chreber] 1793. Beschreibung der Seesonne. *Naturforscher, Halle* 27: 1-6.

Shepherd, S.A. 1967. A review of the starfish genus *Nectria* (Asteroidea: Goniasteridae). *Rec. S. Aust. Mus.* 15(3): 463-482.

Shepherd, S.A. & E.P. Hodgkin 1965. A new species of *Nectria* (Asteroidea: Goniasteridae) from Western Australia. *Jl R. Soc. W Aust.* 48: 119-121.

Sibuet, M. 1977. *Ophidiaster reyssi*, nouvelle espèce d'astéride bathyale de l'Ocean Atlantique. *Bull. Mus. natn. Hist. nat. Paris* (Zool.) No. 343: 1085-1090.

Sladen, W.P. 1889. The Asteroidea. *Rep. scient. Results Voy. Challenger* Zool. 30: 1-935.

Smith, A.B. 1988. To group or not to group: the taxonomic position of *Xyloplax*. In Burke, Mladenov, Lambert & Parsley (eds). *Echinoderm Biology: Proc. Sixth int. Echinoderm Conf. Victoria, 1987*, 17-23. Rotterdam: Balkema.

Smith, G.A. 1927a. On *Asterina burtoni* Gray. *Ann. Mag. nat. Hist.* 19: 641-645.

Smith, G.A. 1927b. A collection of echinoderms from China. *Ann. Mag. nat. Hist.* 20: 272-279.

Stuxberg, A. 1878. Echinodermer fran Novaja Zemljashaf samlade under Nordenskioldska Expeditionerna, 1875 och 1876. *Ofvers. K. VetenskAkad. Forh. Stockholm* 1878(3): 27-40.

Sukarno & M. Jangoux 1977. Révision du genre *Archaster* Müller et Troschel (Echinodermata: Asteroidea: Archasteridae). *Rev. Zool. afr.* 91(4): 817-844.

Tablado, A. 1982. Asteroideos Argentinos. Familia Poraniidae. *Com. Mus. argent. Cienc. nat. Bernardino Rivadavia.* (Hidrobiol.) 2(8): 86-106.

Tortonese, E. 1937b. Descrizione di una nuova stella di mare *(Goniodiscaster australiae* n. sp.) *Boll. Musei Zool. Anat. comp. R. Univ. Torino* (3)45(69): 293-297.

Tortonese, E. 1952b. Studio comparativo di *Asterina gibbosa* Penn. e *A. panceri* Gasco (Echinodermi Asteroidi). *Pubbl. Staz. zool. Napoli* 23: 163-172.

Tortonese, E. 1960b. Il neotipo di *Asterina panceri* (Gasco) (Asteroidea). Doriana 3(108): 1-3.

Tortonese, E. 1962. Un asteroide nuovo per il Mediterraneo: *Asterina stellifera* (Moeb.). *Doriana* 3(118): 1-7.

Tortonese, E. 1976. Researches on the coast of Somalia. Seastars of the genus *Monachaster* (Echinodermata Asteroidea). *Monit. zool. ital.* (N.S.) 7(6): 271-276.

Tortonese, E. 1980. Echinoderms collected along the eastern shore of the Red Sea (Saudi Arabia). *Atti Soc. ital. Sci. nat.* [1979] 120(3-4): 314-319.

Tortonese, E. 1984. Notes on the mediterranean sea-star *Peltaster placenta* (M. Tr.) (Echinodermata, Asteroidea). *Boll. Mus. civ. Stor. nat. Verona* 11: 99-112.

Tortonese, E. & A.M. Clark 1956. On the generic position of the asteroid *Goniodiscus placenta* Müller & Troschel. *Ann. Mag. nat. Hist.* 9: 347-352.

Verrill, A.E. 1909. Description of new genera and species of starfishes from the North Pacific coast of America. *Amer. J. Sci.* 28: 59-70.

Verrill, A.E. 1913. Revision of the genera of starfishes of the subfamily Asterininae. *Amer. J. Sci.* (4)35: 477-485.

Verrill, A.E. 1914c. Nomenclature of certain starfishes: *Asterina. Amer. J. Sci.* (4)37: 483-484.

Zeidler, W. & F.W.E. Rowe 1986. A revision of the southern Australian starfish genus *Nectria* (Asteroidea: Oreasteridae). *Rec. S. Aust. Mus.* 19(9): 117-138.

Zeidler, W. & S.A. Shepherd 1982. Sea stars (Class Asteroidea). In S.A. Shepherd & I.M. Thomas (eds) *Marine Invertebrates of Southern Australia.* Adelaide. 400-418. [Not seen]

Ziesenhenne, F.C. 1964. A new sea-star from Easter Island. *Ann. Mag. nat. Hist.* 6: 461-464.

Note from the Editors

A guiding principle of *Echinoderm studies* has been to cover all aspects of echinoderm biology in order to promote a better comprehension of this remarkable group of animals. The current volume is the fourth in the series which began in 1983. Twenty-seven echinoderm specialists have contributed to twenty-four reviews which deal with various fields of echinoderm research, from palaeontology to molecular biology. Interest in all aspects of echinoderm biology is continuing to increase, and the series seems to have reached its major goals: To be a tribune and an authoritative reference source for echinoderm specialists and biologists in general.

367

CONTENTS OF VOLUME 2 (issued 1987)

CONTENTS OF VOLUME 3 (issued 1989)